Empowerment of Women through Science and Technology Interventions

Empowerment of Women through Science and Technology Interventions

– Editors –

Nirupama Prakash

Betty McLellan

Barbara Wejnert

CENTRE FOR SCIENCE & TECHNOLOGY OF THE NON-ALIGNED AND OTHER DEVELOPING COUNTRIES (NAM S&T CENTRE)

2010

REGENCY PUBLICATIONS

Delhi - 110 035

ISBN 9789351240150

Centre for Science and Technology of the Non-Aligned and Other Developing Countries (NAM S&T Centre)
Core-6A, 2nd Floor, India Habitat Centre, Lodhi Road,
New Delhi-110 003 (India)
Phone: +91-11-24644974, 24645134, Fax: +91-11-24644973
E-mail: namstct@gmail.com
Website: www.namstct.org

Published by : **REGENCY PUBLICATIONS**
(*Division of Daya Publishing House*)
4760-61/23, Ansari Road, Darya Ganj,
New Delhi - 110 002
Phone: 23245578, 23244987
Fax: (011) 23260116
e-mail : info@regency-books.com
website : www.dayabooks.com
website : www.regencybooks.com

Laser Typesetting : **Classic Computer Services**
Delhi - 110 035

Printed at : **Chawla Offset Printers**
Delhi - 110 052

Printed in India

Foreword

The NAM S&T Centre sponsored volume "Empowerment of Women through Science and Technology Interventions" by Nirupama Prakash, Betty McLellan and Barbara Wejnert presents an important contribution to the existing literature on gender and global development issues. Its' focus on using modern technology and science to empower women globally is particularly relevant and pertinent within current initiatives of the Millennium Development Goals undertaken by the United Nations aimed to bring woman into the forth runners of the modern world development.

The modern stage of the world development and rapidly changing technologies demand from the people a stale improvement of their communicative and technical competencies. This demand is especially important for the women from the developing countries as the tool helping them to form their own future and the future of their countries and regions. That is why it is so important to educate women in the field of the technical sciences, fundamental sciences, biology and communication. The importance lays also in fact that women in many regions compose very dynamic, community oriented groups and are conscious of the needs of the new skills education. The group-solidarity will facilitate women's success and gained education to be shared with the relatives, tribe, village and local community.

It is very important, thus, to convince women to make efforts to achieve new life goals especially in those regions where women do not believe in their opportunities. Extremely important becomes also support of girls who decide to study in traditionally men fields as biological, technical, mathematics and informatics science. It is very important at the same time to struggle against the stereotypes which limit the women's role to the homemaker and family care provider to, uninvolved in the highly professional and scientific activities. As it is underlined in the manuscript, all of these goals could be achieved through science and technology interventions.

The important role in the mentioned above processes have to play state authorities, which need to be interested in using women's abilities to conduct academic and research work in the fundamental and humanistic sciences. The modern women, using the processes of the traditional (family and community base) and formal education, due to their natural ability to multitasking are capable to concentrate on the implementation of the fundamental sciences on the level of a family or community. The psychological strength and flexibility of woman equals if not surpassing that of man, therefore, it is absolutely essential to not lose such an enormous potential which have the half of the mankind and can improve societal quality of life.

Moreover the development of e-learning may support the additional educational possibilities without necessity of the additional transport and textbooks costs, or withdraw from family duties. The Internet development in the communication processes widens the possibility to actively participate in the exchange of the world scientific information, as well as collection of information and presentation of the women's own researches and scientific results.

The globalization and democratization processes prove that there are no dreams which cannot be realized. The new technologies such as biotechnology, mobile phone, satellite TV and computers are necessary in the same scale in Africa, Asia or Europe, and there is no difference who – a man or a woman – is able to programme, to repair or to start equipment. The important thing is that the equipment works and makes possible the development of humankind.

Elzbieta Sawa-Czajka

Provost, University at Torun, Poland

Foreword

The NAM S&T sponsored volume "Empowerment of Women through Science and Technology Interventions" by Nirupama Prakash, Betty McLellan and Barbara Wejnert is a very interesting, well written and timely book which is coming out at a time in which women empowerment issues are at the forefront of the academic discourse. The global Women Empowerment Initiatives launched by United Nations as a part of Millennium Development Goals, and the World Conferences on Women are aimed at reducing the severe global burden of women. These initiatives still remains in an early phase of development, hampered by difficulties in four areas: the strength of the actors involved in the initiative, the power of the ideas they use to portray the issue, the nature of the political contexts in which they operate, and characteristics of the issue itself. Therefore the editors of this volume concludes that women "are an under-utilized pool of talent and resource as they comprise more than half of the population of any society and could contribute immensely towards the social and economic development of societies through participation in science and technology programmes."

In my opinion, Empowerment of Women through Science and Technology Interventions not only appears at a crucial moment to help refocus attention on women issues including women health and representation among health professionals, but the book addresses the main areas that have been identified as hampering the success of women advancement in scientific and professional fields.

It is always challenging to organize and edit a volume with diverse contributors into a cohesive unit. Yet, Prakash, McLellan and Wejnert are successful at including a number of significant contributions representing several academic disciplines and number of practitioners into a very cohesive manuscript presenting a multidisciplinary and rigorous analysis of problems of women empowerment through professional advancement and modern technology. What is unique about their approach is that they link a broad analysis of processes involved in empowerment of women, while also recognizing the impact of specific policies, infrastructures and effects of global economic and political situation on women access and utilization of technology and science.

A major contribution of this collection of essays is the analysis of recognition of women's achievement in science in various developed and developing countries. All in all, Empowerment of Women through Science and Technology Interventions offers a serious discussion and rigorous presentation of evidence that could be of interest both to academic and practitioner audiences. A very welcome and necessary addition to the existing literature.

Eunice Rodriguez, Dr PH, ***Associate Professor (Teaching)***

Department of Pediatrics and Centre for Education in Family and Community Medicine,
Stanford University School of Medicine,
Modular, G, Stanford, CA 94305-5208

Preface

The gender dimension of science and technology has become one of the most important and debated issues worldwide. Over the past 30 years, the United Nations General Assembly and UN Economic and Social Commission (ECOSOC) have emphasized issues related to inequalities, insufficiencies and disparities in the access of women to education, training and the labour market. Various major international initiatives have been undertaken on the subject; including the United Nations Decades on Women and Development (1975-1995), and special attention has been directed towards the role of women in science and technology. Gender equality is one of the eight United Nations Millennium Development Goals, which clearly call for action related to science, technology and gender.

Women are identified as seriously under-represented in science, mathematics and engineering fields. They are an under-utilized pool of talent and resource as they comprise more than half of the population of any society and could contribute immensely towards the social and economic development of societies through participation in science and technology programmes.

The challenge is to find ways of changing the situation such that women can be used as a resource for science and technology. There is no doubt that worldwide, countries need to open up opportunities to bring more women to science and technology, thereby producing a critical mass of scientifically skilled women. It is important for ordinary women to appreciate and to access the findings of science and technology to improve the living conditions of their families. It is also important that women scientists take up this agenda and develop projects to address the under-representation of women.

There is need to remove structural obstacles and barriers that continue to exclude girls and women from the study of science and technology. There is urgent need for the development and mobilization of all segments of populations across cultures to contribute to the eradication of poverty, fighting diseases, stemming environmental degradation and improving global economic competitiveness through the application and development of science and technology.

Science and Technology have been an integral part of Indian civilization and culture. Over the years Indian women have overcome the traditional mindsets and have excelled in professions like teaching, medicine and pure sciences. Women have made important contributions in all walks of life and made inroads into new fields like engineering, information technology, biotechnology, nuclear science, space science and many such specialized fields in the domain of science and technology.

While these developments have been highly satisfying, constraints in the form of socio-cultural factors, such as discrimination, lack of self-confidence and gender disparity continue to affect Indian women and their choices of career.

In an Indian National Science Academy (INSA) study on "Gendered Science: Trends and Analysis of Contributions of Indian Women Scientists", taken up with a view to focusing on achievements and recognition of Indian Women scientists, it has been pointed out that women reaching to higher positions in the cadre in their science careers are few and far between. Many women have done exceedingly well in their scientific and academic pursuits, however not many are seen being awarded and rewarded. Some of the contributing factors mentioned in the INSA study were: family, motherhood, inadequate support systems, societal/cultural issues that could be due to fixed mind set, restriction on movement, nepotism and sexism.

The study revealed dissatisfaction of most respondents regarding professional growth and career advancement. Reasons cited were lack of time; household responsibilities; lack of encouragement from the organization and ill health. Respondents who were denied promotion attributed it to gender bias and lack of connections. The parameters focus on strong flavor of gendered science. The study suggests that implementation of remedial measures to improve women's participation in science need to become priority in policy making.

Kalpana Sharma has pointed out that in India, as in many other countries, women have had to fight to be accepted as capable of being equal partners with men in science and technology. Within science, there are areas considered suitable for women and others considered outside their realm of capability. Thus it has taken a struggle, for instance, for women to become engineers. Rather than looking at the reason why more women do not pursue careers in a particular branch of science and technology, it is generally concluded that women have no aptitude for that stream. Each time a woman becomes an aeronautical engineer, or a nuclear physicist, or excels in some area previously considered a male preserve, she is applauded and celebrated, but strictly as an exception.

Even before they reach the point of choosing a career in science, women have to make difficult choices. Every year, girls do far better than boys, in science and arts, in the Class X and Class XII examinations. In some institutions, the majority of the top achievers are girls. What happens to them after that? Do they drop out? Are they forced by their families to make choices which are not their own? Do they fail to get through the competitive entrance examinations? Are they forced to make pragmatic choices about the future because they are conditioned to believe that marriage and family come first? Is there no way for them to balance their commitments to family with their desire to follow a career? There are great concerns now being shown in India about careers for women in science and consideration is being given to nurturing their talent by facilitating and providing various options.

The declining number of girls who follow through on their apparent aptitude for science at the school level is evident in the few women scientists at the top of the academic pyramid. Those women, who do manage to pursue a career in science, often have to strategize how to survive and to move ahead.

In the education sector, information and communication technologies (ICTs) hold great potential. Distance education can enable women who don't have physical access to schools to participate through online, radio or other ICTs in education classes and programmes. E-learning, which can involve a wide variety of technologies such as DVD, CD-ROM, or internet, offers students access to a vast amount of information and resources that are not possible in a single instructional setting. Online teaching support programmes for teachers who have minimal teaching background can assist with lesson planning, curriculum design and other learning tools. Training of women in the use of ICTs, media management and content development can enable them to participate in decision-making processes at government, private sector and civil society levels. Moreover, teaching methods and tools must be gender sensitive and responsive to women's and girls' unique need and learning styles.

Education, training and skill development are critical to ICT interventions. These areas represent an entry point for encouraging women to become more involved in ICT applications development, shaping ICT solutions and framing ICT policies according to female-specific needs and experiences.

Scholars argue that women suffer from an unequal attainment in other scientific fields too–from engineering and biology to nuclear science–and the inequality is observed along some important dimensions: recognition, awards, productivity, consulting, and inventions (scientific entrepreneurship). Some recent evidence gives cause for optimism showing that, especially in the field of biological sciences, the gender gap has narrowed. This evidence pertains only to the most-developed societies of Western countries and it does not necessarily represent the situation in a wide range of countries.

Workshop Presentations

To deliberate on and address issues of concern regarding women and science & technology, the Centre for Science & Technology of the Non-Aligned and Other Developing Countries (NAM S&T Centre) organized a 3-days International Workshop on Empowerment of Women through Science & Technology Interventions from 14-16 December, 2008 at Tehran, Iran jointly with Iranian Research Organization for Science & Technology (IROST), Ministry of Science, Research & Technology, Tehran and Iranian Association of Women's Studies (IAWS), Tehran.

The presentations in this volume show that the gender gap persists in a majority of developed and developing countries. The diverse underlying causes of gender disparity in science could be summarized as pertaining to the overall level of societal development determining the level of economic opportunities open to women for employment and education; to the existing political system promoting or inhibiting policies of gender equality; and to gender stereotypes depicting women as incapable of achieving in technical and scientific fields. As the presenters at the Workshop indicated, the gender gap in academic science, in technology and engineering is not only a topic of ongoing policy changes and scholarly debates, but also is of interest to policy makers and governments engaged in initiatives to narrow the gap between women and men in these fields.

The voices gathered in this publication attest to a growing attention among women, scientists and governments to professional achievement of women in fields not typically occupied by females. Governments and policy makers draft policies to empower women through education and employment especially in the fields of science and technology. As discussed by Gloria Tiny Seitei, such initiatives were undertaken, for example, by the Bostwana government developing different policies and programmes to address critical issues affecting women. According to Shokouh Navabinejad, during the last decade the Iranian government has implemented interventions to ensure women's education at all levels of schooling and across various disciplines, including access to science and technology. There has been an increase in the number of technical and vocational schools for girls across the country, as well as an increase in the percentage of girls entering the university in science and technology fields.

According to Aisyiyah Jenie, the government of Indonesia also has greatly addressed gender equality and involvement of women in science and technology. The government commitment has led to policies prioritizing gender equality at various levels of societal development, including the establishment of policies promoting gender equality and the implementing of these policies among institutions. Correspondingly, the government of Malawi, concerned about the education of girls in science and technology, has developed a number of programmes to encourage girls to take science-based subjects. The governmental initiatives are supported by the nongovernmental organizations. As indicated by Symon Osman Mandala, a group of women scientists and technologists from the academia, private sector, civil society and government formed the Women in Science and Technology NETwork (WISTNET) to promote the participation of girls and women in science and technology in Malawi.

A comparable situation is analyzed by Shamima Abdul Carrim Delbar discussing the strategies that the democratic government of Mauritius has designed to assist women in scholarly and scientific achievements. The government created the "empowerment program" within which women can take advantage of training leading to professional employment. The training is mainly addressed to women employees teaching how to effectively participate in social dialogues and consultations.

Science and Technology (S&T) tools are frequently used to promote empowerment of women farm workers and agricultural producers. According to Vinita Sharma, government in India uses S&T to provide information about farm production, managerial capabilities and understanding of the science involved in the processes and products of farm production. As the author argues, rural women have a special understanding of natural resource management and hence they can play a crucial role in re-nurturing and re-greening rural India. Congruently, education about more efficient rural production is disseminated via Science and Technology (S&T) tools and interventions among women dairy producers in the Himalayan region. Kavita Mehra discusses this process using as an example three organisations: a) initiatives of National Dairy Development Board (NDDB) aimed at expanding the participation of women dairy farmers, and strengthening women's capacity by making available the S&T information and practices in dairy; b) application of biotechnology by rural women in Himachal, a totally new endeavour by women directed by the Institute of Himalayan Bioresource Technology (IHBT), a Council of Scientific and Industrial Research (CSIR); and c) a voluntary organization, Self Employed Women Association (SEWA) demonstrating the capacity to deploy ICT tools. The task of empowering women and communities on health, social and development issues through mass media and promotion of ICT is the study by Nirupama Prakash focusing on Rajasthan region in India. Nirupama Prakash is analyzing the activities of the Centre for Women Studies and One World South Asia (OWSA) that uses information and communication technologies as an empowering tool for rural development with prime focus on social equity. This included a series of capacity building programmes for male and female village volunteers on specific issues such as gender violence, child marriage, skewed sex ratio and health related millennium development goals. This rural based project with ICT methodology which has been undertaken in the traditional state of Rajasthan has created a pool of gender sensitized community and empowered women who are now computer literate and techno savvy.

Three authors discussed recognition of women's achievement in science in various developed and developing countries. Pranav N. Desai analyzes the situation in India in comparison to other countries. Barbara Wejnert looks at the problem of women employment, education and achievement in science, engineering and managerial positions in well developed democracies, and democratizing, developing Eastern European states. Felizitas Sagebiel focuses on gender related barriers in achievement of women engineers in Western European societies. In contrast, Choo Poh Sze, identifying issues pertaining to the participation of Malaysian women in science and technology, concentrates on providing role models of women scientists, promoting an image of successful women in science; providing mentoring and fellowship facilities and making the teaching of science more female-friendly. In general, it was found that overall the gender gap has improved over time but remains large in science and technology.

Khin Htike Htike Lwin discussed the performance of Myanmar women in Science and Technology and outlined two major programmes implemented by the Ministry of Science and Technology designed to assist women to succeed. A project to improve the social and economic conditions of rural people in Nepal was described by Kayo Devi Yami. More than 80 per cent of the participants were women farmers and several of those conducting the research were women qualified in Science and Technology. South African Women in Engineering (SA WomEng) was founded to combat the lack of women in senior positions in the engineering industry. Naadiya Moosajee discussed the extensive and exciting programme being undertaken by SA WomEng.

The various mechanisms in place in Sri Lanka aimed at the economic empowerment of rural women are outlined by Jayasamara Gunarathne. Such empowerment will have the effect of lifting women's social status and allowing more women to have input into the development of social policies. Rose Rita Kingamkono made two presentations at the Workshop. One dealt with the issue of childhood malnutrition in Tanzania and described an important initiative designed to bring economic empowerment to women and provide more nutritious food for children. The other provided an analysis of Science and Technology in Tanzania with a special emphasis on gender.

Out of a concern about the low numbers of women involved in Science and Technology in Zambia, Government and Non-government organisations have embarked on projects to empower women in these areas. The situation in Zambia was described by Atridah Mulonga who said that, even though there are a lot of opportunities for the empowerment of women, more are needed.

There were two presentations about the situation in Iran. Akram Ghadimi discusses the position of women and explains that, while great progress has been made in education and health in recent years, there is more to be done in terms of paid work and political participation. Zohreh Nosrat Kharazmi presented a content analysis of the involvement of young Iranian women in weblogs with a focus on their sense of National Identity.

P.R.M.P. Dilrukshi and Seetha I. Wickremasinghe discussed the study they undertook into the status of women scientists in Sri Lanka with very interesting results. In Sri Lanka, as with most other countries represented at the Workshop, much progress has been made toward the empowerment of women through science and technology interventions, but it was agreed that much more needs to be done before women and men can be said to enjoy equal status.

There was general agreement during and after the Workshop that the event was a very valuable experience for all who attended. Much was achieved and the benefits will continue to be felt for years to come.

The three editors of this volume want to express our appreciation to the Centre for Science & Technology of the Non-Aligned and Other Developing Countries and to all contributors. We have been delighted to have had the opportunity to contribute to bringing these papers together into this powerful document. Thanks are also due to Ms.Madhusmita Das, Project Fellow at Centre for Women Studies, BITS, Pilani for helping in initial editing of papers

Professor Nirupama Prakash, *Ph.D*

Chief, Women Studies & Societal Development Unit,
Birla Institute of Technology & Science, Pilani-333 031, Rajasthan, India

Adjunct Associate Professor Betty McLellan, *Ph.D*

James Cook University,
Townsville, Australia

Associate Professor Barbara Wejnert, *Ph.D*

Department of Global Gender Studies
University at Buffalo, Buffalo, New York, USA

Introduction

Women are a great human resource and the role of women in society is absolutely vital for its progress. Arguably, to empower women is to increase their control over the decisions that affect their lives both within and outside the household. There are many developing countries where women are not even listed in any citizen registration database, which makes it very difficult for them, especially the poorer ones in rural areas and of under-privileged class to assert social security, social protection, legacy, land etc. as there is practically no proof of their existence. Through their empowerment women gain greater share of control over resources–material, human and intellectual such as knowledge, information, ideas – and financial resources involving access to money and control over decision making in the home, community, society and nation. It leads them to become a critical and vital stakeholder among the human capital and their strategic presence in the society is widely accepted. The great leader and the Father of the Indian Nation, Mahatma Gandhi, who firmly believed in non-violence as the soul means to achieve peace in the world, also had a deep appreciation of the value of educating women. He had said that: *"when a man is educated, an individual is educated; when a woman is educated, a family and a country are educated."*

Market economy trends in the new era of globalization have widened the gap between education and technology opportunities for men and women. Worldwide the capacity of women to engage in the knowledge society is grossly underutilized. In the knowledge society, they need access not only to new technologies but also to education, entrepreneurship and employment opportunities as well as the ability to participate fully in knowledge based activities. The involvement and engagement of women in the Information Society on an equal footing with men would directly contribute to improving the livelihood of people, making it more sustainable and thereby promoting the social and economic advancement of societies. Governments must design and implement national policies and programmes that promote science and technology education for women and girls and encourage women to enter into high value added ICT career. Programmes should be strengthened to bring about a greater involvement of women in science and technology. These should include measures to motivate girls to take up science and technology for higher education and also ensure that development projects with scientific and technical inputs have a total involvement of women. Efforts to develop a scientific temper and awareness need should be stepped up. Special measures should be adopted for their training in areas where they have special skills like communication and information technology. Efforts to develop appropriate technologies suited to women's needs as well as to reduce their drudgery have to be given a special focus too. It is not the concern of one nation only, but there are many players and stakeholders in the aim to reach this millennium goal. Programmes should be worked out to empower women through innovative scientific activities integrating action oriented literacy, sound micro-finance and micro-enterprise training as well as an understanding of legal rights and advocacy. Apart from the efforts of the United Nations family of organizations, multilateral bodies and civil society, the positive role of women also depends on the supportive attitudes of their local family unit, the local community in each village and town. We just hope that women become empowered at 100 per cent and become equal to men so that both of them work side by side for a better world of today.

In order to deliberate on the emergent need for capacity building in women to economically empower them, the Centre for Science & Technology of the Non-Aligned and Other Developing Countries (NAM S&T Centre) organized a 3-days International Workshop on Empowerment of Women through Science & Technology Interventions from 14-16 December, 2008 at IROST Headquarters in Tehran, Iran jointly with the Iranian Research Organization for Science & Technology (IROST), Ministry of Science, Research & Technology of Iran and Iranian Association of Women's Studies (IAWS), Tehran. The Workshop was attended by about 60 experts and senior professionals from 14 countries including the host country Islamic Republic of Iran. The overseas participants were from Botswana [Dr. (Ms.) Gloria Tiny Seitei] Botswana Technology Centre; Germany [Prof.(Ms.) Felizitas Sagebiel], Associate Professor, Faculty of Educational and Social Sciences, Bergische University of Wuppertal; India [Dr.(Ms.) Vinita Sharma], Scientist-G, Science and Society Division, Department of Science and Technology, Government of India,New Delhi; [Prof. (Ms.) Nirupama Prakash], Chief, Women Studies & Societal Development Unit Group Leader, Humanistic Studies & Coordinator, Hospital & Health Systems Management Programme, Birla Institute of Technology & Science, Pilani, Rajasthan; [Prof. Pranav N. Desai], Chairperson, Centre for Studies in Science Policy, Jawaharlal Nehru University (JNU), New Delhi; [Dr. (Ms.) Kavita Mehra], National Institute of Science, Technology & Development Studies, CSIR, New Delhi; [Mr. Ratan Kumar], Vice-President, Mekaster Engineering &Equipments Ltd., New Delhi; and [Mr. Kapil Vermani], General Manager, Mekaster Engineering& Equipments Ltd., Gujarat; Indonesia [Mrs. Siti Nurul Aisyiyah Jenie], Researcher, Research Centre for Chemistry, Indonesian Institute of Sciences, Jakarta; Malawi [Mr. Symon Osman Mandala], Principal Science and Technology Officer, Department of Science and Technology, Lilongwe; Malaysia [Mrs. Choo Poh Sze], Science and Policy Specialist, The World Fish Centre,Penang; Mauritius [Mrs. Shamima Abdul Carrim Delbar], Industrial Analyst, Ministry of Industry, Science and Research, Port Louis; Myanmar [Dr. (Ms.) Khin Htike Htike Lwin],Principal,

Technological University, Kyaukse; Nepal [Dr. (Ms.) Kayo Devi Yami] Tuladhar,Chief, Science Faculty, Nepal Academy of Science and Technology; South Africa [Ms. NaadiyaMoosajee], Co-founder and Director SAWomEng (South African Women in Engineering),Department of Civil Engineering, University of Cape Town; Sri Lanka [Ms. Jayasamara Gunaratne], Coordinator, Technology Transfer Unit, Ministry of Science & Technology; Tanzania [Dr. Rose Rita Kingamkono], Director, Research Coordination and Promotion, Tanzania Commission for Science and Technology, Dar es Salaam; and Zambia [Ms. Atridah Mulonga], Programme Officer–Promotion and Grants, National Science and Technology Council, Lusaka. The Iranian scholars in the workshop included Dr. Nasrin Mosaffa, Head, IAWS; Dr. Mahnaz Mazaheri Assadi from Environmental Biotechnology Group, Biotechnology Department, IROST; Prof. Shokouh Navabinejad, Psychologist and President, Women NGO's Network in Iran; Dr. Masoumeh Simbar, Assistant Professor in Reproductive Health, Shahid Beheshti Medical Science University; Dr. Mina Mehrnoush, Vice President, Institute of Development and Economic Research, Tehran; and Dr. Zahira Arasty and Dr. Parvaneh Gelard from the Faculty of Enterprises, Tehran University.

The Workshop was conducted in seven technical sessions broadly categorized under the themes 'National Policies and Programmes', 'Role of Information and Communication Technologies' and 'Women and New Technologies', six of which were held at the IROST Headquarters in Tehran. The 7th special session on 'Development of Entrepreneurship and Women's Empowerment' was organized during a study tour to the Faculty of Entrepreneurship in the University of Tehran. The overall technical programme of the conference was coordinated by Dr. Nasrin Mosaffa, Head, IAWS and Prof. A.P. Kulshreshtha, Director, NAM S&T Centre. A Tehran Resolution on 'Empowerment of Women through Science and Technology Interventions' was unanimously adopted at the conclusion of the workshop.

The present publication comprises 22 scientific articles and review papers primarily based on the deliberations of the above workshop, but a number of articles have been contributed by other eminent experts on the subject. I would like to express my deepest gratitude to Dr. Nirupama Prakash, Chief, Women Studies & Societal Development, Birla Institute of Technology & Science (BITS), Pilani, India; Dr. Betty McLellan, James Cook University Townsville, Australia; and Dr. Barbara Wejnert, Department of Global Gender Studies University at Buffalo in taking pains to edit the manuscripts for this publication and their comprehensive support and guidance in this venture. I am also indebted to Mr. Elzbieta Sawa-Czajka of the University at Torun, Poland and Dr. Eunice Rodriguez of the Stanford University School of Medicine in USA for sparing their valuable time in writing their 'Foreword' for this publication.

The valuable services provided by the entire team of the NAM S&T Centre, particularly by Mr. M. Bandyopadhyay, Dr. V. P. Kharbanda, Mrs. Pooja Kapoor, Mrs. Sudipta Chakraborty and Mr. Pankaj Buttan in compiling the articles and giving a shape to this volume are also deeply appreciated.

I hope that this publication would be a valuable reference material for the entire scientific community especially those working in the field of capacity building of women so that they are able to take part effectively in the national decision making process. This would also help the developing countries in executing their plans and strategies on empowerment of women through science & technology as well as in their economic growth though enrichment of science & technology with deeper involvement of women.

Arun P. Kulshreshtha

Director, NAM S&T Centre

Contents

Foreword v
Prof. Elzbieta Sawa-Czajka

Foreword vii
Eunice Rodriguez

Preface ix
Prof. Nirupama Prakash, Prof. Betty McLellan and Prof. Barbara Wejnert

Introduction xvii
Arun P. Kulshreshtha

Part I: Gender Perspectives in Science and Technology: Issues and Challenges

1. Areas of Weaknesses, Obstacles and Challenges to Economic Empowerment of Women in Developing Countries: Botswana's Current Status 3
Gloria Tiny Seitei

2. Career Barriers for Women Engineers in Europe: Transferability for Developing Countries 15
Felizitas Sagebiel

3. Gender in Indian Science and Technology: Recognition of Excellence 30
Pranav N. Desai

4. Women, Science and Technology: An Overview from Indonesia 44
Aisyiyah Jenie

5. The Position of Women in Contemporary Iran 61
Akram Ghadimi

6. Taking Action to Empower Women Scientists and Technologists through Women in Science and Technology Network (WISTNET) in Malawi 72
Meya Patricia Kalindekafe & Symon Osman Mandala

7. The Status of Women in Engineering in South Africa and the Efforts of SAWomEng to Facilitate the Absorption and Retention of Females in the Engineering Sector 81
Naadiya Moosajee

8. The Current Status of Women Scientists in Sri Lanka 89
P.R.M.P. Dilrukshi & Seetha I. Wickremasinghe

9. Gender, Science and Technology in Tanzania: A Situation Analysis 101
Rose Rita Kingamkono

10. Gender Challenges in Science and Technology in Democracies and Transitional Democracies 134
Barbara Wejnert

11. Country Report on Mauritius 144
Shamima Abdul Carrim Delbar

Part II: Women Development: Technology as a Tool for Empowering Women

12. ICT as a Tool for Empowering Women: A Case Study of Rural Rajasthan 157
Nirupama Prakash

13. Reaching the Un-reached: Technology Model for Women's Empowerment and Sustainable Development 169
Vinita Sharma

14. Women Empowerment to Women Capabilities: Bridging the Gap 189
Kavita Mehra

15. Science and Women Empowerment in Iran 202
Shokouh Navabinejad

16. Content Analysis of Web logs: Perspectives of Young Iranian Females 208
Zohreh Nosrat Kharazmi

17. **Economic Empowerment of Malaysian Women through Science and Technology: Identifying Issues, Providing Solutions** 221
Poh Sze Choo

18. **Empowerment of Women through Science and Technology Interventions in Myanmar** 231
Khin Htike Htike Lwin

19. **Empowerment of Nepalese Women through Science and Technology Intervention in Nepal** 246
Kayo Devi Yami

20. **Various Mechanisms Used by "Vidatha" Technology Transfer Programme for Economic Empowerment of Rural Women through S&T Interventions** 258
Jayasamara Gunarathne

21. **Production of Low-Cost Complementary Foods Using Locally Grown Foods in Tanzania: An Economic Empowerment of Women** 283
Rose Rita Kingamkono

22. **Empowering Women through Science and Technology Interventions: The Case for Zambia** 292
Atridah Mulonga

Part I
Gender Perspectives in Science and Technology: Issues and Challenges

Chapter 1

Areas of Weaknesses, Obstacles and Challenges to Economic Empowerment of Women in Developing Countries: Botswana's Current Status

Gloria Tiny Seitei

Senior Communication Officer, Botswana Technology Centre, P/Bag 0082, Gaborone, Botswana

E-mail: gloria@botec.bw, stinytots@yahoo.com

ABSTRACT

The paper focuses on the areas of weaknesses, obstacles and challenges to economic empowerment of women in Botswana. In particular, it addresses the issues faced by women such as land and infrastructure acquisition, economic empowerment and under representation in science and technology; education in science and technology; the girl child; health; discrimination and violence from male counterparts. It discusses the measures taken by the Botswana Government to empower women by developing different policies and programmes to address critical issues affecting women. Issues of women in the employment sector, rural areas and the education are also highlighted. It reveals the disparities in the fields of science and technology where few women participate and deals with recommendations on actions to empower women through science and technology interventions.

Keywords: *Women, Gender, Science, Technology, Equality, Empower.*

1.0 About Botswana

Botswana is a landlocked country located in Southern Africa. The country is bordered to the south and east by South Africa, to the northeast by Zimbabwe, to the north and west by Namibia and touches Zambia just west of the Victoria Falls.

Figure 1.1: Map of Southern Africa

The tableland of the Kalahari Desert covers about ¾ of the country. National parks cover 17 per cent of the country, with 38 per cent of the country dedicated to wildlife areas. Botswana has the misfortune of being a drought –prone country and this is a phenomenon that the nation has accepted as its misfortune. The rapid economic development of the country is attributable to direct revenues from the diamond mining and the cattle industry. Given this scenario, diversification could be an answer, but it is imperative to apply science and technology to come up with solutions for food self sufficiency in drought situations.

2.0 Introduction

Empowerment of women through science and technology interventions is one of the fundamental issues extensively discussed in the 21st Century and Botswana is not an exception.

While gender is a variable that cuts across all sectors of the economy, gender implications of science and technology must be assessed with every sector of development. Issues of under representation of women in areas such as science and technology, economic, unemployment, education, health, discrimination and violence from male counterparts poses a serious challenge to women.

The National Development Plan 7 (NDP 7) indicates that some of the major problems for the rural people more particularly women is lack of relevant technologies, land, labour, credit capital and so on. Although the NDP 7 report covers the year's between1991 to 1999, these problems still persists in most of the African countries including Botswana.

The Botswana Government also aligns itself with the empowerment of women through the different policies that have been developed to address critical issues that affect women. For instance, the Botswana National Gender Programme gives consideration to issues that affect women like poverty, economic empowerment, decision making, education and training, health, the girl child and gender based violence including human rights. In addition the Government of Botswana established the Women's Affairs Department in 1981 geared towards addressing women's issues in the country.

2.1 The Science and Technology Policy for Botswana

In July 1998, a policy was approved by parliament, has strategies aimed at "Empowering Women in all sectors of the economy in order to ensure that the thrust of S&T impacts positively within the context of economic development and to facilitate women's participation and involvement in science and technology areas".

2.2 International Conventions

Botswana has signed International Conventions in the areas of gender. In addition the Government aligns itself with the Southern African Development Community (SADC) Declaration on Gender and Development and the Commonwealth Plan of Action that stipulates a minimum target of 30 per cent of women in positions of political decision making (NDP 9). Botswana has not yet achieved this target since the current representation of women in political positions stands at 18.18 per cent. It is worth noting that even though the SADC region has made strides in the development of policies geared towards women empowerment, practically there is very little reflection of the gender focus in budget allocations for advocacy, education or awareness initiatives (Mmegi, 2008).

The country is confidently certain that while considerable progress has been made in recognizing women's involvement in issues that could improve their livelihoods, there are still areas of weaknesses and challenges to the economic empowerment of women. For instance, most government department policies are gender blind and issues of gender are mentioned in isolation (United Nations Development Programme, 2003).

3.0 Formal Employment Sector

The United Nations University –Institute of Advanced Studies (UNU-IAS) Report describes the impact of science and technology on women as a tool that could offer opportunities to empower women through learning and employment. It also could help in the reduction of the market demand of unskilled labour particularly in developing countries.

The relationship between science and technology, sustainable development and gender has not been explored in depth in Botswana and participation rates for women in research and development is still relatively low as indicated in the chart 1.1.

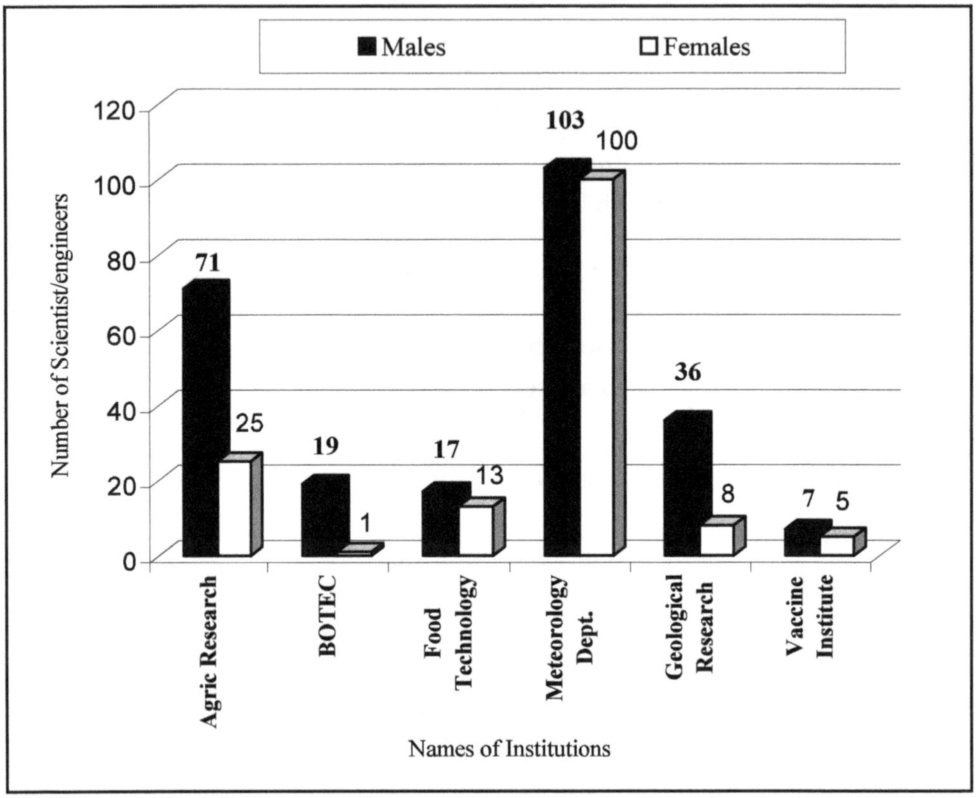

Chart 1.1: Number of Scientists/Engineers by Gender in R&D Institutions

The chart 1.1 shows the dominance of males at 62 per cent as compared to female (38 per cent) scientists and technologists in the research and development institutions. The same can be said for the decision making positions and societal institutions that use various technologies. The missing link in this chain is under representation of the women scientists, engineers who can best reflect the needs and aspirations of women.

There are many factors that contribute to the under-representation of women in the scientific careers.

One factor is combining family responsibilities with professional careers where household responsibilities are not equitably shared and as such women perform several simultaneous roles as wives, child bearers and farmers and the drop out of girls at school due to pregnancy. The Botswana Policy on women in Development cites this as an obstacle that places heavy demands on women's time and energy and incapacitates women to meet their potential, putting them at a disadvantage in many science and technology related jobs that are dynamic and competitive in nature (Taeb, M *et al.*, 2005).

The graph below shows predominance of men as opposed to women in all the fields of the employment sector such as industry, central and local government as well as parastatals and the private sector as shown in the Table 1.1.

The Table 1.1 indicates that Central Government (including Local Government) was the major employer of females accounting for 43.7 per cent compared to 39.3 per cent in the Private and Parastatal Sectors as at March 2008.

At Industry the highest percentage of females were found in Health and social work constituting 64.4 per cent of total employees. Other Female dominated industries were financial intermediaries, hotel and restaurants and education accounting for 62.0 per cent, 59.1 per cent and 53.7 per cent respectively. The least female employees were recorded in mining and quarrying at 12.4 per cent as shown in the table (Central Statistics Office, 2008).

The Chart 1.2 indicates that there are more males than females in all levels of qualification particularly for PhD and masters professionals in research areas. The qualification of females with PhD, Masters and Bachelors degrees recorded for 2004/05 was much fewer than their male counterparts.

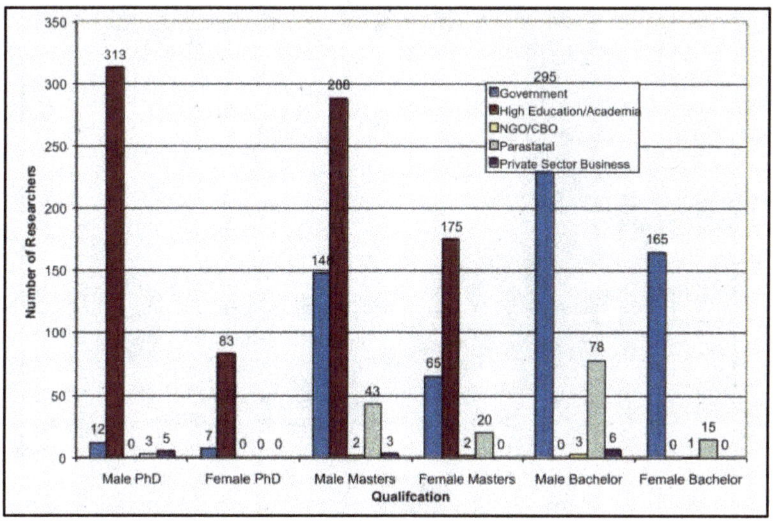

Chart 1.2: Research by Gender and Qualification

The disparity was more pronounced for the PhD's where male researchers with PhD's stands at 79 per cent as opposed to 21 per cent enjoyed by female PhD holders. In the case of masters and Bachelor Degrees, males were double their female counterparts. The academia has the largest number of researchers with PhD followed by the government. Parastatals that do most of the research and development for the government have fewer PhD's and none were recorded for the NGO's. The majority of researchers with bachelor's degree were found in the Botswana Government National Innovation Capabilities (NIC, 2007)

Table 1.1: Estimated Number of Paid Employees by Industry, Sex and Citizenship in March 2008

Industry	*Citizens*			*Non-Citizens*			*All Employees*			*Sep-07*	*Mar-07*	*Annual Percentage Change*
	Male	*Female*	*Total*	*Male*	*Female*	*Total*	*Male*	*Female*	*Total*			
Agriculture	3.171	1,939	5,110	360	94	454	3,531	2,033	5,563	5,505	5,479	1.5
Mining and Quarrying	9,802	1,398	11,199	427	47	474	10,229	1,444	11,673	11.495	11,736	-9.5
Manufacturing	16,937	16,591	33,528	1,881	479	2,360	18,819	17,070	35,888	35.555	35,204	1.9
Water and Electricity	2,200	589	2.789	11	-	11	2,211	589	2,800	2,844	2,835	-1.2
Construction	16,748	2,138	18.886	2,542	382	2,924	19,290	2,520	21,810	21,760	21,789	0.1
Wholesale and Retail	22,278	19,803	42,081	2,333	381	2,713	24,611	20,184	44,794	44,393	44,126	1.5
Hotels and Restaurants	5,829	8,691	14,520	287	158	445	6,113	8,849	14,965	14,996	14,988	-0.2
Transport and Communication	8,361	3,189	11,550	655	89	744	9,016	3,278	12,294	12.566	12,579	-2.3
Financial Intermediaries	2,471	4,485	6,955	334	91	424	2,804	4,575	7,379	7,288	6.813	S.3
Real Estate	11,502	5,475	16.977	868	129	997	12,371	5,603	17,974	18,216	17.942	0.2
Education	2,856	3,838	6,694	1.478	1,194	2,672	4,334	5,032	9,366	9,128	8,789	6.6
Health and Social Work	587	1,216	1,803	219	241	460	806	1,457	2,264	2,234	2,159	4.9
Other Community Serv.	1,186	1,963	3,149	112	118	230	1,298	2,081	3,379	3,105	3,150	7.3
Private and Parastatal	103,926	71,314	175.240	11.507	3.400	14.908	115.433	74,714	190.148	189,085	187,588	1.4
Private	95,631	65,894	161,525	11.040	3,302	14,343	106,671	69,196	175.868	174.436	172,882	1.7
Parastatal	8,295	5,420	13,715	467	98	565	8,762	5,518	14,280	14,649	14,706	-2.9
Central Government	59,666	45,051	104,717	1,095	437	1,532	60,761	45,488	106^49	90,186	88,521	20.0
Local Government	13,977	12,633	26,610	239	165	404	14,216	12,798	27.014	26,774	25.869	4.4
All Sectors	177,569	128,998	306,567	12,841	4.002	16.844	190,410	133,000	323,411	306.045	301,978	7.1

3.1 Actions to be Taken

- An effective human resource programme in science engineering and technology should be regarded as a vital consideration to address the imbalance of under representation of women in science and technology fields
- Developing countries should make efforts to empower women by introducing gender mainstreaming in order to achieve gender equality. The gender mainstreaming system will put women and men, boys and girls at the heart of policy making. This will go a long way in making the gender equality issues visible in the mainstreaming of the society. (Gazette 2008)

4.0 Women in Rural Areas

The Science and Technology Policy for Botswana indicate that the country has experienced rapid socio-economic changes that are affecting the socio-cultural fabric of the society. These changes are said to be dynamic and impact on the traditional activities of women and men. However, despite the level of economic development, the traditional gender division of labour still dominates and women continue to be marginalized within the society.

Among the challenges and obstacles faced by rural women is long working hours under difficult circumstances, such as heat and rain, and without proper technologies to ease work load. Particularly in the rural areas where employment (in the informal sector) is possible, the jobs offer particularly low wages and poor working conditions. World leaders are currently anxious about poverty in developing countries. Rural areas are the mostly affected by poverty, the solution to this phenomenon lies in empowerment of women in science and technology.

Research has shown that in most African countries including Botswana, women perform 60-80 per cent of the agricultural activities and over 50 per cent of the entrepreneurs and employees in the small scale and micro business sector are women. NDP 7 also indicates that 35 per cent of all farming households are female headed and this is largely concentrated among poor farmers. Major obstacles faced by women in this sector are as follows:

- Inadequate technology that could contribute to increased productivity and self sufficiency in the production of food.
- Lack of land which results in limited access to credit and economic resources such as draught relief and labour.
- Lack of proper infrastructure such as roads, electricity and water is a major obstacle that contributes to minimal output in the production of food. As a matter of fact increasing women's access to electricity, roads and water would lead to greater agricultural output and self sufficiency in the availability of food a catastrophe that affects women and children.

In order to empower women in all sectors of the economy and to ensure that the thrust of science and technology impact positively on women within the context of economic development and to facilitate their participation and involvement in science and technical areas, the following actions should be considered.

4.1 Actions to be taken

- Develop technologies that are gender neutral and address the need of women. Such a development will go a long way in empowering women, improving production levels, save time and curtail environmental damage.
- Promote women's participation in the development of science and technology and provide advice and information to women's Non Governmental Organizations on the acquisition and use of technologies.
- Provide unlimited access of technologies to rural and urban women that will ease their work load so as to enable them to spend time on educational and income generating activities.
- Encourage the participation of women in policy formulation, resource allocation and implementation structures at all levels.

5.0 Education Sector

It has to be noted that the Government of Botswana has invested on resources for science and technology education at all levels of the education system. In spite of this, the impact still does not match with the investment since the country is still dependant on foreign human resources for science and technology expertise.

In Botswana, unlike a number of other African countries, both boys and girls have equal opportunities to attend school. Factors which may inhibit school attendance are not gender based, but are rather attributable to social ills such as teenage pregnancies; attitudes towards science and technology subjects by the female students; cultural attitudes and possibly inability of parents to pay tuition for their children. The government has however, initiated plans to address such problems particularly for children from low income families. These include payment of full tuition, uniform and the introduction of feeding schemes by the government for the under privileged and children who's parents are within the low income bracket

Research done when drafting the National Policy on Women in Development indicates that "In Botswana, female enrollment is generally greater than that of males at primary and junior secondary schools. However at senior secondary level, female enrollment begins to decrease and by Form 5 males predominate". The reasons for this could be attributed to better performance by boys at examinations which form the basis of selection into science courses at higher level of education. This also explains why females fail to reach the top scientific positions.

There is also a noticeable gender difference in the selection of academic fields. For example, women's choice is more inclined towards the Arts and Social Sciences than the Natural Sciences and Engineering as indicated in the Chart 1.3 compiled by the University of Botswana.

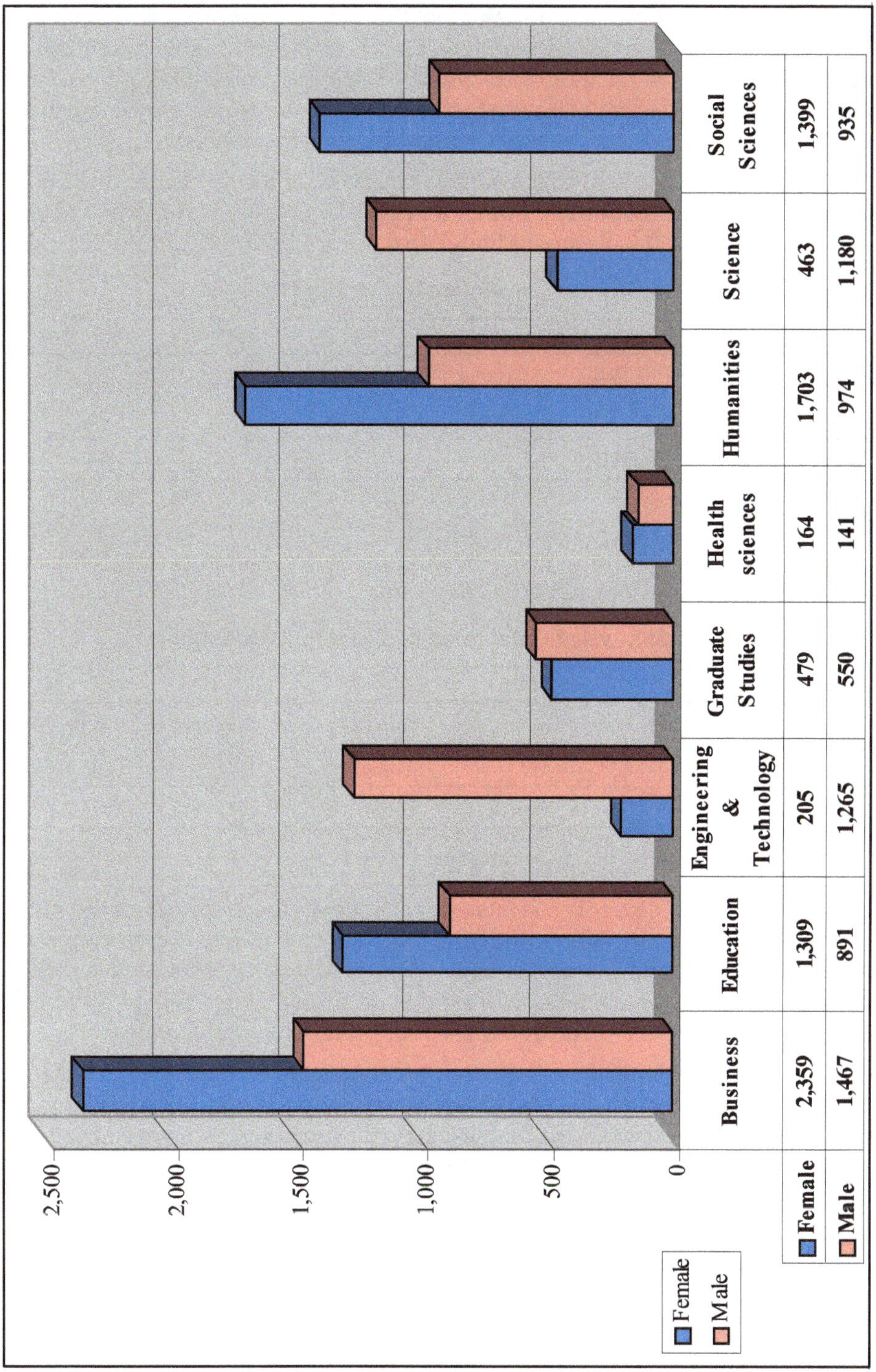

	Business	Education	Engineering & Technology	Graduate Studies	Health sciences	Humanities	Science	Social Sciences
Female	2,359	1,309	205	479	164	1,703	463	1,399
Male	1,467	891	1,265	550	141	974	1,180	935

Chart 1.3: Student Population by Faculty/School and Gender

Despite the fact that there is no discrimination in admission procedures in educational institutions, Chart 1.3 indicates that the participation of women in Science, Engineering and Technology is still very low. The chart shows a higher enrollment of males in Engineering and Technology as well as in Science courses at 86 per cent and 72 per cent respectively. Females have a higher enrollment rate in the Social sciences, Business and Humanities at 60 per cent, 62 and 64 per cent respectively.

The Table 1.2 below qualifies the above statement on the disparities in student enrollment in the faculty of Science by gender at the University of Botswana during the Botswana National Development Plan 7 from years 1990/1991 to 1995/1996.

Table 1.2: Student Enrollment by Gender (1990-1996)

Gender	*1990/91*	*1991/92*	*1992/93*	*1993/94*	*1994/95*	*1995/1996*
Male	315	296	308	422	548	688
Female	82	96	91	146	207	257
Total	397	392	399	568	755	945

The Table 1.2 shows growth in the number of students in the faculty of Science during NDP 7. The number of female enrolling for Science subjects stood at 25 per cent while that of males is the highest at 75 per cent. (NDP 7).

Table 1.3: Student Enrollment by Gender (2004-2008)

Gender	*2003/04*	*2004/05*	*2005/06*	*2006/07*	*2007/08*
Male	1108	1053	1117	1155	1180
Female	361	343	377	427	463
Total	1469	1396	1494	1582	1643

Table 1.3 shows growth in the enrollment of students by gender within the Science faculty at the University of Botswana. However the disparity is still noted between male and females who enroll for the Science subjects. On average, the enrollment of males still stands high at around 74 per cent while female enrollment is low at 26 per cent. The two tables therefore shows an enrollment growth of 38 per cent between the years 1990/1996 and 2003/2008 for both males and females.

An explanation to lack of participation by females in Science and Technology fields was also drawn from the stakeholder meeting that was organized by the Botswana Technology Centre on 31 May 2001. At that meeting it was concluded that girls are marginalized by their male counterparts when they want to enroll for Science subjects because of the cultural stereotypes that regard women as the weaker sex.

5.1 Actions to be Taken

- Introduce science and technology education and training at all levels of schooling and incorporate special efforts to facilitate the participation of women and girls in Science and Technical fields.

- The education programmes or the curriculum should enable students to learn about the poverty alleviation and employment skills for utilization in the future.
- The government relies heavily on imported skilled manpower. This should be regarded only as a short term solution. The proposal would therefore be to intensify education programmes that support continued participation of girls in the field of Science and Technology secondary and tertiary levels to counteract negative factors like teenage pregnancy and the traditional gender stereotyped attitudes
- Introduce interactive science centres for education and fun in order to de-stigmatise Science and to create public understanding and awareness of Science and Technology.
- Improve access for and retention of girls and women in education and vocational training in such fields as Science, mathematics, Engineering, Environmental sciences and Technology.
- Formulate policies enabling higher educational institutions to increase female enrollment in Maths, Natural Sciences, Engineering and Technology. As part of this policy, higher educational institutions should also develop strategies to improve graduation rates in these areas.
- Induce the use of successful female role models who have taken up careers in Science and Technology

Conclusions

It is crucial that the gender issues particularly on Science and Technology be appraised within the various sectors of development. It has been proved in this report that presently in Botswana and possibly in most developing countries, females have limited access to Science and Technology career than males. This results in women being few or absent in Science and Technology areas because of stereotypes, stigma, social aspects, expectations, the biological make up and fear that science is abstract.

It is crucial that the cultural practices and stereotypes that have taken place for a longer period of time resulting in prejudices and discrimination should be done away with. Botswana was amongst the countries that participated in the World Women Conference in Beijing, China in 1995 thus committing it to promoting the goals of equality, development and peace for women. Although there are noticeable developments, a lot still need to be done to promote gender equality particularly in the areas of Science and Technology.

A people centred approach geared towards men and women will go a long way in facilitating the development and transfer of technology that is culturally, socially, politically and economically viable and appropriate to the needs of the citizens in every sphere of life.

Acknowledgements

I would like to extent my gratitude to the following. The library team at the Women's Affairs Department on providing information on the different policies available within their department. I also extend my appreciation to the human resource units at the Botswana Technology Centre, Department of Meteorology, Geological Surveys, Botswana Vaccine Institute, the National Food Technology Research Service, Agricultural Research and the Faculty of Science at the University of Botswana for providing the available statistics on employment and enrollment by gender on Science, Engineering and Technology. Last but not least I thank the Acting Director of Technology Information Department at the Botswana Technology Centre for providing guidance, without his support, this presentation could have not been able to take the shape it has. I hope and trust that Botswana will ensure that there is gender equality in all spheres of live in the future and will also realize the importance of empowering women through Science and Technology interventions.

References

Alexander, E.M., 1996. National Seminar on S&T Policy Proceedings – "Gender issues and Concerns in Science and Technology", Gaborone, Botswana, pp.164 – 165.

Botswana Gazette, 2008. "Government Kick-Starts Gender Equality Programme", Weekly Newspaper, 19-25 November, pp. 29-30.

Botswana Government., 1999.National Gender Programme Report, Women's Affairs Department, Gaborone, pp. 4-5

Botswana Government, 1995. Policy on Women in Development, Women's Affairs Department, Gaborone, pp. 8-11

Botswana Government, 1991. National Development Plan 7, Ministry of Finance and Development Planning, Gaborone, pp. 352-353

Botswana Government, 2003. National Development Plan 9, Ministry of Finance and Development Planning, Gaborone, pp. 348-359

Botswana Government, 1998. Science and Technology Policy, Ministry of finance and Development Planning, Gaborone, pp. 32–33

Central Statistics Office. 2008. Botswana Government. Stats Brief. Gaborone, pp. 5.

Factbook. 2007/2008. Department of Institutional Planning – University of Botswana, pp. 16, 17, 20. http://www.ub/ip/factbooks.cfm?pid=749

Mmegi., 2008., "Financing Media for Gender Equality", Weekly Newspaper 21 November 2008, pp. 46.

NIC., 2007. Botswana Government. Summary of NIC Survey Results "The Botswana National Innovation Capabilities Database for Science & Technology, Ministry of Communications, Science and Technology, Gaborone, pp. 4.

Taeb, M *et al.*, 2005. UNU-IAS Report on Revisiting Women's Participation in Science and Technology: Emerging Challenges and Agenda Reform, pp. 6, 9, 18. http://www.ias.unu.edu/binaries2/womenST_final.pdf

UNDP, 2003. Report. Gender Mainstreaming Project with Four Pilot Ministries, Gaborone, Botswana, pp. 10,11, 133, 134.

Chapter 2

Career Barriers for Women Engineers in Europe: Transferability for Developing Countries

Felizitas Sagebiel
Associate Professor, Faculty of Educational and Social Sciences,
University of Wuppertal, Germany
E-mail: sagebiel@uni-wuppertal.de

ABSTRACT

Two European Commission projects out of 5th and 6th Framework Programme Womeng "Creating Cultures of Success for Women engineering" (website: www.womeng.net) (2002-2005) and from PROMETEA. "Empowering Women Engineers in Industrial and Academic Research" (website: www.prometa.info) (2005-2007), are subject of the paper. Hypotheses about gender aspect of professional culture and career for women engineers and a short description of methodology of Womeng project form the basis for presenting results on: Dominance of men's working culture (masculine culture, minority situation, coping strategies), women engineers' career definition and barriers, women engineers' priorities (part-time working culture, work-life-balance with children, women managers' leadership style), and men's networks. In PROMETEA the question was in what ways and to what amount social structures and cultures of engineering research organisations are traditionally male oriented dominated by hegemonic masculinity and men's networks. The qualitative methodological design included interviewing women engineers and organising focus group discussions with men and women engineers separately. Results focus first on gender aspects of organisational cultures, analysed by gendered division of labour, gender stereotypes and gender awareness. Second, networks will be described from the perception of women and men engineers in research. It will be argued that most prominent career barriers are not relying on women's characteristics, but on organisational cultures and men's networks in their work environments. Having been work package leader in these issues in both projects aim is to give an overview about state of the art, hypotheses, methodology and results of the special work package, dealing with effects of gendered organisational cultures on careers in engineering research. At the end transferability of results to developing countries will be tried.

Keywords: *Women engineering, Gender aspects, Career, Research.*

1.0 State of the Art

Several theoretical fields can be useful for interpretation of Women results: gender studies, gender in academia, critical men studies, studies on feminist technology, organisational studies and studies on gendered career and profession (Sagebiel, 2006; 2006a; 2005c).

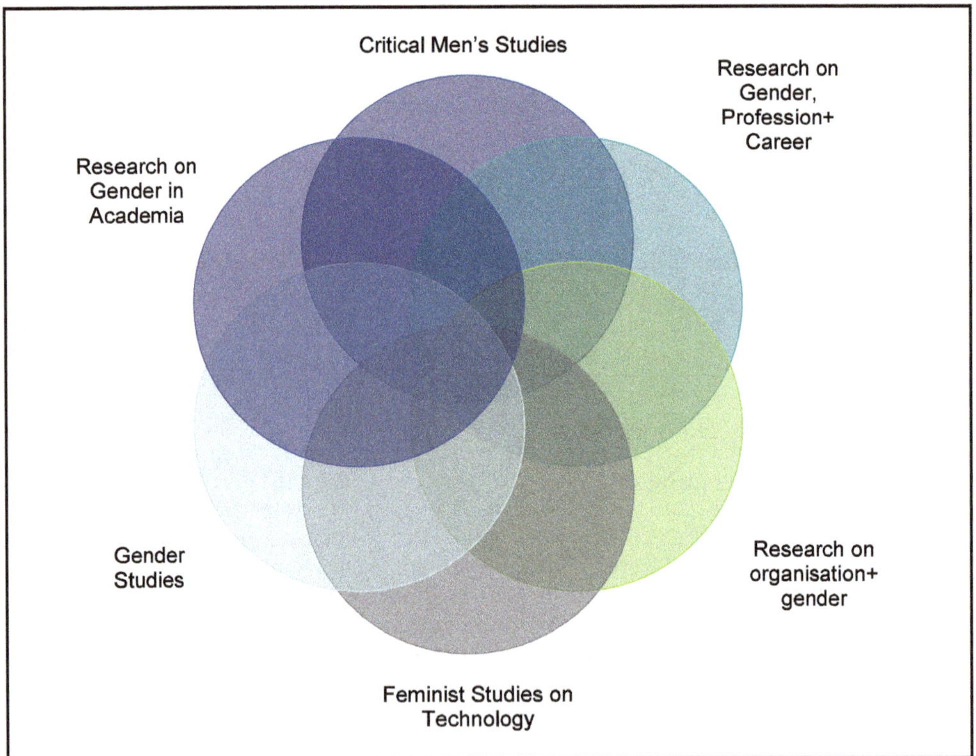

Figure 2.1: Theoretical and Research Fields

2.0 Hypotheses[1] and Methodology for Women

Living in a two gendered society one with gender segregation in professional sphere and the other in a gender segregated educational situation women engineers as professionals are embedded in a male domain. Working in a minority situation

1 For more details and structures of hypotheses see Sagebiel (2006b).

can lead to marginalisation and exclusion or integration through adaptation of male working culture, which is characterised by strong competition and less cooperation. Informal cultural pattern include men's networks on one hand and men's jokes and stories on the other hand.

Women engineers, having taken over a gendered division of labour tend to give so called work-life-balance priority, which lead to preference of part-time work, both strategies influencing their career aspiration. As managers on basis of gendered working styles they develop a new leadership style, which differs from traditional male leadership style. So social constructed gender differences can be weakened or strengthened through women engineers behavioural patterns.

Expert-interview, focus group discussions and website analyses were chosen as investigation tools (Pourrat, 2005; Sagebiel, 2005a), besides official statistics from companies and Statistical Office of the European Communities. For sampling in each country two companies had to be chosen, one from production sector and one from energy sector.

3.0 Results on Professional Experiences and Barriers for Career of Women Engineers (WOMENG)

Starting from description of men's dominating working culture women engineer's career concept, professional priorities and men's networks as career barriers will be analysed (Sagebiel 2005; 2006).

3.1 Dominance of Men's Working Culture

Masculine Culture and Minority Situation of Women

Working in a male domain for some of the women has been a big challenge. Women agree that it is not easy to assert oneself and gain acceptance if you are a woman. In view of an Austrian female engineer, *"It is a men's world and women have to accept that it is a men's world"*. Women engineers have to prove that they are competent, work hard, know what they are doing and want to be taken seriously (In the words of a French woman engineer *"I had to fight to convince the company that as a woman I could make it"*). The Austrian women see men having a conservative worldview and so they cannot handle assertive competent women. Many men are not aware of their old-fashioned gender stereotyping. They want to behave and think modern, but their traditional upbringing manifests in a macho behaviour.

The women at a Slovak company talked openly about facing completely different problems in comparison to their male colleagues. The explanation for the above mentioned discussion lies in traditional positions of women in family and society. Also the general approach to the problems or failures caused by women or men is different: if a woman 'spoils something' the reaction is usually as follows: *"well, she is just a woman, what else could we expect from her (woman has a 'hen brain')?"* If there is a problem caused by a man the reaction is: *"well, it could happen to anybody."*

Even though some women engineers told about dominant traditional masculinity at the same time most of them told that they don't have any problems with this culture. This result can be explained differently:

- First, the awareness of traditional masculinity is confirmed, but this working culture is really not a problem,
- Second, the expected tough image does not allow being weak and having problems,
- Third, it is the product of balancing controversial perceptions of self and inner emotional processes (Sagebiel, 2005b; Sagebiel and Dahmen, 2006).

Coping Strategies between Demonstrating Equality and Difference

How do women engineers cope with their minority situation in the engineering working culture? The respondents comprising of Austrian women clearly stated that they still feel as being a minority and being dominated by a traditional masculinity culture, in which one characteristic part is the culture of sexist jokes. They think that only with great self-assurance they can cope with this working culture.

So one of the coping strategies is to demonstrate great self-assurance, but most of the time men engineers show much more self assurance. In the words, of one Austrian women engineer *"Men have a gigantic self-assurance. Even if they know nothing they open their mouths. Women open their mouths only if they really know something. That is where we do not match at all"*

The opinions of women managers about gender differences in profession, especially in leadership style vary from country to country. While respondents comprising of engineers in France and UK explain differences in leadership behaviour with personality traits, where as in Germany, Austria, Greece, Finland and Slovakia tell about different styles by gender. When asked about what is important for management, most of the respondents comprising of managers see team work, working atmosphere and fewer overtimes as central, in contrast to male colleagues in same position. Overtimes versus part-time are central elements of masculine versus feminine organisational cultures.

A special focus of coping seems to lie on dealing with sexist or stupid jokes. Regarding to the statements in the Austrian focus group it is best to find a joking way to react on sexist jokes, which does not turn male colleagues into enemies. Women need a lot of sensitiveness and delicacy but on the other hand men do not need sensitiveness while communicating with women. Nevertheless one younger manager in Germany uses to behave the same way like *"...I send the jokes back, but better!"*, she always tried to be better and straighter than her male colleagues, a women engineer from Finland said: *"I'm working like a man!"* and some French women adopt that behaviour too.

Respondents comprising of Austrian women think there is no need to change into a man either. *"You should remain a woman. That is okay. But, if you are oversensitive that is of course a problem."*

3.2 Women Engineers, Career Definition and Barriers

Many women do not follow the traditional career definition as vertical progression, they have found their own values describing career. All over the samples and the nations job satisfaction depends on work contents, including widening the individual horizon by doing a horizontal career path, and work atmosphere. These characteristics have a higher importance than vertical career steps. *"Career efforts to pay a certain price"*, was a main statement in both German and Austrian discussion groups. Feel the same: *"A career you can only have if you completely sacrifice yourself. Those who are prepared to have no private life, no hobbies, nothing but the company from dawn till dusk, those are the ones who really make a career...As soon as somebody confesses to a family life, their work engagement will not be unbounded any longer."*

In Austria and Greece respondents comprising of women engineers talked about not planning the next career step, *"it just happened"*, on the contrary to U.K. and Finland where career is noticed as important to all of the women engineers. In Slovakia when women were interviewed they showed no personal interest in leading positions, for which they would have to be supported and urged. But they think women who are interested in leading positions have a good chance in the company. They consider themselves as being competent for their positions. A traditional gendered division of labour in Slovakian society (Hudec and Urbancikova, 2005), which mostly gives the burden of homework to women builds the background of this argumentation.

French women engineers, who were interviewed came up with the explanation: *"Why so few women top managers? Well do they really want it? It implicates so many sacrifices, at the personal level, and familiar, why the top management compel people to spend their lives working? I think that women are not ready to pay such a price!"* The price would be an unbalanced working and private life, which has a high priority for the women.

And the explanation of a German female engineer supplements: *"It's not desirable for women because the female concept of life and relationships does not fit together with managing positions so women can not identify themselves with the way of leadership. It would be much effort necessary to create a female work environment."*

Respondents comprising of women engineer managers told about societal and company internal reasons for less career progression of women; these are gender stereotypes, traditional ideas about gendered division of labour, and traditional role concepts on one hand and overtimes, men's network and restricted entrance of women on the other hand.

3.3 Women Engineers' Priorities

Women engineers mostly see themselves as having different priorities in comparison to men in combining work and private life and those who had climbed a management position seemed mostly to have a different leadership concept.

Objection to Male Culture of Overtimes

Overtimes versus part-time are central elements of masculine versus feminine organisational cultures. Most of women managers, practicing their own leadership concept with focus on a good working atmosphere, are conscious of the importance of reducing overtimes to urgent and necessary cases.

Concerning the working hours and doing overtimes there is a big cultural difference in Europe. In Germany, France, Slovakia, and Austria and in the U.K. it is quite normal to do overtimes and show all time availability. Overtimes do not seem to be a big problem for women engineers in Austria, being in the position when requested. But, here in one company the culture of working overtime and late night leads to strange practice. The women engineers in the focus group told that some men go on hour-long coffee breaks and in the evening they work and send e-mails late at night with the intention to impress the others or some cheat by setting the computer clock ahead of time. The majority of superiors are unable to assess the work of the employees.

The fear of overtimes in a management position is very high in Germany preventing women engineers to fully develop professionally.

Women's Culture of Part-Time Working

Part-time working does not have the same standing all over Europe. In some countries like Germany, Austria and France part-time working is quite normal and accepted and offered at all companies especially for women having children. But in other countries it is offered rarely like in the U.K. and in Slovakia. Having no possibility of working part-time causes quite different decisions concerning family and career and can lead women engineers to drop out.

Being an employee working part-time is a way of work-life-balance for women especially when having children. But like the example at a German company shows, reducing working time can go along with some negative effects. Through reducing the working time the visibility shrinks and also the intensity of social contacts, which are necessary for networking and information spread. A French women engineer who quit talked about her experiences after reducing her weekly working time: *"At the beginning, yes, some people tried to transform me into a secretary, and I said: no, I can't do that. It was the job of his assistant!"*, so women are, if they take a part-time job, sometimes considered as an assistant.

Work-Life-Balance with Children

Respondents comprising of women managers all over Europe show and live the possibility of combining family and career. But, they are unknown role models as women engineers who are not working at this career level being in a minority situation normally do not have contact to women managers.

Especially younger women engineers still fear a career break when having children and looking for a work-life-balance (voices of German and Greek women engineers). Women managers, having experiences in combining family and being a manager, know the difficulties but found their personal way of combat problems.

The respondents comprising of German women engineers in focus groups talked about a company' phenomena which might be transferable to other companies as well: before getting pregnant most of the women were aspirants for reaching the next career level, the promotion of superior's stopped through pregnancy and taking parental leave, and career promotion did not start again after returning back to the work.

Eventually the superiors still had the old-fashioned opinion that a combination of career and family is not possible or the superiors assumed that the women don't want to be further promoted, because they will concentrate on their child.

Women in Germany, Austria, Slovakia, France and Finland do not necessarily drop out from job when having children. In France, there is a great difference between good practice company practicing maternity leave as a kind of diversity and finding solutions to support and the women and the companies seeing pregnancy and maternity just as a problem. Company internal child care facilities are not offered everywhere; this should be improved to take pressure especially from younger women engineers who fear career breaks or problems as parents.

Some women in the U.K. do not want to have the focus on childcare and work-life-balance, because not all women want to have children. Superiors in all nations are still assuming that women will have children. But the reality shows that some women do not have and do not want to have children. Those women always feel the prejudice, the opinion of all women having children. And this prejudice causes the loss of support for women or more support for men.

Women Managers' Leadership Style

When asked about what is important for management, most of the asked managers see team work, working atmosphere and fewer overtimes as central, in contrast to male colleagues in same position. But, the opinions of women managers about gender differences in profession, especially in leadership style vary from country to country. Hence, women-engineers in France and UK explain differences in leadership behaviour with personality traits, many of those asked in Germany, Austria, Greece, Finland and Slovakia tell about different styles by gender.

3.4 Men's Networks as Career Barrier for Women Engineers

The importance and influence of networks in career making processes was confirmed in all project partner countries. Most of women engineers are aware of the male networks and their importance for career except in Finland, where men's networks seem not being so influential for career. Often women engineers told that they have only restricted access to the men's networks, which to change seems to be very difficult. Only one respondent French female engineer denied the fact that one could stay outside of the networks because of his/her sex: *'you enter networks only because of your competencies'*.

The appearance of men's network is quite different. It starts with the informal meetings while drinking a coffee or smoking a cigarette, continuing to so called informal meeting after work. In Finland men have their own 'sauna meetings', where they can communicate also about the professions.

Women networks exist but the understanding of women and men networks is split into two components, one concerning private and the other concerning professional sphere. Professional sphere is related to having a formal women's network, *e.g.* raised in an engineering association or especially for entrepreneurships. In Austria women networks are just starting. And in France already women networks exist, but they are associations, they seem to induce more a reflection on subjects like work-life-balance, dual career than to help women to find a job, so it's more an advising function than some concrete support.

The informal network like men have it has nowhere been the point of discussion. Especially for German women engineers who participated in focus group discussion it was clear that networking is necessary for career progress, but how to do networking was not clear for all. A recommendation was taking a mentor in a higher position, which can help with the men's network. *'You need someone to push and you need someone to pull'* (a German woman manager told). Because of the few women engineers and fewer women engineer managers, they are not as powerful as male networks.

Women engineers need time to build up their own networks including men and women at high positions and they need cooperative men, not fitting in dominant masculinity. Besides women engineers discrimination also new masculinities experience the same, if they do not work in a company with a living diversity culture.

4.0 Hypotheses and Methodology for PROMETEA

In PROMETEA, the question was in what ways and to what amount social structures and cultures of engineering research organisations are traditionally male oriented dominated by hegemonic masculinity and men's networks. Hypotheses from the former European project Womeng (European Commission, 2006) about gendered organisational cultures and networks in professional sphere of engineering (Sagebiel 2005b; 2006a; 2006c) were transferred to the field of engineering research. Further hypotheses were taken from literature related to research work about investigations of gendered dimensions of science and career (Sagebiel, 2006b).[2]

Women engineers in research, working in a traditional masculine organisational culture, are excluded from men's networks and cannot rely on a powerful women's network to compensate the exclusion from men's networks.

In PROMETEA gender aspect of networks was analysed by definition, functions, processes, activities and evaluation and women's integration or exclusion in different engineering research organisations (industrial, academic and governmental). Characterisation and function of men's networks were presented in the perception of women and men engineers in research. They were described and analysed in a comparative perspective of different cases (industrial, academic and governmental) as well as different partner countries. Norms and values of gendered organisational cultures were characterized as they influence careers of women engineers.

Qualitative data on gendered organisational cultures and networks in academic and industrial settings of engineering and technological research have been gathered through semi-structured interviews with women and focus groups with men and women separately. Focus group discussion was the privileged method to get known the more tacit elements and was taken from women project (Godfroy-Genin and Sagebiel, 2007; Sagebiel, 2005a).

---

2 For more hypotheses used in this work package see PROMETEA final report (www.prometea.info) and Sagebiel (2008).

5.0 Gendered Networks and Networking in Engineering Research Organisations

Even though women were asked individually and in focus groups men talked much more about networks, so they produced as much material.

5.1 Understanding of Networks and Networking

Understanding of networks differs depending on awareness and perception of networks and networking. Some interviewees have only a vague impression and can't really tell their definitions, others refuse to talk about informal networks and only refer to professional formal ones. Overall men could describe different kinds of networks with different functions.

Male respondents in Sweden talked about three kinds of networks: networks with former fellow students, networks with people one has worked together with (in different projects) and external networks with people with the same (technical) area of interest. Respondents of other countries also referred to this kind of differentiation. Swedish academic informants also talked about the researcher networks which are built up in conferences and, increasingly, by international research projects. However, there was a general opinion among women and men engineers that becoming a network member is gender independent.

5.2 Manifest and Latent Functions of Networks and Networking

In summary manifest function of networking are getting to know people generally or relevant people, fresh information exchange, sharing material, acting collectively and getting job perspectives. External networks have manifest functions as initiating projects, especially international ones. Publication is another manifest function, but at the same time the information channels for publication are most of all not very transparent. In higher education networks are seen as necessary for career at a certain point of qualification with the aim to selling oneself and enforcing one's visibility. Internal networks build the unofficial organ gram of the organisation which allows information exchange. In some countries networks are depending on hierarchical structure and decision of superior about the amount of restriction of information. From men focus group participants some take advantage from tacit knowledge they get by networking, others think it is not always career relevant.

A male Austrian engineer at university explained in detail how he networks and how it supports his work: *"If you take a look, you need networking to initiate projects. You need external partners, you have international partners and without networks this doesn't work. And yes, if you do your PhD you can also use the network of a colleague. If you know someone who works at the institute who works with specific equipment some things work out much easier. Getting any analyses, getting any chemicals, borrowing equipment, all that works will much easier if you walk over and say hello"*.

Even if the Finish saying goes that "good work always sells itself", it's just not enough, because actually the knowledge and information "between the book covers" is silent. One of the male engineers, interviewed in Finland, said that the network helps you to *"'peel the cream of the top' before anyone else gets the chance to do it. That is not necessarily always fair but it helps you to know that if you had not done it, somebody else would have".*

5.3 Process of Networking: From Initiation to Patronisation

Networks work with sympathy, personality, good performance, same professional and private interests, fraternisation after getting drunk together and luck. Several interviewees told that networking would start with common education in school or later during study time or even later during in service, for example, training in companies. Commitment often would go back to these roots.

In a German discussion round, at a governmental institute, men explained that first personal links would be already made during study time and can sometimes last a long period in someone's professional live. *"Networking already starts during study time and continues on conferences or while preparing together a project proposal."* Additionally, this institute was a good example for prolonged commitment within a specific community. When the institute was founded by some professors (mostly coming from a certain university nearby), they supported their close research assistants and offered jobs at the institute to them. The connections to this university would be still strong now-a-days, and the new personnel were likely to be recruited out of the university department of the current institute's leader, like one interviewed women engineer confirmed: *"You always can see where the institute leader has his connections; there is a commitment which obliges him to some kind of patronisation of his roots!"*.

5.4 Evaluation and Devaluation of Network and Networking

Male respondents are aware of the importance that networks have for daily work life and particularly career advancement. Difficulties for becoming a member of networks were not reported. Some respondents stated that network's impact might be overvalued, because at least professional performance counts for climbing up the career ladder and not connections. But some male discussants see this to be totally different, for them network contacts are highly influencing organisational cultures and decision makers; for career progress knowing the right people in decisive positions is essentially.

Women engineers perceived that they were not part of powerful networks, even though many of them felt or knew the importance of networks. Some of them under evaluated the importance of networks and networking for their careers. They spent too less time for informal opportunities to contact these networks outside the narrow defined work environment. Men in contrast, even though they told that becoming a member was seen as something not being reflected. They realised that meeting and knowing the right people at the right time and place, opens many opportunities.

British men engineers in a focus group discussion in business sector (B_FGM_O) saw networking as vital: *"I think it's the hardest part of the scientists job"* and his colleague adds: *"I think you ignore it at your peril, I would say!"* A woman engineer in Germany consciously stated: *"Networks are the nuts and bolts–privately and professionally. Without (private) networks I could not have worked in that way."*

On the contrary male researchers in France and Slovakia show a distant approach towards informal networks, *"I prefer the term: "scientific community", that is very different from the influential networks, which are very insane, chaotic and perverse"* (male engineer France). And one Austrian engineer from university personally feels *that "scientists are more defined by their performance than, like a strategic lawyer. The later can tell you anything and you can tell a lot, but here we have numbers, data and facts. Here you can't sugar-coat anything".*

5.5 Rituals and Activities of Networking

Many rituals and activities especially of informal networking, take place without women. Networks are knitted and tightened after work. In most of the partner countries women engineers feel more responsible for family duties and they will not join these activities as much. Besides for joining special sports (extreme) and sauna meetings the sex difference counts and, even if women are not excluded in these activities and rituals, men don't seem to look for alternatives.

In a German men focus group discussion in governmental research it was clear for all that important contacts to other people are mostly made after work, while going out for dinner or some drinks. *"My first superior told me, if you want to be successful in raising funds or getting project partners, you have to get once drunken with your future cooperation partner*!" More experienced participants even agreed on this story. *"Inhibitions get weak and you can talk more frank and free about cooperation structures".*

5.6 Women's Exclusion from or Integration in Men's Networks

For answering hypotheses about women's lower integration in men's networks, two questions were analysed on basis of interviews with women engineers and of men and women focus groups.

Women were asked about access to men's network (extent, areas). Men and women were asked about barriers against networking (lack of time, restricted access) and gender differences.

Men asked about exclusion of women generally answer that they see no access barriers for women in joining networks. On the other hand in some situations having women included could damage reputation of men's network, men and women realised. One interviewed German female engineer, working in industry, had a very clear opinion about how male networks work, but this was outstanding: *"Networks function due to the fact that they exchange information and informally a "non-aggression pact" exist and may be supports itself maybe. The men are afraid to damage their reputation with a woman as a network partner. And men probably see there more common characteristics among themselves. Are women defined as the 'others'?"*. Another male German engineer in industry argued *"Fostering networks with women is more inhibited"*, because it easily gets a kind of sexual touch.

Family responsibilities are often used arguments or prejudices of women themselves and moreover of men against women joining men's networks. But, some women didn't realise the usefulness of networking for their careers and this might have influenced their time argument. Less time resources (except in Sweden, where only one woman saw problems in timing) and less mobility are factors, besides special activities and rituals by men, in which women engineers are not interested in.

One of the Chilean respondents said: *"Networks are hard to get in since it often implies going for drinks someplace you don't particularly care about..."* While asking Swedish informants in academic research about networking with men, one informant said in a humorous manner: *"No, you can't network with them" [everybody laughed]*. And she continued explaining what it takes to network with men: *"Then you have to play football with them and I have no desire to do that"*.

One other female discussant from Sweden said: *"I was never invited to the sauna evenings. One of the other women with a management position higher up in the hierarchy was invited but I think she turned down the invitation because it did not feel right for her"*.

A special phenomenon found in PROMETEA results was that partly women did not perceive being excluded from men's networks, and this seems to be combined with non perception of gender differences and discrimination.

The hypothesis focussed too on women's networks, but about these PROMETEA results showed hardly any engagement of women engineers in research. Moreover these *women's networks* overall were not estimated as helpful.[3]

6.0 Summary of Results of WOMENG and PROMETEA and Conclusions

Engineering is still defined as a men's field based on a masculine image of engineering in society.

Women engineers are conscious of working in a male domain and talked about discriminating situations connected with dominant masculinity culture, but those in jobs in the company, did not define those as problems. Only women engineers who left their prior jobs complained about this working culture. Most Respondents comprising of women engineers were satisfied with their job, but sceptical about combination of career progression and preferred work-life-balance, viewing this as a decision between two not compatible choices. Their restricted entrance to men's networks is seen as an important career barrier by women engineers, especially in management positions. Diversity programmes seem to help to weaken the power of traditional men's networks.

The results of PROMETEA show interdependence between existing gender equality in society and organisational culture in engineering research. Gendered differences in networking exist everywhere. But, discrimination and excluding processes are more open (manifest) in very traditional societies and more tacit (latent) in more equal societies. Future research should further develop concepts of gendered organisational cultures and networks. Also, men's networks as instrument for social regulation should be more focussed, and, relationship between men's networks' power and gender awareness should be more investigated.

---

3 Only some Swedish female focus group discussants referred to networks with other women, when asked about networks.

As globalisation tends to transfer organisational cultures and men's networks to companies all over the world, women's career barriers will be globally transferred too. How special cultures of countries (developed and developing) will impact global companies' cultures, can be hypothesized by looking at different results from different European countries. Overall countries with a more traditional gendered society and less gender awareness tend to have more manifestation of gender discrimination whereas gender bias in more equal societies is more latent.

7.0 Acknowledgements

The author has coordinated this work package in both European projects Women and PROMETEA.

I would like to thank our project partners from PROMETEA in the participating countries for the great input they provided. This article is based on their results and analysis. In Austria: Birgit Hofstätter, Anita Thaler and Christine Waechter, in Chile: Dámaris Fernández Donoso, Claudia Paz and Sonia Yánez, in Finland: Liisa Husu and Paula Koskinen, in France: André Béraud, Anne-Sophie Godfroy-Genin, Cloé Pinault, Yvonne Pourrat and Jean Soubrier, in Germany: Jennifer Dahmen and Gaby Hoeborn, in Greece: Nikitas Nikitakos and Maria Lambrou, in Lithuania: Ala Kovieriene, Diana Saparniene and Virginija Sidlauskiene, in Russia: Elena Myasina and Vera Uvarova, in Serbia: Jovan Dudukovic, Jelena Jovanovic and Sanja Vranes, in Slovakia: Oto Hudec and Natasa Urbancikova, in Spain: Carme Alemany, in Sweden: Helen Peterson and Minna Salminen-Karlson, in UK: Wendy Faulkner, Lisa Lee and James Stewart, at Schlumberger: Pierre Bismuth.

References

Collinson, David L., Hearn, Jeff, 2000.'Men' at 'work': multiple masculinities/multiple workplaces" in Understanding Masculinities. Social relations and cultural arenas, ed. Máirtín Mac and Ghaill, (Open University Press: Buckingham, Philadelphia), pp. 61-76.

European Commission, 2006. *EU RESEARCH ON SOCIAL SCIENCES AND HUMANITIES. Creating Cultures of Success for Women Engineers, Womeng*, Final report.

Godfroy-Genin, Anne-Sophie, 2008. (Methodological issues in the PROMETEA project: classifications, fuzzy borders of technology, gender in *Prometea International Conference Proceedings Women in Engineering and Technology Research*«, ed. Anne-Sophie Godroy-Genin in press) (LIT Verlag).

Godfroy-Genin, Anne-Sophie, Sagebiel, Felizitas, 2007. Möglichkeiten und Schwierigkeiten internationaler und multimethodologischer Forschung über Gender in den Ingenieurwissenschaften« *Zeitschrift für Frauenstudien und Geschlechterforschung*.pp.27-43.

Hudec, Oto/ Urbancikova, Natasa, 2005.Human Resource Development in the Area of Engineering in Slovakia« in *Creating Cultures of Success for Women Engineers*, ed. Anita Thaler and Christine Waechter *'Conference Proceedings of Final International Workshop of the WomEng Project*, Graz, pp. 117-123.

Sagebiel, Felizitas, 2005a. Using a Mixed International Comparable Methodological Approach in a European Commission Project on Gender and Engineering in *Methodological Aspects in Cross-National Research,* ed. Hoffmeyer-Zlotnik, Jürgen H.P. / Harkness, Janet, *ZUMA-Nachrichten spezial* 10, pp. 47-64.

Sagebiel, Felizitas, 2005b. Organisationskultur und Geschlecht in den Ingenieurwissenschaften Europas« in *IFF Info, Zeitschrift des Interdisziplinaeren Frauenforschungs-Zentrum, Vol.* 22, Nr. 30, pp. 48-60.

Sagebiel, Felizitas, 2005c. Gendered organisational cultures in engineering. Theoretical reflections on WomEng results and future research perspectives« in *Creating Cultures of Success for Women Engineers,* ed. Anita Thaler and Christine Waechter '*Conference Proceedings of Final International Workshop of the WomEng Project,* Graz, pp. 143-156.

Sagebiel, Felizitas, 2006a. Ingenieurinnen in Europa. Karrieren und Barrieren« in *Journal Netzwerk Frauenforschung NRW* Nr. 20, pp. 42-50.

Sagebiel, Felizitas, 2006b. Men's network and other barriers for women's career in engineering. Results from EU-project Womeng and perspectives of EU-project PROMETEA« in *Proceedings of Australian Technology Network Women's Executive Development Conference "Change in Climate? Prospects of gender equity in universities"*, Adelaide, (peer reviewed CD).

Sagebiel, Felizitas, 2006c. Men's network and other barriers for women's career in engineering. Results from European Project WomEng« in *Breaking the glass ceiling,* ed. Gul Gelkan, Proceedings from Second International Conference on Women's Studies, Eastern Mediterranean University (EMU), Center for Women's Studies, Famagusta, Turkish Republic of Northern Cyprus, (CD and in press).

Sagebiel, Felizitas, 2007.Gendered organisational engineering cultures in Europe« in *Gender in Engineering – Problems and Possibilities* ed. Welpe, I. / Reschka, B. / Larkin, J. /,pp. 149-173. (Peter Lang Verlag).

Sagebiel, Felizitas, 2008.Gendered Organisational Cultures, Men's Networks and Women Engineers' Career Chances: Results from two European Research Projects. In: Engendering Leadership Conference, Perth, Online conference proceedings.

Sagebiel, Felizitas, 2008. Motivation of young people for studying SET. The gender perspective. In: Proceedings of Sefi Conference, Aalborg, (CD) (Co-authors: Dahmen, Jennifer, Davidsson, Bodil, Godfroy-Genin, Anne-Sophie, Rommes, Els, Thaler, Anita, Urbancíková, Natasa).

Sagebiel, Felizitas, 2008.Gender, Career and Family Life: Women Engineers in Europe. The International Journal of Interdisciplinary Social Science 2 (5).

Sagebiel, Felizitas, 2008. Gendered organisational cultures and networks in engineering research« in *Prometea International Conference Proceedings "Women in Engineering and Technology Research"*, ed. Anne-Sophie Godroy-Genin (in press) (LIT Verlag).

Sagebiel, Felizitas, Dahmen, Jennifer, 2005. Maennlichkeiten in der europaeischen Ingenieurkultur. Barrieren oder Aufforderung zur Anpassung für Frauen« in *Soziale Technik*, Vol. 1, No. 15, pp.19-21.

Sagebiel, Felizitas, Dahmen, Jennifer, 2006.Masculinities in organisational cultures in engineering education in Europe. Results of European project WomEng« in *European Journal of Engineering Education, Vol.* 31, No. 1, pp. 5-14.

Sagebiel, Felizitas, 2006. ENGAGE – Distance Learning of Gender Awareness in Engineering. A European Idea. In: Proceedings of the International Conference on Digital Media and Learning, Bangkok, Thailand, 13-14 March, 2006. (Together with Anne-Sophie Genin, Natasa Urbancikova and Jennifer Dahmen)

Chapter 3

Gender in Indian Science and Technology: Recognition of Excellence

Pranav N. Desai
Professor, Centre for Studies in Science Policy,
School of Social Sciences, Jawaharlal Nehru University
New Delhi – 110 067, India
E-mail: dpranav@hotmail.com

ABSTRACT

Gender asymmetry, tilted in favour of men or women, has persisted in all human civilizations. However, gender asymmetry, as a major social structure, has not been paid due attention in sociology of science till recently The present paper is an attempt to explore whether this phenomenon paucity could be explained in terms of recent technological change and whether is it simply correlated with the level of economic development. In recent times, the information revolution has led to increasing introduction of large-scale automation in production. This has raised expectations of narrowing down the gender gap in various science and technology enterprises, though in many countries, the gender gap has not disappeared. Women represent only 28 per cent of the total global Research and Development (R&D) workforce and only a few countries have achieved gender parity. There is also a generally held notion that women in the industrially advanced countries enjoy a higher share in Science and Technology (S&T) enterprise compared to those in developing or scientifically less advanced countries. Hence, an attempt has been made to analyse the share of women scientist in different regions and economic groupings of the world. In context of global structure of human resource in R&D, the paper has provided a special focus on women in Indian science and the issues such as the share of women in Indian science and comparison of the same with the gender profile of the western science. Recognition received by women scientists is another yardstick of the social status of women. Thus, an analysis of correlation between the proportion of awards received by women scientists and their share in total R&D workforce may reveal gender discrimination patterns. It is contended that a greater participation of women could be encouraged and ensured by affirmative policy interventions.

Keywords: *Gender, Globalization, India, Innovation, R&D, S&T, Literacy, Scientific productivity.*

1.0 Introduction

Gender asymmetry, tilted in favour of men or women, has persisted in all human civilizations. However, gender asymmetry, as a major social structure, has not been paid due attention in sociology of science[1] till recently. The present paper is an attempt to explore whether this phenomenon could be explained in terms of technological change or is it simply correlated with the level of economic development or that it is more complex. The sickle-based agriculture mainly operated by women had given rise to matrilinear societies as the women played a dominant role in production and hence in family rearing unlike her predecessor of the pastoral or hunting stage economy. The technological change brought about by the invention of plough for cultivating land reassigned the predominant role of men in economic and social spheres. Even during the industrial revolution of the eighteenth and nineteenth centuries, most of the technologies discriminated women in production system due to the very nature of these technologies except in some definite instances where female labour began to emerge. In recent times, the information revolution has led to increasing introduction of large-scale automation in production. This has raised expectations of narrowing down the gender gap in various science and technology enterprises. Though in many countries the gender gap has narrowed down, it has not completely disappeared. There is also a generally held notion that women in the industrially advanced countries enjoy a higher share in Science and Technology (S&T) enterprise compared to those in developing countries. However, women represent only 28 per cent of the total global Research and Development (R&D) workforce and only a few countries have achieved gender parity (UNESCO, 2008). Could it be due to the cultural influence of the dominant patriarchal value system?

The present study assumes significance as in recent times, with the awareness of the predicament of environmental degradation, the patriarchal value system supposedly responsible for the unmindful exploitation of Nature. Hence, an attempt has been made to analyse the share of women scientist in different regions and economic groupings of the world. In context of global structure of human resource in R&D, the paper has provided a special focus on women in Indian science and the issues such as the share of women in Indian science and comparison of the same with the gender profile of the western science. Recognition received by women scientists is another yardstick of the social status of women. Thus, an analysis of

---

1 "Sociology of science deals with the social conditions and effects of science, and with the social structures and processes of scientific activity". See for details Joseph Ben-David and Teresa A. Sullivan (1975). "Sociology of Science". *Annual Review of Sociology* 1: 203–222.

correlation between the proportion of awards received by women scientists and their share in total R&D workforce may reveal discrimination meted out if any. It is contended that a greater participation of women could be encouraged and ensured by affirmative policy interventions.

Status of women is usually measured in terms of their level of education and participation in economic activity. Another way of measuring their status would *inter alia* depend on their control over knowledge and opportunities. Recognition received by women scientists is another yardstick of the social status of women and so here is an attempt to analyse various awards received by the women scientists in recognition of their work. However, the proportion of their participation would depend on various factors such as their total literacy rate, masculinity ratios etc. As far as trend of sex ratios (Visaria, 1992) in India is concerned, the proportion of women has constantly declined in the twentieth century, from 972 females per 1000 males in 1901 to 927 in 1991and only slightly improved to 933 in the census year 2001. India is one of the few countries in the world with higher masculinity ratio as the gender gap in survival is greatest during the first five years of life. The biological imbalance is eliminated by the age of 1 year through the higher mortality of male child world over. In contrast to this phenomenon, in India a higher mortality rate of female child (43.0 per cent) has been reported (1984) compared to male child (39.5 per cent) during 0-4 years (World Bank, 1991). Every sixth infant death is due to gender discrimination as there is a greater neglect of girl-children during the first five years of life in terms of nutrition and medical care. The fact that the incidence of diseases due to malnutrition is higher among women and particularly girl-children is documented in various National Sample Surveys (1971, 1981, and 1991). Moreover, with the access to modern reproductive technologies in urban areas, there is an increasing trend of abortion of female foetuses in India as well.

The proportion of female literacy remains quite low, even though it has increased from 32 per cent according to 1991 census (Visaria *et al.*, 1992) to 54 per cent in the census year 2001 (Census of India, 2001). Notwithstanding the regional imbalances and rural-urban disparities that exist in the country, the average literacy rate for rural areas is only 46 per cent against 73 per cent in urban areas.

2.0 Status of Women in Science

It will be pertinent here to discuss some of the feminist perspectives on science while analyzing status of women in science. A Marxist-feminist scholars viewed science and technology as integral not only to the systems of domination of the late capitalism and its new forms of imperialism, but also to one of patriarchal domination. Even the old Marxist social theorists who treated science as an integral part of the capitalist production system are alleged to have assumed gender division as a natural and not a social structure. According to some scholars "science is neither race less, sexless nor classless; she is a white man, bourgeois, and infected too" (Rose, 1994).

There is another trend amongst Marxist thinking gaining currency is ecofeminism. The ecofeminists (Shiva, 1989; Mies, 1986) view scientific rationality as intrinsically the rationality of violence, desire and domination. They treat science and scientific rationality both as the core constituent and the driving motors of capitalist accumulation through exploitation of women/nature and colonies (as in the past) and particularly the biogenetic technologies are treated as mechanisms which convert nature into a genetic resource to be engineered, patented and owned for corporate profits. Thus from the ecofeminist standpoint, S&T and rationality by definition are violent and exploitative against timeless categories such as women and nature. According to this viewpoint, no change or alternative model of science is possible and according to conventional Marxist viewpoint no change against the oppression of women or women scientists is possible without the overthrow of the capitalist system or class society where patriarchal value system is an essential and integral part of the same. Hence, change has no place in either trend of thinking. With this trend of thinking some radical feminists have disagreements, which revolve around the question of whether the major determinants of gender inequalities are to be found in direct power relations between men and women or somewhere else. The Marxists and liberal feminist theories lacked this focus on power or culturally conditioned relationship.

A study (Ruizo, 1987) on the relationship between the participation of women in academic production and the level of technological and economic development has suggested that the reason behind higher participation of women in science in semi-industrialized countries like their counterparts in the developed countries is that scientific production is still a part of cultural production and is not yet fully locked into technological and economic growth. This view, thereby, implies that most of the technologies generated so far have discriminated against women. However, this situation is expected to change with the changing nature of technologies and with the increasing diffusion of automation and computerized operations wherein the share of women's employment is known to have increased over the years.

Some of the Marxist-feminist scholars have also held that the family values, domesticity and unpaid labour of women have benefited the class interest. As against these views an empirical study carried out in India, where there is a greater complexity of social stratification, has revealed otherwise. By contrast, where a more class-based system of social stratification operates, change has been far more dramatic leading to high level of education and freedom to choose careers and opportunities (Seymour, 1995). Thus, there are intertwined factors at play and it is increasingly being realized by scholars of all persuasions that gender asymmetry requires to be tackled from a holistic approach.

An attempt has also been made here to explore whether the phenomenon that women are underrepresented as compared to men in science is universal or is it a function of income levels, and whether other factors, such as, cultural and political factors, play a more dominant role. An international comparison of share of women (Tables 3.1 and 3.2) in science in countries at different national income levels[1] reveals the following. In the low-income economies data are not available for many countries but it is estimated that the share of women ranged from about 5 to 40 per cent. In the lower-middle-income countries this figure ranged from 20 to as high as 63 per cent. In this category of countries the erstwhile communist bloc countries could exhibit a proportion exceeding 50 per cent of the total S&T personnel. Some of the Asian countries like Philippines and Thailand are estimated to have a figure of around 50 per cent. For the upper-middle income economies, the Figure ranged from 3 to 61 per cent. The category of high-income countries included mostly the western European countries.

Table 3.1: Global Share of Women Scientists

Economic Grouping of Countries	*Share of Women Scientists*
Low income	5-40
Lower Middle Income	20-63
Upper Middle Income	3-61
High Income	30-40

Source: UNESCO Institute for Statistics "Statistics on Research and Development".

Contrary to expectations and despite the high levels of female literacy and economic well being the proportion of women scientist ranged only from 30 to 40 per cent and exceptions were a few like Sweden and Denmark where the figure touched the 50 per cent mark. Even a country like the USA and India, which are at opposite extremes of economic development share in common the same proportion of women S&T personnel. Thus the data available do not suggest a definite relationship between national income levels and share of women in science.

The data on female literacy suggest that though the high-income countries have achieved almost 100 per cent female literacy, there is some degree of discrimination against women at the level of career opportunities in S&T organizations. This discrimination becomes quite visible in high-income countries and especially in the western European countries, Japan, the USA, as they have achieved almost 100 per cent female literacy. Especially in the case of USA, the share of women in S&T personnel is significantly lower than their share (41 per cent) in the total labour force (WDR, 1994).

Another region where the share of women scientists is significantly lower than the level of female literacy is the west Asian countries. It is possible that political and cultural factors have greater roles to play than simply the economic factors alone in under representation of women in S&T.

2.1 Share of Women Scientists in India

The aggregate share of women in S&T personnel[2] in India was only about 15 per cent. Women in engineering accounted for 2 per cent and in science branches it was

---

2 The economies are classified on the basis of gross national product (GNP) per capital, as the criterion used by the World Bank and are listed in ascending order of income. Low-income economies are those with a GNP per Capital of 675 but less than 8,356 are classified as middle-income and a further classification has been made for upper-middle-income economies with a GNP per Capital of more than 2,695 and lower-middle-income with less than that. The high-income-economies are those with a GNP per capita of 8,356 or more. See World Development Report, *1994* (Oxford University Press, 1994), ix and 162-163.

Table 3.2: Region Wise Total R&D Personnel by Sex (FTE)

Country	Year	Total	Female	Per cent Female	Country	Year	Total	Female	Per cent Female
Algeria	2005	7,331	2,449	33.4 per cent	Cambodia	2002	494	112	22.6 per cent
Burkina Faso	2005	942	196	20.8 per cent	India	2000	3,18,443	44,264	13.9 per cent
Ethiopia	2005	5,112	443	15.5 per cent	Kuwait	2005	489	198	40.4 per cent
Gambia	2005	84	4	4.8 per cent	Malaysia	2004	17,887	6,467	36.2 per cent
Lesotho	2004	51	18	50.0 per cent	Pakistan	2005	53,159	3,102	5.8 per cent
Madagascar	2005	1,477	458	31.0 per cent	Republic of Korea	1998	1,28,669	17,069	13.3 per cent
Seychelles	2005	180	30	34.9 per cent	Sri Lanka	2004	5,475	1,820	33.2 per cent
South Africa	2005	28,798	10,949	38.0 per cent	Turkey	2005	49,251	15,347	31.2 per cent
Tunisia	2005	16,289	7,416	45.5 per cent	Kazakhstan	2000	12,829	6,466	50.4 per cent
Macao, China	2005	409	94	22.9 per cent					
Austria	2004	42,891	10,112	23.6 per cent					
Belgium	2005	53,517	17,057	31.9 per cent					
Denmark	2005	43,499	16,011	36.8 per cent	Bulgaria	2005	15,853	8,278	52.2 per cent
Germany	2005	4,80,758	1,23,428	25.7 per cent	Croatia	2004	11,162	5,533	49.6 per cent
Cyprus	2005	1,157	442	38.2 per cent	Czech Republic	2006	47,729	15,056	31.5 per cent
Greece	2005	33,958	12,215	36.0 per cent	Estonia	2006	4,740	2,013	42.5 per cent
Greenland	2004	48	11	21.9 per cent	Latvia	2006	6,520	3,202	49.1 per cent
Iceland	2005	3,226	1,265	39.2 per cent	Romania	2005	33,222	15,890	47.8 per cent

Contd...

Table 3.2–Contd...

Country	*Year*	*Total*	*Female*	*Per cent Female*	*Country*	*Year*	*Total*	*Female*	*Per cent Female*
Ireland	2005	16,690	4,819	28.9 per cent	Slovakia	2005	14,404	6,372	44.2 per cent
Italy	2005	1,75,248	61,048	34.8 per cent	Slovenia	2005	8,994	3,345	37.2 per cent
Lithuania	2005	11,002	5,925	53.9 per cent	The former Yugoslav Rep. of Macedonia	2005	1,434	750	52.3 per cent
Luxembourg	2005	4,392	900	20.5 per cent					
Malta	2006	752	150	19.9 per cent					
Portugal	2005	25,728	11,661	45.3 per cent	Country	Year			
Spain	2005	1,74,773	66,020	37.8 per cent	Cambodia	2002			
Sweden	2005	77,704	22,574	29.1 per cent	India	2000			

Note: Data from USA and Japan are not available in the table hhttp://stats.uis.unesco.org/ (Accessed December 1, 2008).

about 19 per cent. Nevertheless, if disciplinewise analysis is carried out the picture that emerges is not too depressing. This figure varies substantially from one specialization to other. There is greater representation of women in branches like life sciences and medical science (21 and 16 per cent) followed by branches like mathematical and physical sciences. The proportion of women in engineering sciences is negligible and this proportion is much lower compared to many developed countries. However, if comparisons are made in specific situations such as women faculties in physical sciences in higher education sector (during the year 1990) then the proportionate share of women faculties (10 per cent) as well as the absolute number are more favourable and surpass those of developed countries like the USA, the UK and Germany (less than 5 per cent) (Barinaga, 1994). It is also important to note that not only that the share of women scientists is much lower compared to men but women also report a higher unemployment rate than do men. In the USA (NSF,1986), the unemployment rate for women scientists (2.7 per cent) was more than twice that for men (1.3 per cent) and particularly in physical sciences the unemployment rate reported for women was more than 3 per cent implying certain level of discrimination.

More surprises are in store if the social structure of human resources in R&D is analyzed in terms of percentage share of women (DST, 1993). The share of women scientists engaged in R&D is only 9 per cent which much lower than their share of 15 per cent of the total S&T personnel. If the branchwise distribution of women scientists in R&D is analyzed, following picture emerges. Though the share of women scientists in engineering branches is negligibly small, their share in engineering R&D is 6 per cent. This due to the fact that the higher proportions of women engineers have preferred R&D career and their share in this branch is 33 per cent out of the total women S&T personnel in R&D. This figure is as high as in natural science R&D (36 per cent). Their share in medical R&D (27 per cent) is also higher than their share in the total number of medical personnel. Another fact which requires to be noted about women in Indian R&D is that there is a higher concentration of women in the middle and lower salary bracket than at higher salary brackets particularly as compared to men (DST, 1994). This analysis is helpful in probing whether there is any underrepresentation or discrimination of women in rewarding their excellence in scientific work.

2.2 Scientific Productivity and Innovation

Gender parity deteriorates while ascending the ladder from elementary education to higher education, employment and other opportunities in scientific careers of women (Tables 3.3 and 3.4). This process referred as 'leaky pipes' or 'crystal glass ceiling' by many ends up losing productive human resource.

This also results into underrepresentation of women scientists at various levels of decision making spheres that constitute boards, governing councils, funding, fellowships and award granting agencies or S&T policy making agencies. Hence, some invisible barriers obstruct the full potential of innovation capabilities of women scientists. As the awareness regarding this phenomenon is of recent origin, academic circles have not paid due attention to gendered regimes of advantages and disadvantages of the innovation process. What is commonly accessible is the literature on scientific input like gender parity index in enrollment at various educational and employment level and there is a dearth of sex disaggregated data on scientific productivity like published, papers, patents design,copyright etc.

Table 3.3: Percentage Enrollment of Women at Various Levels in Different Faculties (2000-01)

	Percentage of Women		
	Graduation	*Post-graduation*	*Ph.D.*
Arts	44.2	44.7	38.6
Science	39	42.5	37.2
Engineering/Technology	21.8	15.8	16.5
Medicine	45.5	34.4	29.3
Agriculture	17.2	18.8	14.6
Veterinary Science	21.6	18.6	14.5

Table 3.4: Share of Women Scientists and Technicians in Government R&D Institutions and Some Universities

Organisation	*Grade*	*Total Scientists*	*Per cent Women*	*Total Technicians*	*Per cent Women*
CSIR	–	5030	13	3250	14
DBT	–	179	31.8	277	23.1
ICMR	–	615	27.3	1939	20.1
DAE	–	436	16.5	–	–
DOD	–	127	8.7	–	–
ICAR	Assistant Professor	12750	10.4	–	–
ICAR	Associate Professor	4750	6.2	–	–
ICAR	Professor	2500	3.5	–	–
IISc	Academic	316	6.6	34	14.7
IISc	Scientific	113	9.7		
Hyderabad University (Science Departments)		101	15.8 (Only one professor)		
Jawaharlal Nehru University (Science Schools)		82	16	20	0

Source: INSA Report, 2004. Science Career for Indian Women: An examination of Indian women's access to and retention in scientific careers. Indian National Science Academy, New Delhi, p. 9.)

As study (Lemoine, 1992) conducted on the productivity pattern in terms of published papers and patents of the Council of Scientific and Industrial Research

(CSIR) scientists in India reveals that "there is no divergence in the productivity of male and female scientists who have written 10 articles or less". On the other hand, the different values of the parameter for the whole population of authors and for those men and women researchers with 11 or more papers provide some evidence of the existence of inequalities in the distribution of scientific productivity. The proportion of less productive scientists seems to be larger among male than among female scientists". Thus, these results are contrary to the generally held opinion or the studies conducted in the western countries (Naldi *et al.*, 2005) that women scientists are less productive than their male colleagues. Similarly in the patenting activity, the proportion of less productive innovators is bigger among males than females. "Such results can be expressed in more general terms according to which, for every 100 male inventors with 1 patent there will be 26 with 2; 12 with 3 and so on. By contrast, for every 100 CSIR female inventors with 1 patent there will be 38 with 2; 21 with 3 and so on".

According to an estimate, women own only 1 per cent of the total intellectual property in the world and that it is feared the unfolding of the globalization will further complicate the situation. The TRIPS agreement does not recognise that women are not in themselves a unified category and that the divisions along a "North–South axis are important, as are divisions along lines of ethnicity and class" (Barwa and Rai, 2002).

Women do not have the same level of access to scientific resources and educational opportunities in the South as do the women in the North. Moreover, women are the custodian of traditional knowledge (TK), but TK is still a hotly debated issue for its consideration under intellectual property protection (Desai, 2007). Further, there are also studies showing obstacles at innovation level or for commercializing the inventions. This is due to the fact that women do not enjoy equal access to markets as markets are also embedded in social relations. Moreover, the assessment and recognition of excellence is considered to be important element in many opportunities in the careers of scientists. However, it is debatable whether intrinsic or extrinsic incentives are more important and for all branches of S&T. Thus, any gender bias in the assessment and reward system might hinder productivity and opportunities in the careers of women scientists.

The ability of scientific institutions to respond to socio-economic changes might differ from country to country or from one branch of S&T to another. The following discussion provides an analysis of the situation that prevailed in the reward system in Indian science.

3.0 Recognition of Excellence

Measurement of scientific excellence at various levels from publication of papers, granting of patents and promotion in careers is an essential process in any scientific institution. Objectivity and universalism are considered to be core of scientific ethos or Mertonian values of CUDOS (Communalism, Universality, Disinterestedness, Originality and Skepticism). It is this perception of science that is being challenged by many. Prevalence of Gender bias if any in scientific institutions can only endorse this perception. It is now increasingly being accepted that "Merit and talent are not sufficient conditions to become a successful scientist. Resources, time, social networks, encouragement – unevenly distributed between the sexes – are necessary prerequisites" (European commission, 2004). In the preceding context the system of recognition of excellence prevailing in Indian S&T is discussed to see whether there is any correspondence between the share of the awards granted to women and their share in the total S&T human resource.

Table 3.5: Awards for Excellence in Scientific Research in India

Sl.No.	*Name of the Awards*	*Disciplines Covered*	*Year of Establishment*	*Prize Amount (US $)***	*Total No. of Awardees*		
					M	*F*	*Total*
1.	Shanti Swarup Bhatnagar	Maths Sci, Phys Sci Engg Sci, Earth Sci Chem Sci, Bio Sci	1958	2127	275	7	282
2.	Rafi Ahmed Kidwai	Science Agricultural	1956	213	188	6	194
3.	Jawaharlal Nehru	Agri Sci Bio Sci	1969	106	125	7	132
4.	Om Prakash Bhasin	Science and Technology	1985	106 to 2127	349	2	51
5.	Hari Ohm Ashram	Phys Sci Applied Sciences Theoretical Sci Life Sciences Sciences and Society Interaction between	1974	106	67	3	70
6.	Federation of Indian Chamber of Commerce	Life Sciences Physical Sciences (Mathematics) including Agricultural	1967	1062	74	2	76
7.	Indian National (Award) Science Academy (Medal)	All Basic Sciences including Medical Sciences	1969	213 to 531	220 250	3 19	223 269
8.	Indian Council of Medical Research*	Medical Sciences	1953	21 to 213	325	100	425
	Total	1573	149	1722			

*: Includes 37 award instituted in different years starting from 1954 in different fields of medical sciences.

These awards include 11 oration awards for research or medical practice and one Kshanika Oration Award exclusively for women.

The rest of the 25 research awards are in the field of biomedical research, clinical research, nutritional sciences, cancer research, leprosy, mental health, community medicine and 14 awards for young scientists below the age of 40 years.

**US $ 1 = Indian Rs.47.02

There are eight major agencies, scientific as well as industrial, which recognize excellence in scientific research by announcing awards (Table 3.5). The awards are given in different branches of science, engineering and medicine in cash or as medals or research grants. Most of the awards were established in the early fifties and more were added in the subsequent years. Out of the total number of 1722 awards presented so far, the share of women constitutes only 9 per cent. This proportion corresponds fairly well with the proportion of total women scientists engaged in R&D.

Compared to other areas; there is greater representation of women as high as 24 per cent in the Indian Council of Medical Research awards in medical and life sciences. It has to be noted that the share of women in R&D in this sector constitutes 27 per cent compared to men with only 12 per cent holding Ph.D. and 24 per cent master's degree out of the total women personnel in medical science R&D.

4.0 Concluding Observations

A global comparison of the social structure of human resources in terms of gender and national income levels of different countries does not reveal any definite or marked relationship between the proportion of women in science and income levels of countries. Some of the low and middle-income countries possess higher share of women scientists than the most advanced western countries and in some of the specific branches the proportion of women is much higher than that of the developed western countries. Thus, the share of women varies considerably in different branches and sectors.

Moreover, though most of the high-income countries have achieved almost 100 per cent female literacy there is considerable under representation of women in S&T organizations implying discrimination at that level. Hence, it seems that cultural and political factors together play a more dominant role than national income levels and that a holistic perspective is required to study this phenomenon as several intertwined factors are at play.

In Indian science, under representation of women seems to be a function of general literacy level and the status of women in society rather than any discrimination against them by scientific enterprise *per se*. This could be a result of the fact that women do not constitute a critical mass in R&D organizations. In these kinds of organizations a tendency of protecting women cannot be ruled out.

This is also reflected in many of the special schemes for women announced by the Government of India. Representation of women in different branches of S&T varies considerably as also their performance and recognition of excellence. The latter aspect, of course, requires further in-depth studies.

---

3 It requires to be noted that though data as share of women scientists in the total S&T personnel are available only for the year 1981 and not beyond, there are indications to suggest that their share since then has increased and certainly not reduced. This may be inferred from the data available on faculty wise enrollment of women in higher education during the period 1981 to 1992. The share of women in the enrollment in science has increased from 29 to 34 per cent. In medical sciences the share went up from 26 to 33 per cent and that of engineering increased from 5 to only 8 per cent. See for details *Research and Development Statistics, 1992-93* (Government of India, Department of Science and Technology, New Delhi, 1992), 11 and 91.

Finally, it appears that a greater participation of women could be encouraged and ensured by affirmative policy interventions or in other words empowerment of science and technology could be ensured through interventions of women scientists and engineers.

References

Barinaga, M., 1994. "Surprise across the Cultural Divide', Science, (63), pp.1468-1472.

Barwa, Sharmishta and Shirin Rai, 2002. "The Political Economy of Intellectual Property Rights: A

Gender Perspective" in Newell, P *et al.* (Eds), Development and the Challenge of Globalization,

Glasgow, Great Britain: ITDG Publishing, pp.42.

Census of India, 2001, http://www.censusindia.gov.in/Census_Data_2001/India_at_glance/literates1.aspx (accessed 1 December 2008)

Desai, Pranav N., 2007. "Traditional Knowledge and Intellectual Property Protection: Past and Future", Science and Public Policy, Vol. 34, No.3, pp. 185-197.

DST, 1994. Research and Development statistics, 1992-93, Government of India, Department of Science and Technology, New Delhi, pp.82-85.

European Commission, 2004. Gender and Excellence in the Making, European Commission, Brussels, p12.

INSA Report, 2004. Science Career for Indian Women: An examination of Indian women's access to and retention in scientific careers. Indian National Science Academy, New Delhi.

Lemoine, W., 1992. "The Frequency Distribution of Research Papers and Patents According to Sex. The Case of CSIR, India", Scientometrics, Vol. 24. No. 3, pp.449-469.

Mies, M. 1986, Patriarchy and Accumulation on a World Scale (Zed Press, London).

Naldi, Fulvio, Daniela Luzi, Adriana Valente and Ilaria Vannini Parenti 2005. "Scientific and Technological Performance by Gender" in Henk F. Moed, Wolfgang Glänzel and Ulrich Schmoch (Eds) Handbook of Quantitative Science and Technology Research: The Use of Publication and Patent Statistics in Studies of S&T Systems, Kluwer Academic Publishers, New York

NSF, 1988. Women and Minorities in Science and Engineering, National Science Foundation, Washington, January, 1988, pp. 11.

Rose, H., 1994. Love, Power and Knowledge: Towards a Feminist Transformation of the Sciences, Polity Press, Cambridge, UK, pp.4.

Ruizo, B., 1987. "The Intellectual Labour Markets in Developed and Developing Countries: Women's Representation in Scientific Research", International Journal of Science Education (9), pp.385-91.

Seymour, S., 1995, "Family Structure, Marriage, Caste, Class and Women's Education: Exploring the Linkages in Indian Town", Journal of Gender Studies, Sage Publications, 2 (1), pp.68-85. New Delhi.

Shiva, V., 1989. Staying Alive, Zed Press, London. Statistics on Research and Development 2008 UNESCO Institute for Statistics. http://stats.uis.unesco.org/ (accessed 1 December, 2008).

Visaria, P., 1992. The Sex Ratios of Population of India, Monograph No. 10, Office of the Registrar General, New Delhi.

Visaria, P., A. Gumber and L. Visaria, 1992. School Enrollment, Attendance, Drop-outs and Literacy in Western India: Some Data on levels and Determinants, Working Paper No. 41, Gujarat Institute of Development Research, Ahmedabad.

World Bank, 1991. Gender and poverty in India: Issues and opportunities concerning women in Indian Economy, World Bank, Washington, Report No. 8072 – IN, June 14, pp.123.

World Bank, 1994. World Development Report *1994* (Oxford University Press, New Delhi, pp.219.

World Bank, 2009. World Development Report: Reshaping Economic Geography (The International Bank for Reconstruction and Development / the World Bank, Washington D.C)

Chapter 4

Women, Science and Technology: An Overview from Indonesia

Aisyiyah Jenie
Researcher, Research Centre for Chemistry, Indonesia Institute of Sciences, Kawasan PUSPIPTEK Serpong, Tangerang – Indonesia
E-mail: jenie_sna@yahoo.com, siti045@lipi.go.id

ABSTRACT

With a total population of 222 million, Indonesia is the world's fourth most populous country. Over 85 per cent of Indonesia's citizens are Muslims, making it the most populous Muslim country in the world. Alongside secular laws, Islamic principles and traditional customs affect the lives of Indonesian women and their situation varies between different regions and ethnic groups. Much of Indonesian legislation emphasises the importance of equal opportunities for men and women, including in education, but typical stereotypes and sex roles prevail in rural Indonesia, where the man is seen as the bread winner and the woman as the mother and wife. While 50 per cent of Indonesian women are economically active, only 30 per cent constitute salaried employees outside the agriculture sector. The Indonesian government has paid much concern on the gender equality and the involvement of women in science and technology. The government commitment has led to policies of gender mainstreaming at all levels, including the establishment of the gender mainstreaming mechanism and obligations. In the field of research and development, women researchers or scientist in Indonesia has vast opportunities to develop their interest. However, in their ways as students, professionals, researchers, scientists, and academicians, women involved in science and technology (S&T) in Indonesia are still constrained by a number of factors. This report is intended to provide and overview on the situation of women and gender in science and technology in Indonesia. It includes both information on their involvement, their achievements as well as gender machineries and frameworks in the country.

Keyworsd: *Women, Science and Technology, Indonesia, Research, Education, Report.*

1.0 Introduction

Indonesia, located in South-east Asia, comprises of 17,508 islands, in which only 6,000 of them are inhabited. The country consists of distinct ethnic, linguistic and religious groups.

Indonesia is a republic, with an elected legislature and president. The nation's capital city is Jakarta which also acts as the country's largest commercial centre. The country shares land borders with Papua New Guinea, East Timor and Malaysia. Other neighboring countries include Singapore, the Philippines, Australia, and the Indian territory of the Andaman and Nicobar Islands.

Figure 4.1: Map of Indonesia

Indonesia's estimated Gross Domestic Product (GDP) for 2007 is US$408 billion. In 2007, estimated nominal per capita GDP is US$1,812, and per capita GDP PPP (Purchasing Power Parity) was US$4,616 (International Dollars). The services sector is the economy's largest and accounts for 45.3 per cent of GDP (2005). This is followed by industry (40.7 per cent) and agriculture (14.0 per cent).

However, agriculture employs more people than other sectors, accounting for 44.3 per cent of the 95 million-strong workforce. This is followed by the services sector (36.9 per cent) and industry (18.8 per cent).

Major industries include petroleum and natural gas, textiles, apparel, and mining. Major agricultural products include palm oil, rice, tea, coffee, spices, and rubber. The country has extensive natural resources, including crude oil, natural gas, tin, copper, and gold. Indonesia's main export markets (2005) are Japan (22.3 per cent), the United States (13.9 per cent), China (9.1 per cent), and Singapore (8.9 per cent) (Wikipedia, 2008).

With a total population of 222 million (National Statistics Agency, 2006), Indonesia is the world's fourth most populous country. Over 85 per cent of Indonesia's citizens are Muslims, making it the most populous Muslim country in the world. Alongside secular laws, Islamic principles and traditional customs affect the lives of Indonesian women and their situation varies between different regions and ethnic groups. Much of Indonesian legislation emphasises the importance of equal opportunities for men and women, but typical stereotypes and sex roles prevail in rural Indonesia, where the man is seen as the bread winner and the woman as the mother and wife. While 50 per cent of Indonesian women are economically active, only 30 per cent constitute salaried employees outside the agriculture sector (Wikigender, 2008).

With the annual population growth rate of just over 1.1 per cent per year, demographic data shows that women constitute more than half of the Indonesian population (50.1 per cent). The actual composition of males and females are about 49.87 to 50.13. Below are the country data compiled by the UNDP in 2005-2006 (UNDP, 2008):

Table 4.1: Gender Related Country Data of Indonesia (UNDP, 2008)

	Value
Human Development Index (HDI)	0.728
Gender Related Development Index (GDI)	0.721
Population	222 million
Women Life Expectancy (WLE)	71.6 years
Women Literacy Rate (WLR)	86.8 per cent

2.0 Science and Technology in Indonesia

The development of science and technology in the country has an essential objective to increase prosperity in order to develop a civilized nation. Along with the new paradigm in this globalization era, which is technology based economy; technology becomes an important factor bringing significant contribution to enhance the nation's quality of life (The Ministry of Science and Technology, 2006). The implication of this paradigm is the transition of the global economics which was previously based on resources (Resource Based Economy) into economics based on knowledge (Knowledge Based Economy/KBE). In KBE, the strength of one nation is measured by their S&T capability as the primary economic factor to enhance competitiveness, replacing capital financial, land and energy.

However, the current situation in Indonesia is that the economic growth based on science and technology (measured as total factor productivity/TFP) is still quite low and even it reached negative for around 63 years since the country's independence. The S&T role in the country reached a high mark only during 1991-1995 with TFP of 4 per cent. Outside this period, the national economic growth was based on capital financial and manpower. This is the reason why the country suffered badly from the economic crisis in 1997, whereas for as long as 32 years (New Order) the national economic growth reached its peak (approx. 7 per cent-8 per cent per year) and only within a few months during the crisis in 1997-decreased drastically to 3-2 per cent in 1998 (Firdausy, 2008). Therefore, the only option for this country is to develop and enhance science and technology in the form of sustainable development.

There are many ways to develop and enhance science and technology. Each method is relative from one country to another. According to Firdausy (2008), at least there are five methods that are most likely possible for the situation in Indonesia. First, by providing incentive which are not only limited to conducive and anti-inflation macroeconomics policies but also applied to policies related to trade and local competitiveness. For the later, attention on beating inefficiency and bureaucracy (most likely known as high cost economy) must be conducted. Second is by increasing the quality of manpower or human resources. This does not only mean skills but also their work performance.

Education on S&T must be formally embedded in the national education system.

Third, is to develop and enhance the industry of S&T itself. This includes developing and enhancing R&D institutes and universities (public and private), and also the most important factor is coordination between the institutes and universities. One of the largest government's research based institutions in the country is Indonesia Institute of Sciences (LIPI). Established on August 23rd, 1967, LIPI manages more than 25 research centres of different field work under five deputies.

Together with Ministry of Research and Technology and other governmental research institution, LIPI arranges Indonesian Research and Technology Framework and the policies related. At this moment, Indonesian Research and Technology Framework is focused on six major fields, namely:

1. The strengthening of food resilience,
2. The increase of new findings and exploitation of new sources of energy and renewable energy,
3. The development of transportation technologies and management,
4. The development of information and communication technologies,
5. The development of defense technologies,
6. The development of health technologies and medicine (RESGEST, 2006).

The fourth method is by establishing a financial mechanism and exception budgeting, such as providing credits on purchasing of advanced equipments, technical assistance for industries and financial incentive to motivate industries to perform more R&D activities. The budget for R&D which is currently dominated by the government, should be reduced and shared by the private sector and government organizations. In other words, national S&T development should meet the requirements of investors, society and inventors (market driven technological development). To achieve this, research activities conducted by research institutes and public universities should align with industries, government organization (BUMN) and market demands. The last method proposed by Firdausy (2008) is to establish a stimulating and protective S&T framework for private domestic and private foreign sector in order to enhance S&T nation wide. Therefore corruption as an adverse effect of autonomy and decentralization policies and also long process of bureaucracy should be eliminated since it may have disadvantages towards national S&T development. However, the open door policy for imported technology still has to be applied carefully as so it does not cause drawback to the national S&T development and the society. The same awareness should also be applied to high-technology enhancement. In a way, S&T development is not only a matter of how to increase the economic growth but also on how to form a prosperous nation.

To endorse the importance of science and technology for the national development, the president of Indonesia, Susilo Bambang Yudhoyono, in 2005 has stated the year 2005-2006 as Indonesia Science Year. This was based on two main reasons.

- First, the development of technology of a country will strongly and rapidly flourish if it were based on strong mastery and fully comprehend of basic sciences.
- Second, the country has to sufficiently and consistently focus on the development of science in order to put forward continuous development based technology enhancement.

This statement by the president was a positive response for the Indonesian scientist and researchers demand which was to emerge science mainstreaming as a national movement to strengthen the national development foundation (Jenie, 2005).

3.0 Women, Science and Technology

3.1 Gender Mainstreaming in Science and Technology: Why Needed?

Until March 2008, the total amount of population under poverty line according to the National Statistics Agency is 34.96 million or 15.42 per cent from Indonesia's total population. Half of it are women and women as housewives are the dominant group amongst them. Around 22.2 million of the poor population lives in the rural areas, while 12.76 million lives in the cities. This indicates that women are often the lowest group below poverty line, thus are not able to access the advantages of science and technology.

The assessments conducted by the Gender Working Group UNCSTD (United Nation Commision on Science and Technology for Development) showed that technology change in developing countries, which were intended to increase productivity, has eliminated the productive and economic role which has traditionally been done by women. Research and assessments in Indonesia on changes in rural areas, which are most common caused by the change in technology, has shown that there is a different impact of technology towards men and women (Lukito, 2008).

The above situation has an effect on how further women are involved in science and technology. According to UNESCO (2004), for the last twenty years in many countries in the Asia Pacific region, including in Indonesia demonstrated that women lagged behind men in science and technology, education and training, careers, participation in policy and decision making positions as well as in the enjoyment of benefits emanating from science and technology development and application. Women's access to science and technology is still restricted by gender equality legislations which are very unfortunate and only places women as being the user of S&T (*science for whom?*) instead of being the inventor (*science by whom?*). This condition does not only occur in developing countries, but in all situation, women does not tend to have an important role. While in fact, women could play an important role in the sosial-economic development of a developing country, through their triple-responsibility in society management, reproduction and production. This triple responsibility can be translated as the foundation from food production, energy and water, health treatment, family income and children's first experience to education and nutrition which builds the platform for sustainable development (Siswono, 2004).

United Nations Commission on Science and Technology for Development (UNCSTD) Gender Working Group (1993) identified seven key issues in gender dimension of science and technology that is considered to be particularly important and for which transformative actions were both necessary and feasible and therefore strongly recommended. These issues are:

1. Gender equity in science and technology education
2. Removing obstacles to women scientific and technological careers
3. Making science responsive to the needs of society
4. Making the science and technology decision-making process more "gender aware"
5. Relating better with local knowledge systems
6. Addressing ethical issues in science and technology and
7. Improving the collection of gender disaggregated data for policy makers.

Gender mainstreaming has been globally recognized as the most effective strategy to achieve gender equality and justice in all fields of life and in all sectors of development. This obviously includes science and technology. The strategy aims at both research groups and institutions. Target groups include teachers, lecturers, researchers, technologists, technicians, R&D managers, principal investigators, policy and decision makers, including those who set research priorities, planning, programming and budgeting officers, monitoring and evaluation teams/groups, gender differential impact study teams/groups, statisticians, and S&T reporters. Whereas target institutions include education and training institutions from preschool to higher education institutions, R&D institutions, policy and decision making institutions, commissions, councils, advisory bodies, planning-programming and budgeting institutions and scientific professional organizations.

3.2 Government Frameworks

Considering the importance of the role of science and technology in the nation's social economic development, it requires the whole society to be involved in the development of S&T.

The Indonesian government has paid much concern on the gender equality and the empowerment of women in this sustainable development. The commitment has led to policies of gender mainstreaming at all levels, including the establishment of the gender mainstreaming mechanism and obligations. The country's machineries on gender in science and technology include the ministries and government policies and programmes (RESGEST, 2006).

- Ministry for Research and Technology, was established in 1973, has a main function to coordinate research and technology development activities of all government agencies
- The Ministry for Women Empowerment, established since 1978, with the primary task to formulate policies, initiate, coordinate, monitor and evaluate the implementation of programmes and activities to enhance the status and role of women, in the family and in society. The vision of women empowerment is the achievements of gender equality and justice within the family, community and state.

There is also a strategic and capable National Comission on Violence against women which takes a right based approach to policy influencing and advocacy to eliminate violence against women. A number of effective and well-networked civil society organisations focussing on gender rights are developing nationally and in the region.

- Law No. 18 year 2002 on National System of Research, Development and Application of Science and Technology. The aim of this law is to strengthen the S&T resources in order to accelerate the achievement of international competitiveness for sustainable development.
- Presidential Instruction No. 9 of the year 2000 on Gender Mainstreaming in National Development, which includes to mainstream gender in implementing their planning formulation, implementation, monitoring and evaluation of policies and development programmes. The State Minister of Women Empowerment is tasked with providing technical assistance.
- Presidential Instruction No. 4 year 2003 on the Coordination of Strategic Policy Formulation and Implementation of the National Development of Science and Technology. An important quote from this strategic policy is that *"national development in the field of science and technology also needs to be implemented with gender perspective, in line with international efforts to achieve gender equality"*.

Indonesian government adopts two national objectives to promote gender equality and justice in the five-year National Development Programme 2000-2004 (RESGEST, 2006). The country's Mid-Term National Development Plan (2004-2009) includes gender mainstreaming as one of the strategies which concerns the "Increase of S&T Capabilities and Capacities" in 6 priority areas as mentioned in the Indonesian Research and Technology Framework mentioned above.

More specifically, the national development objectives toward the achievement of the increased quality of life of women, the enhanced role of women, and child protection are compiled into the National Policy of Women Empowerment for 2004-2009, which are:

1. To ensure gender justice in various laws and regulations, development programmes, and public policies;
2. To improve the position of the GDI and the GEM rank of Indonesia, through the enhancement of achievements of gender equality and justice, and the positions of women in public life
3. To eliminate violence against women.
4. To improve the welfare and protection of the rights of the child (RESGEST, 2006).

3.3 On Education, Research and Development

In the field of education, gender stereotyping could be identified in the selection of field of study at vocational schools and universities. Social science is dominated by female high school students who mostly studied in the field of business and management (64.6 per cent) while only 18.8 per cent of them studied industrial engineering. One of the reasons why girls are not very much interested in engineering and technology is caused by the cultural and traditional perception, that engineering and technology are more suited for boys. Gender studies (Fausto-Sterling, 1981; Wittig and Petersen, 1979) also mentioned that women might just not have quite the same mental capacities as men whereas girls in high school take less mathematics and science courses and on the average perform more poorly than boys on standardized mathematics achievement tests.

However, Indonesia has a good story to tell on gender equality in the enrollment in primary, secondary and tertiary education. Boys and girls are enrolled in equal numbers at the primary and secondary level as shown in Table 4.2. It is worth noting that in the generals secondary education sector the number of girls and boys studying maths and science are almost equal. For tertiary education, boys outweigh the girls only by 2 per cent difference. These remarkable improvements can be traced back to the 1970's investments in primary school construction and hiring and deployment of women teachers. Based on the data, educational attainment shows higher performance on female-to-male ratio compared to other gender gap sub indexes, such as economics and even political empowerment.

Over the last decade girls have caught up with boys in higher education. Over the past ten years gender parity in higher education has improved remarkably. Around 15 per cent of both young men and young women are getting a higher education. Gender bias in subjects chosen is evident in choice of vocational subjects. Most girls prefer tourism to technical and economic related trades.

Table 4.2: Gender Equality in Primary, Secondary and Tertiary Educational Attainment (NIEW, 2007)

Gender Gap Subindexes	Rank	Score	Sample average	Female	Male	Female-to-male ratio
Economic Participation and Opportunity	**67**	**0.598**	**0.596**			
Labour force participation	89	0.60	0.69	51%	85%	0.60
Wage equality for similar work (survey)	10	0.79	0.64	—	—	0.79
Income (PPP US$)	61	0.52	0.52	2,289	4,434	0.52
Legislators, senior officials, and managers	77	0.20	0.37	17%	83%	0.20
Professional and technical workers	63	0.72	0.79	42%	58%	0.72
Educational Attainment	**80**	**0.949**	**0.939**			
Literacy rate	77	0.92	0.91	87%	94%	0.92
Enrolment in primary education	88	0.98	0.97	93%	95%	0.98
Enrolment in secondary education	73	0.99	0.94	57%	57%	0.99
Enrolment in tertiary education	83	0.79	0.86	15%	19%	0.79
Health and Survival	**88**	**0.969**	**0.973**			
Sex ratio at birth (female/male)	1	0.94	0.94	49%	51%	0.95
Healthy life expectancy (years)	95	1.02	1.04	58.9	57.4	1.02
Political Empowerment	**63**	**0.101**	**0.138**			
Women in parliament	73	0.13	0.22	11%	89%	0.13
Women in ministerial positions	70	0.12	0.21	11%	89%	0.12
Years with female head of state (last 50)	17	0.07	0.04	3.3	46.7	0.07

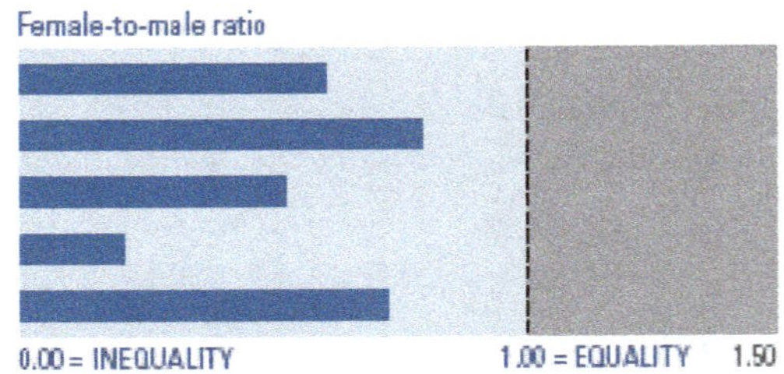

Data from the Department of National Education in 2006–2007 on the number of graduates from both public and private universities is shown in Table 4.3. In average, it is almost equal between female and male graduates both in public and private universities. This shows that in the new age of information and technology in Indonesia respectively, high education is considered important not only for young men but also young women. In particularly, women having higher education have a double impact both in socio-economic development as well as improving the socio-economic status of women themselves (Gallaway and Bernasek, 2004). Many development economists tend to view that in developing countries such as Indonesia, women's education is primarily a means to the end of reducing population growth whereas there is strong correlation between more education for women and lower fertility rates. Although this statement still needs more justification (Angeles *et al.*, 2005). On the other hand, feminists economists tend to view women's education as critical for achieving improvements in the socio-economic status of women and contributing to their empowerment, as well as promoting social development. Women's education is a means to the end of improving the lives of women, which has the further benefit of contributing to socio-economic development (Gallaway and Bernasek, 2004).

Table 4.3: Number of Graduates in Public and Private Universities 2006–2007 (Ministry of National Education, 2008)

	Public		*Private*		*Total*	
	Value	*Per cent*	*Value*	*Per cent*	*Value*	*Per cent*
Male	29,748	51.04	72,309	51.88	102,057	51.64
Female	28,536	48.96	67,057	48.12	95,593	48.36
	58,284.	139,366	197,650			

In the field of research and development, R&D activities in Indonesia are mostly conducted in public R&D Institutions. R&D activities in public sectors are carried out in 17 Ministerial Departments and 33 Non Ministerial Agencies. One of them is the Indonesia Institute of Sciences (LIPI) as mentioned above.

Within year 2002-2003, the number of women researchers in R&D units were 648 women researchers or 28 per cent of the total researchers (2293 person); 128 women technicians or 30.2 per cent of total technicians (424 person) and 1931 (43.75 per cent) other women supporting staff such as administrators, librarians, archivist (4.423 person). In 2003, the number of researchers were 3351, consisting of 991 (29.6 per cent) females. There were 871 technicians, consisting of 239 (27.4 per cent) females, as shown in Figure 4.2. In average, although they have equal enrollment to higher educations the number of women researchers are still outnumbered by the men. According to Fausto-Sterling (1981) one explanation as to why there are so few women researchers or scientists is that they face multiple forms of discrimination – in obtaining jobs, level of salary received and in their ability to gain access to informal networks of communication important to pursuing a career patterned on currently existing success models. More women will want to become scientists if job discrimination is eliminated

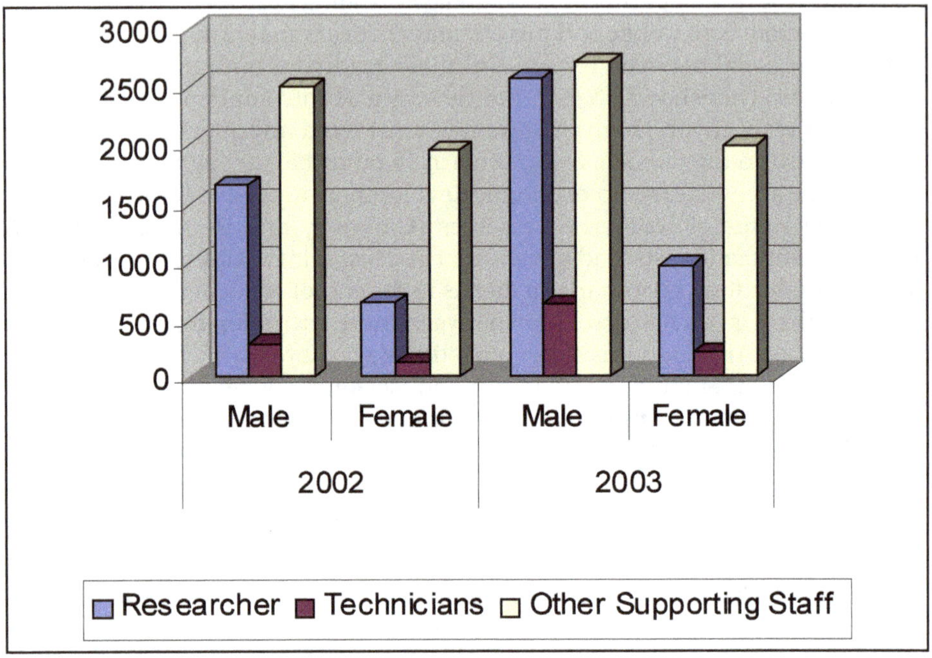

Figure 4.2: Number of Women Researchers in Indonesia 2002-2003 (RESGEST, 2006)

and more efforts made to encourage girls in primary and secondary schools to consider and prepare for scientific careers.

3.4 Success Stories

In Indonesia, looking for a women scientist does not appear to be an easy task. This does not generally mean that there are no women scientists at all; however there is still a lack of available information and database. Moreover some names are missing from journals or textbooks and achievements are not well recognized.

A reward in science is one way to appreciate and socialize the research result. In Indonesia, one of the internationally recognized awards in science for women is the UNESCO-L'Oreal International Fellowship. These fellowships are allocated to young women researchers in Life Sciences at the doctoral and post-doctoral level, whose promising projects has been accepted by a reputable institution. The award is given to three women researchers from each of five UNESCO regions, Africa/Arab States, Asia Pacific, Europe, Latin America and North America. From the Asia Pacific region, Indonesia has accepted the award three times since 2004 represented by the women researchers from institutes and universities. First, Ines Atmosukarto, a microbiologist from LIPI in 2004, followed by Fenny Dwivany, a molecular biologist from Bandung Institute of Technology (ITB) in 2007. The latest in this year, Indonesia once again received the award represented by Made Tri Ari Penia Kresnowati, a Bioprocess Technologist from Bandung Institute of Technology (ITB) (L'OREAL-UNESCO, 2008).

The three scientists gained their education until doctoral degree, and achieved their PhD from various international universities. Ines who earned a doctorate from Australia's Adelaide University in 2002, has done extensive research on endophytes, micro-organisms that live inside plants and which have the potential for novel uses in the pharmaceutical industry. She stated that each plant can harbor different endophytes, even plants from the same species may have different ones, and therefore there is so much diversity and potential in Indonesia with over 250,000 species of plant growing. She added that various papers showed that endophytes could produce antibiotics, fungicides, and possible new cancer treatments. Fenny, on the other hand earned her PhD in molecular biology from the University of Melbourne, Australia in 2003. She began extensive research in gene isolations, rice genes and the latest research on expression profiles of ACS genes in bananas to prevent it from ripening. As for Penia, she earned her doctoral degree from Netherland's Delft University of Technology in bioprocess technology. Finishing her doctoral degree at age 30, she became the youngest lecturer to hold a doctoral degree at the ITB's Chemical Engineering department. All of them are currently in Australia, continuing their research. Ines is at Canberra developing a new type of vaccine. Fenny is conducting her second phase of research which is to silence or surpress the ACO gene of banana using RNA interference which will take place at the Plant Cell and Biology Research Centre of the University of Melbourne, Australia. Penia is conducting stem-cell related research in Monash University's Department of Chemical Engineering and will be working at the University's Bio Engineering Laboratory (BEL), conducting research into the design of bioreactors for creating health products from stem cells. Their results will then be brought back to Indonesia to be applied and socialized in the country (ReferencesView, 2008; VESKI, 2008).

It is worth noting that all three of them, received their doctoral degree and this prestigious award at their early 30s which is a very young age in the world of science and in Indonesia's education system particularly. However, the success in their career is seemingly balanced with their success in the family household. Despite all of them are married and mothers of young children, they have full support from their spouses-which is a very critical factor-and similarly came from well educated families.

Another programme by UNESCO and L'Oreal is the National Fellowships which is awarded to national initiatives, such as national science fellowships and mentoring programmes. Indonesia has also joined and continuously received this award since 2006. Besides financial reward to support their sustainable research, rewards may open new networks for scientists, especially women. With a broad network system, women scientists may exchange information on their research. In the future, hopefully the appreciation is not only by giving certain rewards and recognition, but the most important is to establish an equal culture for women to be involved in science and technology enhancement.

4.0 Future Challenges

Indonesia has made notable improvements in gender mainstreaming in science and technology since it launched in 2000. However, even in this modern globalization era the country still needs to struggle and faces many challenges on gender issues.

Patriarchy culture still holds out in every angle of life which places women as men's subordinate. In their ways as students, professionals, researchers, scientists, and academicians, women involved in science and technology (S&T) in Indonesia are still constrained by a number of factors.

4.1 Cultural Value and Gender Stereotyping

This factor is mostly relevant to women in pursuing their higher education or career. Gender stereotyping still exist around maintaining the cultural value placed upon wifely duty (*kodrat wanita*) of a women. Chief among the socio-cultural constraints is the conflict between the roles of wife/mother and career woman. Wright and Tellei (1993) stated that most managerial women cited the conflict between their family and work roles as significant in their own work histories. Given the strength of the cultural value attached to marriage and motherhood, it is not surprising that female professionals are more likely to be single than their male counterparts. Moreover, stereotypical images of career women such as "Men are better decision makers/ emotionally more stable, "Intellectually superior" or "Women are more interested in husbands and children" occasionally found in popular business periodicals which reflect and reinforce prejudicial attitudes.

4.2 Science and Technology as a Masculine World

The perception that science and technology is a masculine world is a problem that women need to deal with further. Now-a-days modern science in Indonesia is conducted in large groups. Generally research is done by a scientist who obtains a grant in his or her name and then employs a variety of underlings, post-doctoral fellows, graduate students, technicians, analysts, administrators, etc. to aid the work. The hierarchical relationships among the varying types of scientific workers are expressed in terms of prestige, experience, salary and who bosses whom. Not surprisingly, women are found in disproportionate numbers in the lower ranks of the scientific hierarchy (Fausto-Sterling, 1981).

Patriarchy culture still holds out in the field of science and technology. Although teachers in primary and secondary schools are dominantly female, science teachers in high schools tend to be male. Teachers often give differential treatment to their students, such as lack of motivating girls in science. The science and curricula itself are still designed to the interest of boys rather than girls (RESGEST, 2006). Therefore, women cannot be fully and equally integrated into the world of science so long as patriarchy culture still remains and science is practiced and taught in a hierarchical manner.

4.3 Family Support

Socio-economic status in the family background, such as lack of funding is another obstacle that many young girls have to face in order to pursue their education. Many families especially in rural areas cannot meet their budget for their daughters' to further their studies to higher education (RESGEST, 2006). Along with the perception that girls are associated with domestic responsibilities, education is not priority to families in rural areas.

In developing careers in science and technology, Indonesian women are often not permitted to work fully outside the household. The years in which a women scientist have to prove herself in her work coincides with the years in which the majority of women establish a family, namely between the age of twenty-five and thirty five years. Women with young children have difficulties to work long and irregular hours, a situation which clearly does not work in favor of their career prospect.

Women faces double burden to balance between work and family. This situation mostly is impending women for study or work outside or abroad (RESGEST, 2006). The situation gets more difficult when the perception of the husband often slows down the knowledge development of these women scientists. Their belief is that women should deal with domestic matters, instead of conducting research which may take many years. This does not only happen in developing countries, but also in developed and advanced countries.

4.4 General Issues Related with Science and Research

General issues on research are also faced by women scientist and researchers in Indonesia. These issues include access to information, research funding and the total salary or payment received as researchers. Limited access to information, such as international scientific journals, are especially experienced by researchers in rural areas. The rapid development of science and technology requires boundless information access to sustain research and development activities.

Although the government seems to pay a significant attention for the development and enhancement of S&T, the total amount of research budget is only less than 1 per cent (0.07 per cent) of the state's total revenue. This is far too low compared to other neighboring developing countries such as Malaysia and Thailand which has already reached until 2 per cent. Ideally, according to UNESCO, research budget should be at least 3 per cent of the state's total revenue per country (Hanggara, 2008).

This implies to the amount of qualified researchers existing in Indonesia and the salary they accept. As being a scientist is not a wealthy profession, teenagers and fresh graduates from universities tend to choose much 'prosperous' career such as being in bussines, entertainment or working in high profile companies and offices. According to Wright and Tellei (1993) middle-class women typically have a high-school education and remain in the formal labour force (in teaching or sales) up to the time of marriage, after which they engage in full-time domestic responsibilities.

Upper-class women-those from noble, wealthy, or high-ranking military families-are more likely to have paid employment in the professions, civil service, and in family businesses, and to remain in the formal labour force after marriage, although they will take time out when children are born.

5. Conclusions

Indonesia through its government has adopted the gender mainstreaming strategy to achieve equality for women, including in science and technology. To endorse this strategy, policies and programmes have been established which are aligned with the national sustainable development programme. Statistic data on education shows that gender equity is almost equal in the enrollment in primary until tertiary education. However, gender bias still exists in the subjects chosen, where some subjects tend to be 'more suitable' for male. Women researchers are still outnumbered by men as girls are mostly not encouraged to pursue a career as scientists or researchers from their primary education.

As Indonesia faces further challenges to growth and a continued demand for the enhancement of science and technology, educated women should find more doors opening for them. However, obstacles still occurs, including cultural and traditional values of women, S&T being a masculine world, family supports and also general issues related with research in Indonesia. Besides all of the challenges lying ahead of the country, Indonesia is still proud of those who have achieved both international and national recognition. This can be a good role model for girls and young women in this era to be more involved in science and technology and take part in the national development.

Also it can be a reminder for the government on how important science and technology is to the national development, and how it requires the involvement of the whole society, without exception–including women.

Myriam Sarachik, a professor in physics from the City College of New York once simply said that, "The importance for women is to seek the strength within. Believe in your strength and respect it. That is the most important thing for you in choosing to do what you love. Surely, in that way, you will invest your heart in all your work."

References

Angeles, G., Guilkey, D.K. and Mroz, T.A., 2005. The Effects of Education and Family Planning Programmes on Fertility in Indonesia, Economic Development and Cultural Change, 54(1), pp. 165-201.

Department for International Development (DFID), 2007, "Gender Equality in Indonesia", United Kingdom, http://www.dfid.gov.uk/consultations/ indonesia-gender per cent 20.pdf (accessed 1 Nov., 2008)

Dwivany, F.M., 2007. "Don't Be Afraid to Be a Scientist (Jangan Takut Jadi Peneliti)", Indonesia http://www.itb.ac.id/news/1575.xhtml (accessed 4 Nov., 2008).

Fausto-Sterling, A., 1981.Woman and Science, Women's Studies Int. Quart., 4(1), pp. 41-50.

Firdausy, C.M., 2008. "Science and Technology as a Mandatory in Economic Growth (IPTEK sebagai Keharusan Pertumbuhan Ekonomi)", Jakarta, Indonesia. http://www.ristek.go.id/index.php?mod=News&conf=v&id=2959 (accessed 29 Nov., 2008).

Gallaway, J.H and Bernasek, A., 2004. Literacy and Women's Empowerment in Indonesia: Implications for Policy, Journal of Economic Issues, 38(2), pp. 519-525.

Hanggara, R., 2008, "Limited Budget, Indonesia's Research Decreases (Kurang Dana, Riset Indonesia Terpuruk)", Jakarta, Indonesia. http://kampus.okezone.com/index.php/ReadStory/2008/08/19/65/138084/65/dana-kurang-riset-indonesia-terpuruk (accessed 28 Nov., 2008).

Indonesia Institute of Science (LIPI), "About LIPI", Indonesia, http://www.lipi.go.id/ (accessed 28 Nov., 2008)

Jenie, U.A., 2005. "Science and Technology for Development (Iptek untuk Pembangunan)", Indonesia,http://www.tokohindonesia.com/ensiklopedi/u/umar-anggara-jenie/index.shtml (accessed 20 Nov., 2008)

L'OREAL-UNESCO, "For Women in Science: Indonesia", <http://www.loreal.com/_en/_ww/index.aspx?direct1=00008&direct2=00008/00007> (accessed 27 Nov., 2008).

Lukito, H.P., 2008. "MDGs, Gender and S&T in Indonesia (MDGs,, Gender dan Iptek di Indonesia)", Indonesia, http://mediaindonesia.com/index.php?ar_id=MzY3MzI (accessed 25 Nov., 2008)

Ministry of National Education (Depdiknas), 2008. "National Education Statistics (Statistik Pendidikan Nasional) 2006/2007", pp. ii-vi, Jakarta, Indonesia

Ministry of Science and Technology (KNRT), 2006, "Buku Putih Indonesia 2005-2025: Penelitian, Pengembangan dan Penerapan Ilmu Pengetahuan dan Teknologi", Jakarta, Indonesia

NAM Institute for Empowerment of Women (NIEW), 2007. "Project: Indonesia", Malaysia. http://www.niew.gov.my/project/INDONESIA.pdf (accessed 5 Nov., 2008).

ReferencesView, 2008. "Dr. Ines Irene Atmosukarto, Fenny Martha Dwivany", http://www.zoominfo.com/Search/ReferencesView.aspx?PersonID=480265438,http://www.zoominfo.com/Search/ReferencesView.aspx?PersonID=1183018047 (accessed 27 Nov., 2008).

RESGEST, 2006. "Report on Gender, Science and Technology in Five Asian Countries", pp.28-84, 119-129, Jakarta, Indonesia, http://www.resgest.org/publications/resgest per cent 20report.pdf (accessed 5 Nov., 2008).

Siswono, 2004. "Gender Inequity Causing Limited Access of Women in Science and Technology (Ketidaksetaraan Gender Tutup Akses Perempuan di Bidang Iptek), Indonesia, http://www.gizi.net/cgi-in/berita/fullnews.cgi?newsid 1096016756,52964 (accessed 24 Nov., 2008).

UNDP, 2008. "Human Development Reports 2007/2008", <http://hdrstats.undp.org/ indicators/268.html> (accessed 5 Nov., 2008).

UNESCO, 2004. "Guidelines for Gender Mainstreaming in Science and Technology", UNESCO Office, Jakarta, pp. 1-7.

VESKI, 2008. "VESKI Fellow and L'Oreal Collaborate", Australia. http://www.veski.org.au/Default.aspx? Id=38 (accessed 27 Nov., 2008).

Wikigender, "Gender Equality in Indonesia", <http://www.wikigender.org/wiki/index.php?title=Gender_Equality_in_Indonesia> (accessed 6 Nov., 2008).

Wikipedia, "Indonesia", http://en.wikipedia.org/wiki/Indonesia (accessed 6 Nov., 2008).

Wright, L. and Tellei, V.C., 1993.Women in Management in Indonesia, Int. Studies of Mgt. and Org., 23(4), pp. 19-45.

Chapter 5
The Position of Women in Contemporary Iran

Akram Ghadimi
National Research Institute for Science Policy (NRISP), Iran
E-mail: akramghadimi@yahoo.com

ABSTRACT

This paper is a brief overview of women's situation in Iran. In our research, we will delve into issues such as Population, Education, Higher Education, occupation, Women's Cooperatives and Political Participation in order to substantiate our argument. We would also like to look into the relationship between on one hand literacy rate, improve economic situation, political participation in the one hand decrease in fertility rate.

Keywords: *Education, Higher education, Occupation, Political participation, Fertility.*

1.0 Introduction

Women in Iran, today, are struggling to achieve equality in the spheres of employment and political representation. They have made great progress in the educational and health fronts after the Islamic revolution, but lag behind in terms of paid work and political participation. In post-revolution years, the situation of women in Iran, like other groups of society, has witnessed many changes. This paper statistically analyses the situation of Iranian women since 1950's.

2.0 Population

Table 5.1 shows the demographic situation of Iran in 47 years. The gender proportion of the population is 104.This implies that there are 100 females to 104 males.

Table 5.1: The Population Combination of Iran during 1956-2003

Year	*Females*	*Males*
1956	9309760	9644944
1966	12432921	13355801
1976	16352396	17356347
1986	24164049	25280961
1996	29540329	30515159
1997	30030698	30994247
1998	30494482	31473335
1999	31011437	31952423
2000	31501807	32431511
2001	31993352	32913350
2002	32226049	33314175
2003	32935956	34055617

As a result of a highly successful family planning programme, Iran's total fertility rate dropped by 6.7 between 1980 and 2001,. Iran has changed its official position on family planning twice since the 1960s. In 1966, Iran was one of the first countries in the developing world to establish a national family planning programme, and by the mid-1970s, 37 per cent of married women were practicing family planning, with 24 per cent using modern methods.

Soon after the 1979 Islamic revolution, Iran replaced the previous regime's population programme with pro-fertility policies. At the end of the Iran-Iraq war in 1988, the rapid population growth of more than 3 per cent a year (one of the highest rates in the world) was seen as a major obstacle to economic reconstruction and development.

At the end of the Iran-Iraq war in the late 1980s, the Iranian government renewed its interest in family planning amidst deteriorating economic conditions. The country's National Birth Control Policy issued in 1989 had the endorsement of the country's highest Islamic authorities, and religious leaders joined with health and policy experts in a campaign to persuade the public of the need for family planning through newspaper reports, television spots and Friday prayer speeches.

Iran's family planning programme today serves as a model for developing nations and other Muslim countries. Infant, child and maternal mortality rate all have declined, and contraceptive use among married women has risen from 37 per cent in 1976 to 73 per cent in 1997 (The Guttmacher Report on Public Policy, 2001).

In response to a growing demand from an increasingly educated population, Iran used the media to help revitalize its family planning programme and promote awareness about the population growth rate, which was too high and, if left unchecked, could have negative impact on the national economy and the population's welfare. The level and rate of decline in fertility rate have surpassed expectations. The official government target in its first Five-Year development plan was to reduce the total fertility rate to 4.0 births per woman by 2011. By 2000 the rate was already down to half the state goal, at 2.0 births per woman (The Guttmacher Report on Public Policy, 2001). The growth rate of life expectancy from 1997 to 2003 was 4.3 per cent for both males and females (The Guttmacher Report on Public Policy, 2001).

Now, we can turn to factors affecting decline of population growth rate.

3.0 Education

The education variable is one of the highly important social indexes in the evaluation of development strategies of a country. Education improves public awareness of society and this increase could influence the political, economic, occupational, and health situation of different groups in society. (The Guttmacher Report on Public Policy, 2001).

Table 5.2 shows that the education level of women was 7.3 per cent in 1956 and reached 79.7 per cent by 2003, indicating an enormous increase in the education level of Iranian women. In 1986, the difference in the education level of men and women was about 18.9 per cent, which reduced to 9.1 per cent in 2003.

Table 5.2: The Education Rate of Six-year-Old Children and above Population in 1956-2003 (in percentage)

Year	*Males*	*Females*
1956	22.2	7.3
1966	40.1	17.9
1976	58.9	35.5
1986	71	52.1
1996	84. 7	74.2
1997	84.45	74.24
1998	85.3	75.33
1999	85.53	75.7
2000	85.7	76.04
2001	86.07	76.97
2002	87.5	77.8
2003	88.8	79.7

As it can be seen, men have higher education rate; however the 5.3 per cent increase in education among women during 1996 to 2002 compared to the 3.3 per cent increase in men's education in the same years shows the realization of the government's goals aimed at lowering the gap of education index between men and women. (The Guttmacher Report on Public Policy, 2001).

4.0 Higher Education

4.1 College/University Students

According to the statistics, in the academic year 2002-2003, 809,567 students were admitted to the universities and higher education institutes. Of these numbers, 50.99 per cent of women constituted university students.

According to Table 5.3, the percentage of female students in the total number of state universities shows a 47.9 per cent increase in the growth rate, from 36.1 per cent in 1997 to 53.5 per cent in 2004. The rapid increase in the number of female students at state universities has exceeded the growth in the number of male students. During 1997 to 2003, the increase in the number of university students was 16.37 per cent for boys and 135.9 per cent for female students (The Guttmacher Report on Public Policy, 2001).

Table 5.3: Gender Combination of State University Student

Year	*Males*		*Females*	
	Number	*Per cent*	*Number*	*Per cent*
1997	369907	63.88	209163	36.12
1998	386693	61.83	238687	38.17
1999	371263	58.11	267650	41.89
2000	379319	55.89	299333	44.11
2001	387185	52.78	346342	47.22
2002	381505	50.21	378365	49.79
2003	396719	49.00	412848	51.00
2004	430493	46.59	493420	53.41

Table 5.4 shows the share of females in the whole number state universities' students of has increased during the aforementioned period. The highest growth was in the field of agricultural and veterinary sciences with 108.68 per cent and the least was in the field of medicine at 28.04 (The Guttmacher Report on Public Policy, 2001).

5.0 The Islamic Azad University

The proposal for the establishment of the Islamic Azad University was first made by the then Speaker of Parliament, Akbar Hashemi Rafsanjani, in 1982. It was intended to maximize the use of the country's potential in order to expand higher education, promote the level of knowledge and literacy in the society, and secure trained experts without depending on government funds. Driven by the need to meet the increasing demand for training specialized workforce, the University is a public, non-state, non-profit institution that does not use the public budget and is mainly financed through students' fees and public donations (The Guttmacher Report on Public Policy, 2001).

Table 5.4: Gender Combination of State University Students According to Fields of Study

Year	Medicine		Humanities		Natural Science		Technical Engineering		Agricultural and Veterinary		Art	
	Total	Women Per cent	Total	Women Per cent	Total	Women Per cent	Total	Women Per cent	Total	Women Per cent	Total	Women Per cent
1997	94377	52.06	242989	39.55	81916	43.29	113204	11.58	32392	24.35	14192	52.64
1998	98273	53.57	264793	41.96	89631	45.70	118813	10.79	35632	31.36	18338	54.48
1999	97826	56.30	267844	46.68	92379	49.18	120800	12.64	38112	36.58	21952	58.83
2000	95329	58.23	289336	49.52	86566	54.32	141859	16.10	41225	39.32	24337	59.51
2001	100826	60.99	312854	53.80	91278	56.49	154190	17.44	47547	43.69	30428	60.09
2002	93847	63.53	329684	56.66	97095	59.19	154945	18.93	48736	46.46	35563	63.28
2003	93032	64.99	346545	59.03	100495	62.05	183234	20.04	53840	49.38	32421	68.27
2004	91794	66.85	410792	61.95	121509	64.48	204291	21.34	58345	50.81	37182	69.89

The University began its academic activities in 1983 by establishing nine academic branches in the country and admitting 2045 students through a nationwide entrance exam for ten fields of study. Today, with 330 academic branches throughout the country in ten regions and with several overseas branches in Great Britain, Lebanon, the United Arab Emirates, Qatar, Kuwait, Tajikistan and Tanzania, more than one million students study in the university (The Guttmacher Report on Public Policy, 2001). There are 48,616 academics who teach at the University on a part-time and full-time basis, from the post-secondary schools through Ph.D. programmes.

In the academic year of 2002-2003, there were 864,190 students at Islamic Azad University. It is interesting that 50.99 per cent of students of state universities and 50.3 per cent of Islamic Azad University students were female students.

According to Table 5.5, the percentage of female students in Azad universities shows a 21.8 per cent increase in the growth rate, from 40.8 per cent in 1997 to 49.7 per cent in 2004. The rapid increase in the number of female students at state universities has exceeded the growth in the number of male students during 1997 to 2003. The increase in the number of university students was 34.1 per cent for male and 92.2 per cent for female students (The Guttmacher Report on Public Policy, 2001). The most number of female students study in B.A programmes, both in state universities and Open (Azad) universities.

Table 5.5: Gender Combination of Azad University Students

Year	*Males*		*Females*	
	Number	*Per cent*	*Number*	*Per cent*
1997	362872	59.15	250596	40.85
1998	396062	60.07	263226	39.93
1999	368795	55.11	300442	44.89
2000	389588	53.65	336640	46.35
2001	438640	52.45	397609	47.55
2002	416571	51.64	390068	48.36
2003	428755	49.61	435435	50.39
2004	486616	50.26	481590	49.74

6.0 Occupation

Data on percentage of women in the total workforce of Iran between 1956 and 1991 shows that, in 1956, 9.6 per cent of employees were women. This figure increased to 13.2 and 13.7 per cent in 1966 and 1976, respectively. In 1986, this figure dramatically reduced to 8.8 per cent but rose slightly to 9.3 per cent in 1991.

Studies show that in 1993, 25.6 per cent of government employees in the civil services were women. The highest rate of female employment was in the Ministry of Education and the Ministry of Health with 42.1 and 41.3 per cent, respectively, while only 1.6 per cent of employees in other ministries and state organizations were women. (Zanjan and Saber, 1995)

The state of women's employment in Iran shows that, in spite of a significant increase in the total number of literate women as well as women with high educational levels, their share of the employment market has not increased substantially between 1956 and 2003.

According to Table 5.6, the total employment rate of women and men in the years under investigation reduced, and women employees have a slight share in the total figures. Low employment rates and lower income make Iranian women vulnerable to economic depression and stagnation (The Guttmacher Report on Public Policy, 2001). Now-a-days, universal or global economic competition motivates countries to adopt measures to continue their survival and catch up with the necessities of this competition. Empowerment of the economic situation of a country needs the participation of the whole population in the economic sphere and activities. Therefore, empowerment of females and promotion of their economic power, as half of the population, is very important and essential. Women cooperatives are an instrument through which women can represent their abilities in the economic field.

Table 5.6: The Rate of Occupation

Year	*Men*	*Women*
1975	70.8	13.9
1985	68.4	8.2
1991	65.5	8.7
1997	59.22	9.94
1998	60.83	11.6
1999	61.86	11.70
2000	62.09	11.91
2001	61.82	11.69
2002	62.40	12.00
2003	62.80	10.90

7.0 Women's Cooperatives

After the victory of the Islamic Revolution and from the very beginning of executing the plan of job creation for the unemployed, women's cooperatives started their activities. Before establishing the Ministry of Cooperation up to 1990, in sum, there were 65277 women gathered in 5507 cooperatives and since 1991 (when the Cooperative Act and establishment of the Ministry of Cooperation was approved), up to the end of March 1999, there were 1084685 women members out of the total 9627278 actual members of cooperatives across the country (11.4 per cent). During 1991 to 2002, the rate of women's membership of the cooperatives has been showing significant growth.

During 1997 to 2003, the rate of growth in women cooperatives was 187.3 per cent; in other words, number of women cooperatives was doubled during the period of study.

Table 5.7: Number of Women's Cooperatives

Year	*Number*
1997	624
1998	469
1999	929
2000	1124
2001	1793

It should be noted that women cooperatives are those establishments where more than 50 per cent of members are female (The Guttmacher Report on Public Policy, 2001). These women's cooperatives work in areas of production and service.

8.0 Political Participation

Political participation is the basis of political life and in a democracy it is the right of all citizens to be able to participate in politics. Therefore, "the necessity of participation of women from development viewpoint which constitute 50 per cent of the population is crucial for achieving actual and more sustainable development, because today, development is impossible without participation of all the people, and democracy is impossible without recognizing the equal rights of women" (The Guttmacher Report on Public Policy, 2001).

Granting equal voting rights and accepting the presence of women in politics are prerequisites of women's participation in political activities. In this section, we take a brief look at the progression in the presence of Iranian women in politics.

The changes in the presence of women members of parliament in the National Parliament of Iran before the Islamic revolution can be seen in Table 5.8. It shows an increase from 1963 to 1975.

Table 5.8: Relative Distribution of Women Elected to the National Parliament during the Pahlavi Period

Term	*Twenty-first (1963)*	*Twenty-second (1967)*	*Twenty-third (1971)*	*Twenty-fourth (1975)*
Number	6	7	18	20
Percentage	3	3.3	6	7.5

As Table 5.9 a indicator that in the Iranian Senate during the period between 1963, when women were granted voting rights, and the Islamic Revolution, (1979) the number of women members always fluctuated between 3.3 and 5 per cent.

Table 5.9: Relative Distribution of Women Elected at the Senate during the Pahlavi Period

Term	*Fourth (1963)*	*Fifth (1967)*	*Sixth (1971)*	*Seventh (1975)*
Number	2	2	3	1
Percentage	3.3	3.3	5	3.3

Table 5.10 shows the increase in percentage of female candidates from the first session of the Islamic Legislative Assembly until the seventh session (1979-2003). The number of women candidates and elected women members of parliament (MPs) shows an increase during these years; however, the percentage of women among candidates is greater than the percentage of women MPs. It shows that more women are interested in political posts but they do not have support among voters, which could mean that voters lack confidence in women MPs.

Table 5.10: Relative Distribution of Female Candidates and Females Elected at Islamic Legislative Assembly

Term	*Candidates*		*Elected*	
	Females	*Per cent*	*Females*	*Per cent*
First (1979)	90	3.02	4	1.66
Second (1983)	32	1.98	4	1.66
Third (1987)	47	1.99	5	1. 6
Fourth (1991)	86	3	9	3.6
Fifth (1995)	351	6.43	14	5.6
Sixth (1999)	504	7. 3	13	4.4
Seventh (2003)	806	9.89	12	4.4

8.1 Islamic Councils of Cities and Villages

Councils are one of the efficient tools of political participation in today's world and perform some of the duties of the central government in dealing with local and regional problems of different countries.

According to the law of 'Organization, duties and election of Islamic Councils' (passed in May 1996 by the Islamic Legislative Assembly), Councils known as Islamic Councils of villages, towns, cities and districts were created for the fast progress of social, economic, and civil projects. On March 9, 1999, the election of Councils was held in Iran for the first time.

Table 5.11: Gender Combination of Members of Islamic Councils of City and Village

Term	*Gender Combination of Members of Islamic Councils of City and Village*			
	Men		*Women*	
	Number	*Per cent*	*Number*	*Per cent*
First (1999)	162649	99.16	1373	0.84
Second (2003)	152974	98.49	2338	1.54

As Tables 5.11 and 5.12 indicates that the number of candidates for Islamic city and village Councils elections has decreased; and also, their political participation as members of the mentioned councils has increased. The rate of growth in per cent of female members of the Islamic Council of Cities and Villages during 1997 to 2003 has been 79.76 per cent. The percentage of participants in the elections is another factor of political participation of women. Regarding these statistics, the number of women who participated in the sixth presidency election was 6884031 (43.99 per cent) (The Guttmacher Report on Public Policy, 2001).

Table 5.12: Gender Combination of Candidates to be elected to Islamic Councils of Cities and Villages

Term	*Gender Combination of Members of Islamic Councils of Cities and Villages*			
	Men		*Women*	
	Number	*Per cent*	*Number*	*Per cent*
First (1999)	328862	97.84	7276	2.16
Second (2000)	213091	97.32	5866	2.88

9.0 Conclusions

As indicators in this paper demonstrate, Iranian women's participation especially in academic fields is widespread. Iranian women have been working hard to improve their status in society. Statistics show there is a relationship between decrease in fertility rate, increase in literacy rate and increase in number of female students. Although Iranian Women are confronted with many problems, change in their position has been positive. According to gender justice, Iranian Women will achieve more success in the future.

References

Farhikhteghan Magazine, 24th Year Book, Azad University, 23/5/2006.

Mousavi, Beladi, Sadreddin, Seyed., 1997. Social and economic indexes of women in the Islamic Republic of Iran, statistics and informatics section, Women Participation Affairs Centre, Tehran, p.1.

Mohebati, Mohammad Goli., 2003. 'Azad University' Encyclopedia of Higher Education, Tehran, Bonyad Daneshnamei Bozorg Farsi, pp.387-391.

National Report on Women's Situation in the Islamic Republic of Iran, 2001. Women Participation Affairs Centre, Tehran, p. 24.

Report of Changes in Women's Situation during 1997 to 2003, Women Participation Affairs Centre, Soureh Publication, Tehran, 2005, p.40.

Report of Changes in Women's Situation during 1997 to 2003, Women Participation Affairs Centre, Soureh Publication, Tehran, 2005, p.41.

Report of Changes in Women's Situation during 1997 to 2003, Women Participation Affairs Centre, Soureh Publication, Tehran, 2005, p.42.

Report of Changes in Women's Situation during 1997 to 2003 , Women Participation Affairs Centre, Soureh Publication, Tehran, 2005, pp.49-50.

Report of Changes in Women's Situation during 1997 to 2003, Women Participation Affairs Centre, Soureh Publication, Tehran, 2005, p.58.

Report of Changes in Women's Situation during 1997 to 2003 , Women Participation Affairs Centre, Soureh Publication, Tehran, 2005, pp.25-26.

Shamim, Alim, 1994. 'Participation of Women in Political Power', Translated by Heydarzadeh, Hamshahri Newspaper, April 3, quoted from Petros Ghali.

The Guttmacher Report on Public Policy, Islam, Women and Family Planning: A Primer, December 2001, pp. 6.

The World Bank. 2003. Gender and Development in the Middle East and North Africa Region: Women in the Public Sphere, Washington DC, p.36.

Women Participation Affairs Centre. 2005. Report of Changes in Women's Situation during 1997 to 2003, Soureh Publication, Tehran, pp.19-20.

Zanjan, Susan and Saber, Firouzeh., 1995. Study of the Role and Place of Women in Legislation, Policy Making, Planning and Execution in the Country, Series of Papers at Congress on Women's Role in Science, Industry and Development, Office of Women's Affairs, Tehran, pp.70-72.

Chapter 6

Taking Action to Empower Women Scientists and Technologists through Women in Science and Technology Network (WISTNET) in Malawi

Meya Patricia Kalindekafe[1] & Symon Osman Mandala[2]

[1]*Chairperson of WISTNET Board of Trustees and Senior Lecturer in Ecology, University of Malawi, Chancellor College, Department of Biology, Zomba, Malawi*

E-mail: meykalinde@yahoo.co.uk, lkalindekafe@chanco.unima.mw

[2]*Principal Science and Technology Officer, Ministry of Education, Science and Technology, Department of Science and Technology Magetsi House, Lilongwe 3, Malawi*

E-mail: mandalamphatso@yahoo.com, smandala@mistmw.org

ABSTRACT

Women scientists and technologists in Malawi like in many other parts of the world hold the main family responsibilities and are expected to combine career and family commitments, often putting the needs of their spouse's career ahead of theirs. They also have to work harder than their male counterparts to prove themselves, sacrifice family priorities or face risky situations in order to be treated at par with their male colleagues. They also face other challenges common to women in other countries such as societal attitudes, age differences, inadequate leadership training and negotiations skills, inadequate role models and mentors and lack of networks. The government of Malawi and nongovernmental organizations have developed a number of programmes to encourage girls to take science-based subjects. Although, women in Malawi are as free as men to pursue and advance in the scientific careers of their choice, in some cases women are up to now still not rewarded on equal terms to their male counterparts with similar abilities. In recognition of the challenges faced by women in science and the potential role that women scientists and technologists play in socio-economic development of Malawi, a group of women scientists and technologists from the academia, private sectors, civil societies and government formed the Women In Science and Technology NETwork (WISTNET) in order to promote the participation of girls and women in science and technology. The network includes women who are information technologists, biologists, chemists, engineers, agriculturalists, environmentalists, and those in the life sciences. These women are in government, universities, the private sectors, non-governmental organizations, community based organizations, faith-based organizations, primary and secondary institutions and some are self-employed. The paper outlines the specific challenges faced by women scientists and technologists in Malawi and presents the main goal and objectives of the network. It also presents the specific actions planned by WISTNET to encourage and empower women who choose science and technology college programmes to stay in science or take up careers in science and encourage girls in primary and tertiary to take science and technology subjects. The paper also outlines proposals on formation of linkages with other networks, development of mentorship programmes, development of innovative programmes to promote uptake of science and technology, ideas for supporting women researchers and professional development of women scientists and technologists.

Keywords: *Mentorship, Linkage, Professional development, Role modeling, Empowerment scientist, Technologist.*

1.0 Introduction

The role of Science and Technology (S&T) in the socio-economic development of any country cannot be over-emphasized. Science and Technology has contributed to the discovery and application of various solutions to problems the world has faced from time to time. These vary from food security, disease control, agricultural productivity and improved health delivery services, among others.

The Government of Malawi recognizes the role that Science and Technology can play in industrial development in the country. The Department of Science and Technology was established with a mission of promoting rapid industrialization through utilization of science and technology so as to contribute towards the efforts of transforming Malawi from being a predominantly importing and consuming nation to a predominantly producing and exporting country.

The role of women in the socio-economic development of any country also cannot be overemphasized. However, realizing women's potential has been a challenge. Malawi recognizes the significant role women play in the development of the country. As a result of this recognition, the government is promoting the meaningful participation of women scientists in its quest to transform the country. The active engagement of women scientists has been as a result of important indicators such as population dynamics among others. For instance, statistics from the population census of 2008 indicates that 51 per cent of Malawi's population is female (National Statistics Office, 2008). But the participation in and utilization of S&T by women has not received the attention it deserves (Government of Malawi, 2002). Not many women are motivated to study S&T subjects at secondary and tertiary education system. The development and transfer of technology often does not focus on the needs of women. Malawi has noted that the participation of women in the labour market and their equal access to employment requires, among other actions improving women's access to technologies that facilitate their occupational and domestic work, encourage self support, generate income, transform gender prescribed roles within the productive process and enable them to move out of stereo-typed, low paying jobs.

But the S&T information is not readily available to the majority of the Malawi population and more seriously to women. This has been aggravated by low levels of literacy of the majority of the population, limited S&T information centres, inappropriate curriculum on S&T subjects in schools, inability to fully implement the S&T Policy, among others, above all, lack of a body to champion issues of science and technology among women and girls.

2.0 Specific Challenges Faced by Women Scientists and Technologists in Malawi

Like most countries, women in Malawi are faced with a number of challenges including:

- Lack of appreciation of the potential role women can play in S&T and the related contribution in development of the country
- Lack of resources for women to further pursue advanced training opportunities and also advocate for female participation in S&T through role modelling, mentoring and community awareness
- Brain drain of professional women in S&T due to lack of financial motivation that would create opportunities for professional development for female Scientists and technologists.
- Lack of a database for female scientists and technologists in Malawi for networking.
- Lack of follow up to past and present efforts in promoting female participation in S&T.
- The teaching of S&T at the primary and secondary school levels in Malawi has little regard for the needs of girls and preparation of girls for careers in science.
- Lack of investment in infrastructure for training institutions, research and human capital
- Careers in S&T are not perceived as being prestigious and fashionable by the society because the nature of the work is often unattractive and too demanding
- Dropping out of talented women out of science instead of moving up the career ladder and that those who remain are less likely to attain leadership positions.
- Lack of role modelling and mentorship for girls.

3.0 Formation of Women in Science and Technology Network (WISTNET)

In recognition of the challenges faced by women in science outlined above and the potential role that women scientists and technologists play in socio-economic development of Malawi and the impact that science and technology in Malawi have on society and their contribution to the discovery and application of various solutions to problems faced from time to time such as food insecurity, disease control, agricultural productivity to improved health delivery services, among others; a group of women scientists and technologists from the academia, private sectors, civil societies and government formed Women In Science and Technology NETwork (WISTNET) in August 2006 in order to promote the participation of girls and women in science and technology.

The network includes women who are information technologists, biologists, chemists, engineers, agriculturalists, environmentalists, and those in the life sciences. These women are in government, universities, the private sectors, non-governmental organizations, community based organizations, faith-based organizations, primary and secondary institutions and some are self-employed.

The Women in Science and Technology Network is viewed as a launch pad for active and meaningful participation of female scientists and technologists in the development and application of science and technology. The network intends to publicise the role of science and technology, mobilize the women and girls to participate in science initiatives, and lobby for greater recognition of women scientists roles in socio-economic development of the country. Although the network has this enormous challenge and noble task to achieve, implementation is in most cases hampered by unavailability of adequate funds. The network has so far been supported financially by the Department of Science and Technology, and this support has been on activity basis. WISTNET has three types of members (WISTNET, 2008[1]).

Full member are women scientist or technologist who have completed their first degree in natural and / or exact sciences, both pure and applied (including Information technologists, Biologists, Chemists, Engineers, Agriculturalists, Medical scientists, Environmentalists and all those in the life sciences) or scientific institutions committed to the objectives of the network. Associate members are individuals who shall be non-scientists and non-technologists or institutions that are non-scientific bodies but interested in furthering the objectives of WISTNET. Candidate members on the other hand are girls or women who have not yet completed their first degree in science and / or technology. Currently about 150 women have completed forms of registration for membership.

3.1 Main Goal

The overall goal of WISTNET is to contribute to the socio-economic development of Malawi and make science more responsive to the needs of society, by promoting science and technology among women and girls through scientific and technological dialogue, capacity building and research.

3.2 Specific Objectives

- Enhance partnerships amongst women in science and technology through linkages at national and international levels;
- Promote increased participation of female scientists and technologists in science and technology;
- Advocate for the participation of women scientists and technologists in decision making and in addressing problems that require scientific and technological intervention;

Proposed Actions/Outputs to Empower Girls and Women in Science and Technology

In order to achieve its goal and objectives, a number of actions have been proposed. These include launching and strengthening WISTNET, establishing networks, conducting a baseline survey, role modelling and mentoring for young girls and enhancing significant recognition of women scientists' contribution to national development (WISTNET, 2008[2]).

Output 1: WISTNET Launched and Strengthened

WISTNET plans to organise a launching ceremony at the beginning of 2009 where renowned women scientists would be invited and make presentations. Once the network is launched, panel discussions will be held and strategic plan developed as a way of strengthening the network.

Output 2: Networks Established

WISTNET will strive to establish networks with various institutions with similar objectives with the aim of sharing information and collaborating in various activities. This will be through hosting science and technology meetings, attending both local and international conferences on science and technology and online networking.

Output 3: Establishment and Maintenance of a Database on Women Scientists and Technologists

Currently there is no disaggregated data on women scientists and technologists. WISTNET will initially conduct a baseline survey. The survey will make a documentation of all women scientists and technologists and publish disaggregated data on a regular basis. Challenges hindering women participation in science and technology will also be documented.

Output 4: Role Modelling and Mentoring Programme Conducted

This programme should provide mentoring for young women in schools, colleges and universities as well as those in science and technology careers. It will be important to ensure that the programme includes provision of training to mentors to ensure that the mentor-mentee relationship provides meaningful and lasting results.

Women scientists and technologists in Malawi will actively tell their success stories and act as champions and role models for girls in schools and other women who were either trained as scientists but are not practicing science. Through

WISTNET, we would like to encourage training of a lot of girls in science and technology, retain the trained women and call back those who left science.

Women scientists and technologists in Malawi will strive to encourage girls to take science by separating some of the common myths from reality. WISTNET will for example actively tell girls and other women that aspiring for a scientific career should not entail a compromise in a girl's or woman's appearance or family life. Girls will be told that it is perfectly possible for a good scientist to be a stylish career woman as well as a good mother and that it is not appearance that counts, but competence.

The following activities are proposed in order to achieve output 4.

(*a*) Identify role models in S&T to make presentations to the youth.

(*b*) Conduct Career talks in S&T in primary and secondary schools.

(*c*) Identify women achievers in S&T career and present their experience to fellow women.

(*d*) Organize College Science Camps for Secondary School Students. The purpose of the College Camp would be to engage secondary school students in science and technology subjects by hosting them at a college campus for a period of time and giving them a taster of college life, the kinds of subjects they would have the opportunity to learn and the types of careers they could branch into. It is hoped that such a programme would broaden their minds and help them to make more informed choices about their careers.

(*e*) Produce and distribute Brochures and Career guides to primary and secondary schools. WISTNET will produce brochures describing the different science and technology offerings within the tertiary education system in Malawi as well as career guides with living examples of the career options available through a science and technology education.

Output 5: Introduce a TV/Radio Programme on Women in Science and Technology

Output 6: Capacity Building

WISTNET will promote capacity building amongst girls and women through:

(*a*) *Advocacy for Bridging Programmes*: WISTNET plans to advocate for bridging programmes to enable women with deficiencies in mathematics and science subjects to access science and technology courses at college. The bridging programmes would provide an extra year or more of tuition to help capable young girls to overcome deficiencies in mathematics and science that may have been caused by external factors beyond their control

(*b*) *Support for girls in science and technology courses.* WISTNET needs to provide support for girls taking classes in science and technology related subjects at college. Such support could include emotional support, advice and coping strategies from those that have been through the system and extra tutoring to help those facing difficulties. This would help to increase the retention of those girls that take science and technology courses.

(c) *Support for Female Researchers such* as:

(*i*) *Scientific writing retreats*: The advancement of women in science and technology requires women to actively engage in research. Research activities require the writing of proposals to solicit funding and lead to publication of papers for conferences and journals. Women need to be encouraged to conduct research and publish and one way of doing this is to have writing retreats. The writing retreats would invite participation from women that have ideas for research but are failing to articulate them into a proposal, or those that are conducting or have conducted research and that need to publish their work. Each retreat would be structured such that at the end of the retreat, every participant would have a tangible product – proposal or paper – ready for submission.

(*ii*) *Peer-reviewed publication*: A peer-reviewed publication would encourage women to write about their research work and provide the much-needed experience and confidence to produce work for international publication.

(*iii*) *Funding for presentations at conferences*: Women engaged in research should be encouraged to present their work at national, regional and international conferences.

(*d*) *Professional Development Workshops*. WISTNET should conduct professional development workshops for women in science and technology to ensure that women stay current within their chosen professions. The workshops should also address other issues of relevance to women and that have a bearing on their professional development such as health and wellness, financial management, work-life balance.

5.0 Workshop Expectations and Conclusion

Expectations

WISTNET expects to create partnerships and linkages with other women and women's networks with interests similar to WISTNET at this workshop.

Traditionally, women's networks tend to revolve around socio-cultural activities such as weddings, funerals, church and this does not provide much space for conversation about issues related to professional development (Nyirenda-Jere, 2008). WISTNET would like to through this workshop to find how it and other similar stakeholders can find space within traditional planned women's activities for members to have the opportunity to engage meaningfully with each other and other stakeholders on issues of interest and relevance to the network's objectives.

Mentoring and role modelling are critical elements of any strategy aimed at improving the access to and participation of women in any aspect of socio-economic development, and for WISTNET, the promotion and uptake of science and technology by women in Malawi. WISTNET wishes to learn from other scientists how they have run the mentorship programmes that were responsive to particular needs, that clearly articulated the role of the mentor and mentee, that had well prescribed mechanisms for providing feedback and follow-ups, that had adequate support structures for both the mentor and mentee – such as training, financial and other resources.

6.0 Conclusions

Women need to understand that advancement on any front requires strategic thinking to clearly identify the objective(s), define the resources that will be required to achieve the objectives, map out a plan of action and put the plan in motion. Often times, as women we yearn for advancement yet we fail to articulate what that means and determine how we will achieve it. This kind of thinking is critical for any network and it is therefore very important for WISTNET and other women's organisations to inculculate in its members the idea of strategic thinking as a means of advancing in the science and technology sector.

The gender conversation among women needs to change from a "them and us" attitude to a "them with us" one. Women need to recognise that there is much to be learned and gained from forging partnerships with male colleagues and professionals rather than seeing them as the enemy.

It will thus be important for women in science organizations to ensure that the male voice or perspective is not lost in the activities that are conducted. Promoting participation of girls and women should not translate into alienating and sidelining boys and men.

7.0 Acknowledgements

We wish to acknowledge the Centre for Science and Technology of Non-Aligned and Other Developing Countries (NAM S&T Centre) for the financial support to attend this workshop which is very relevant to the goals of WISTNET and efforts by the Department of Science and Technology in Malawi in meaningfully engaging women scientists. Finally we wish to thank the Malawi government through the Department of Science and Technology for nominating a representative to attend this important workshop.

References

Government of Malawi (2002). National Science and Technology Policy for Malawi.

National Statistical Office (2008). 2008 Population and Housing Census. Preliminary Report. NSO. Zomba. Malawi

Nyirenda-Jere Towela (2008). Report on the HERS-SA Academy 2008. Submitted to UNICEF Malawi and WISTNET Board of Trustees.

WISTNET (2008[1]). The Constitution of the Registered Trustees of the Women in Science and Technology Network.

WISTNET (2008[2]). Promotion of Women in Science and Technology Development and Application for Wealth Creation and Socio-Economic Development. Proposal for Funding.

Abbreviations

WISTNET:	Women in Science and Technology Network
UNICEF:	United Nations Children Fund
NSO:	National Statistical Office
S&T:	Science and Technology

Chapter 7

The Status of Women in Engineering in South Africa and the Efforts of SAWomEng to Facilitate the Absorption and Retention of Females in the Engineering Sector

Naadiya Moosajee
SAWomEng (South African Women in Engineering),
Department of Civil Engineering, University of Cape Town, South Africa
E-mail: naadiya@sawomeng.org.za

ABSTRACT

Over the last ten years, there has been a marked increase in the number of women entering the engineering industry, but women are still grossly under represented. There is explicit lack of women in senior positions within the engineering industry in South Africa. There also seems to be bottlenecks at a tertiary level, where female engineering students are opting to join alternate industries. To combat this issue, amongst others, South African Women in Engineering (SAWomEng) was founded. It is a unique organisation in that it was founded by students at the University of Cape Town, South Africa, and promotes engineering at a tertiary level, and recently at a high school level, through an annual National Conference. The organisation hopes to make a positive contribution to the plight of women engineers and share lessons learnt in founding a unique organisation at a National Level.

Keywords: *Women in Engineering, Retention, SAWomEng*

1.0 Introduction

"Our country requires engineers, in different categories, given the major infrastructure development that will take place in the country. We need to have young women actively participate in these activities." Education Minister Naledi Pandor at the Pretoria Girls' High School Valedictory Function, 5 October, 2006.

The shortage of engineers is a global issue. The number of women in engineering has been of particular concern as women are still grossly under-represented, with the percentage of women graduates in engineering still below 20 per cent in many countries. According to many commentators, growth in the areas of science, engineering and technology could be the catalyst for job creation, social upliftment and economic development. Perhaps even more significantly, there has been a failure to capitalise effectively on our most under-utilised resource: women.

Issues surrounding gender equality and the emancipation of women have become highly publicised in recent years. Significant changes favouring the upliftment of women in the workplace have begun to influence industries across the board and are reflective of the policies and structures being executed at a national level. However, it is evident that many more practical measures need to be implemented in the male-dominated sectors of our economy. Research and statistics indicate clearly that transformation in the engineering industry has been particularly slow and therefore the sector should become much more aggressive in its stance to empower females in the field.

In this dynamic environment, the most successful industry players will be those who are able to position themselves strategically. This involves utilising opportunities stemming from the long-term movement towards country-wide gender equality. If support and advocacy for initiatives of this nature are encouraged at all levels, a more positive attitude will be cultivated which recognises the contributions made by females towards the economic development of South Africa. It should be realised that a corporate culture which advances the careers of women will also foster profitability.

In order to address the massive skill shortages experienced, and simultaneously promote progressive thinking in this arena, it is vital that tomorrow's future captains of industry are at the core of this process. While many women's networks and support groups have been established in recent years and months, few focus on the plight of the young women entering the specialised field of engineering.

2.0 Status of Women in Engineering in South Africa

In the last ten years, there has been a marked increase in women's entry into the higher education system of South Africa., however, within the SET (Science, Engineering and Technology) sectors, women are still largely underrepresented. The bottleneck in the education system gets even smaller at postgraduate degree levels, particularly in the field of engineering. This results in only a small proportion of South African women end up working in the SET sector. As available literature suggests, an even smaller number ends up being retained and moving up the promotion ranks within the sector (Moletsane, 2008).

This trend can be seen in the Corporate Leadership Census (Business Women Association and Catalyst, 2004), where approximately 60 per cent of companies in the SET industry had no women board directors. The study also showed the stark imbalance in gender and racial representation within the SET industry. Table 1 shows the results of a study, Assessment on the participation of Women in SET, prepared for the Department of Science and Technology on the participation of women in all levels within the SET industry. It can be noted that there is an increase in the number of women at junior levels.

Table 7.1: Participation of Women in the SET Industry (Moletsane, 2008)

Level of Seniority	*Mean*
% share of direct shareholding by women in the company (direct shareholding refers to employee share schemes	0%
% share of indirect shareholding by women in the company (ownership of women outside the organisation)	0%
% of women members of boards of directors	10.88%
% of women in Executive Management	11.63%
% of women in Middle Management	26.10%
% of women supervisors	29.52%
% of highly skilled women who are not in management/supervisory positions	18.03%
% of women in the junior professional category	29.04%

One of the strategies suggested within the study to address the gender imbalances, is to increase the recruitment, retention and participation of women in the economy, particularly in the SET sector. Increasing the number of women in the national system of innovation is both a human right as well as a development issue. As the Department of Science and Technology (DST) acknowledges access to Science and Technology (S&T) for women is necessary to transfer patterns of productivity, contribute to job creation and new ways of working and promote the establishment of a knowledge-based society resulting in wealth creation.

Access by women to Science and Technology also has the potential to positively contribute to improving the quality of life of women, their livelihoods and households (http://www.naci.org.za/accessed November, 2008). Against this backdrop, SAWomEng presents an invaluable platform for the advancement and education of female engineering students.

3.0 SAWomEng Background

SAWomEng (South African Women in Engineering) is a non-profit, student-run body founded at the University of Cape Town. Its mission is to act as an anchor and support base for young female students of all backgrounds and engineering disciplines. Using a proactive and holistic approach, the organisation seeks to both identify the brightest female minds in engineering and provide them with the crucial soft skills necessary to succeed in the South African engineering industry. SAWomEng is committed to supporting UN Millennium Development Goal 3 to promote gender equality and women empowerment.

The organisation was the brainchild of two then third-year University of Cape Town civil engineering students, Naadiya Moosajee and Mabohlale Mampuru, who envisioned a platform through which young women in the field, like themselves, would be able to fast-track their careers. This inspired the creation of SAWomEng, with the annual national conference serving as the foundation for the meeting of the aims and objectives of the organisation in order to promote the retention of women in engineering.

4.0 The Concept

Revolutions have changed fortunes and redirected history. Much revolutionary spirit has stemmed from the hearts and minds of women. SAWomEng hopes to sustain this spark and effect change by challenging mindsets, fostering supportive structures and empowering aspirant female engineers before they enter the world of work. The current skill shortage crisis in the engineering industry is traceable to a number of factors.

These include, amongst others, the previous depression in the industry as well as marketing of the engineering field, especially to university students. Because female engineering graduates lack insight into possible career prospects and opportunities in engineering, it results in the infamous brain drain of future engineers into the more alluring commercial arena for instance.

5.0 The Five Founding Pillars

5.1 Igniting the Engineering Flame

SAWomEng aims to remind students about the exciting, innovative, groundbreaking work that is being done in the engineering industry. This is to prevent the migration of students from engineering into other industries. There is a lack of awareness amongst students, about the industry and industry related issues, such as current projects and career opportunities. This makes engineering seem like a dull and non-lucrative industry as opposed to the more alluring or rather, the better-marketed commerce industry. Thus, we believe that by creating an opportunity for companies to market themselves and inviting industry leaders to share their experiences, we are able to 'reignite the flame'. Students can get excited about their choice of study, and have a window into their field of engineering.

5.2 Mentorship

Being one of the minorities in the engineering industry, female students often lack support. Thus we believe that it is absolutely crucial that links are formed between engineers who are practising in industry and students who aspire to work in industry. SAWomEng will create the necessary platform to foster these ties. This will in turn facilitate the transfer of knowledge and skills.

5.3 Incubating the Engineering Mind

SAWomEng is a safe heaven for the brightest female minds in engineering. Our main aim is to showcase not only the technical brilliance of these minds but also the refreshing perspectives these minds bring to the engineering industry. A technical project will thus form part of the conference. The project will see the students tackling a topical and relevant engineering problems in a multi-disciplinary fashion.

5.4 Meeting of the Minds between Old and New Engineers

Young engineers contribute in great numbers to the exodus of engineers from the engineering industry. We feel that young engineers are misunderstood by industry. Industry is out-of-sync with the needs and priorities of this rapidly evolving species. It is thus necessary to create a meeting ground for old and young engineers. The conference will thus help female engineering students engage with stakeholders in the engineering industry in a relaxed and social environment so as to promote dialogue and networking.

5.5 Embracing the Multi-faceted Nature of Female Engineers

The engineering woman can be likened to a multi-faceted diamond. SAWomEng aims to nurture the facet of women most neglected by the industry: their femininity.

6.0 SAWomEng Vision

The SAWomEng philosophy is inextricably linked to its five founding pillars. Whilst all the pillars are equally important, the mandate of the organisation is to increase the population of women engineers. Currently, the organisation is pitched at female engineering students in an aim to retain them in the engineering industry. It is however equally important to attract women engineers into the engineering field. Our goal is to attract high school girls into the field of engineering.

South Africa is plagued not only by a skills shortage, but also a skills gap. Graduates entering industry face a double-edged sword in the form of fast-tracking and poor mentorship. As a result of the skills shortage, young engineers are forced to accept huge responsibilities early in their careers. However, industry's capacity to provide adequate skills transfer and adequate mentorship has been eroded.

7.0 Girl-Eng

7.1 What is Girl-Eng?

Girl-Eng is an extension of SAWomEng. It is aimed at female students in Grade 11 and Grade 12, promoting maths and science as well as career options in engineering. Girl-Eng advances two of the five SAWomEng pillars, namely: Mentorship and Igniting the Engineering Flame.

7.2 Mentorship

The SAWomEng delegates have been an under-utilised resource thus far. Delegates will now deliver speeches at high schools in their provinces post the conference. Each delegate will also mentor at least one female high school student. We believe that through this mentorship, school girls will be inspired to study engineering and will excel in industry. This is because challenges in both the academic and working worlds will not be insurmountable, as they will have SAWomEng 'ex-patriots' to counsel them.

7.3 Igniting the Engineering Flame

Interaction between the school girls and SAWomEng delegates will result in mutual excitement about the engineering industry. This will aid both the attraction and retention of female engineers, hence 'igniting and fuelling the engineering flame'. The SAWomEng website will play a pivotal role in advancing 'Girl-Eng'.

7.4 Girl-Eng Event 2009

Six schools in each of the four provinces, Western Cape, Gauteng, Kwa-Zulu Natal and Northern Cape will be identified. Sixty female students, excelling in maths and science will be selected to participate in a day long Girl-Eng event in each of the provinces. One of the aims of this event is to bring students from disadvantaged communities in particular, which have no access to information or insight in to the field of engineering to these events. During these events, students will meet top women in industry, participate in exciting activities with sponsor company representatives and get insights into the field of engineering.

8.0 Creating Positive Change through a National Conference

The conference on the theme "Educate, Empower and Celebrate female engineering students" held on the 26th August, 2006 in Cape Town, culminated in a networking dinner. The Conference encouraged 100 top female UCT engineering students to rise to the challenges currently being experienced in industry. Top women in the engineering industry, educational institutions as well as captains of industry addressed the delegates. This addressed another fundamental goal of SAWomEng – which is to bridge the gap between students and industry, by facilitating valuable networking opportunities and promoting mentorship programmes.

2007 saw the launch of the first SAWomEng National Conference, held in Cape Town from the 9th–13th July. South Africa's top fifty female engineering students assembled in Cape Town. Delegates were faced with the challenge of identifying solutions to engineering problems surrounding the 2010 Soccer World Cup. The Conference being over five days enabled valuable workshops and sessions to be held, with leading women in the SET sector. Delegates were also assigned mentors from the engineering industry. The mentorship programme allows for informal interaction between women in the field with women wanting to enter the field of engineering. One of the important session was a round-table discussion with students and women in industry discussing the main barriers to entry for female students.

The 2nd National SAWomEng Conference saw the organisation grow from strength to strength. 60 delegates, from South Africa's top Universities and Universities of Technology were selected to participate in this 5-day conference. The technical project is a vital component of the SAWomEng Conference, which distinguishes the conference from other events. The 2008 technical project comprised a multi-disciplinary expose of South Africa's energy crisis. Delegates were exposed to pertinent issues surrounding the energy crisis and had to find innovative solutions. The aim of the projects is to promote critical thinking within the next generation of engineers and to expose the delegates to critical issues facing engineers.

9.0 Lessons Learnt

Over the last three years, SAWomEng has grown into a National organisation educating and promoting the retention of women in engineering. During this time, 210 students have gone through the Conference. Every year, the organisation grows and becomes stronger, but we still have a long way to go. There are some lessons which we have learnt which other organisations can perhaps learn from. These include:

- Never be afraid to ask for help. When SAWomEng was started, there was a lot of consultation with senior female lecturers at the University of Cape Town. A lot of these women referred us to other prominent women in industry and shared an unofficial network of women in engineering. A vision of SAWomEng is to bring this network together as a non-executive council, to provide experience and support.
- Work with passionate people. At SAWomEng, we realise how many women and men there are, who are sympathetic to the plight of women in engineering. We have many people who apply to be part of the executive, from both genders. The way to change the mindset of women in engineering is not to form "an old girls club" but to include anyone who is passionate about the work that we are doing
- Building the network. The power of networking is often understated. Many of the contacts and funding opportunities came from talking to as many people as possible and attending many events. At SAWomEng we realise the importance of networking and try and impart this advice on the delegates attending the conference
- Funding is always a battle, but persistence is the key. SAWomEng is entirely funded by corporate engineering industry in South Africa. The way the organisation is marketed makes it not only an excellent recruitment opportunity for companies but also an excellent corporate social investment initiative. We deliver a good product and in that way, keep sponsors coming back each year. It is important to be persistent, professional and have a good proposal.
- Never give up. There are many times when you feel like you are fighting a losing battle. When starting a new organisation, it is difficult to get people to trust you. It is important to keep a good reputation and believe in yourself enough to bring everything together. Listen to advice, but trust your instincts and work hard to achieve your goal.

SAWomEng is the first of its kind in Africa and the world and there are very few similar organisations particularly targeting women currently in tertiary institutions. We have found that there is an increase in promoting engineering, but the bottleneck is at tertiary level. The competition for engineers is not just global one, but also an industry wide one. Female engineers are particularly sought after by the Commerce and Banking related industry, and as the SET industry, we need to position ourselves so as to attract female students into the SET environment.

10.0 Conclusions

Women in the SET sector are still grossly under-represented. There are definite bottlenecks at tertiary levels and there are still very few women in senior positions in the SET industry. An increase in the number of women at junior levels, where they make up approximately 30 per cent, can be noted, but there is a definite problem with absorption and retention of women in the SET sector. Against this backdrop, a student founded and run South African Women in Engineering (NPO), was formed to promote the absorption of women into the engineering industry.

SAWomEng has five core principles, which have been identified as possible bridges for the gaps between female engineering students and the engineering industry. These principles are applied at an annual national conference. Since the inception of SAWomEng, 210 female engineering students have attended the conference. SAWomEng has also launched a second wind called Girl-Eng, which focuses on educating high school students about engineering, with particular focus on female students from impoverished and former disadvantaged communities. Important lessons learnt by the founders, when starting an organisation of this kind, is to build a network, and ask for advice from senior women in the field, as well as be persistent and professional at all times.

References

National Advisory Council on Innovation (NACi), *Women in Science, Engineering and Technology: Exploring the facts*, Http://www.naci.org.za/readmore/summary3.html, accessed 20 November 2008

National Advisory Council on Innovation (NACi), *Tracking women and men in the SET sector*, http://www.naci.org.za/readmore/summary2.html, accessed 20 November 2008

National Advisory Council on Innovation (NACi), Annual *Report*, 2007 – 2008

Mampuru, M. and Moosajee, N (2008) *SAWomEng Proposal*

Moletsane, R. and Reddy, V., 2008 *Report: an assessment of the participation of women in set industry for Department of Science and Technology.* (Commissioned by NACI, October)

Chapter 8

The Current Status of Women Scientists in Sri Lanka

P.R.M.P. Dilrukshi* & Seetha I. Wickremasinghe

Science and Technology Policy Research Division,
National Science Foundation,
Maitland Place, Colombo 7, Sri Lanka
**E-mail: dilrukshi@nsf.ac.lk*

ABSTRACT

A study was conducted to investigate the current status of women scientists in Sri Lanka along with and the problems and constraints faced by them in their career development. A postal questionnaire survey and the direct interview method were used to collect the necessary information. The study sample consisted of 1,628 women scientists and 65 per cent of the sample responded (1,058) to the survey. 26 per cent of the sample investigated had PhDs as their highest academic qualification. There was a significant ($p<0.001$) association between the age and the highest academic qualification of the women scientists. The most productive age duration for achieving the Ph.D. qualification was within 41-50 years followed by 31-40 years. The research productivity in terms of the number of publications was significantly ($p<0.001$) high in the women scientists of age group 31-40 followed by the age group 41-50. The designation level of women scientists also showed a significant association ($p<0.01$) with number of publications they produced. 68 per cent of the women scientists work under a male as their immediate superior while 81 per cent of the S&T sector institutions are headed by a male CEO. The analysis also showed that there is no gender bias or discrimination when selecting for rewards and recognition or when selecting for higher administrative positions in the S&T sector. However, indirect discrimination seemed to appear in the recruitment of women for science careers ($p<0.001$). Further, the respondent sample of the Sri Lankan women scientists who have participated more in training programmes, seminars, and other extra curricular activities and produced more publications well acknowledged the fact that the support of the spouse in equally sharing family responsibilities has assisted them towards their career development and achieve its goals.

Keywords: *Women scientist, Science and technology, Research and development, Science careers, Gender gap*

1.0 Introduction

The gender gap reduction in scientific career and achievement is an ongoing policy and a scholarly debate. Nevertheless, the occurrence of gender gap in the employment of Science and Technology (S&T) careers may be seen due to several reasons. The gap in the educational attainment of women compared to men in the area of S&T could be one of the main reasons for this disparity. Another aspect can be the lack of suitable career prospects for women in the field of S&T. The studies conducted in fields as diverse as engineering and biology have found that women scientists suffer from an attainment gap along with at least three important dimensions: productivity, recognition, and reward (Cole and Cole, 1973; Long and Fox, 1995). The study done on differences in promotion according to sex have had mixed results. Some studies have found that the promotion of women takes longer than that of men (Tesch *et al.*, 1995; Wallis *et al.*, 1981; Kaplan *et al.*, 1996; Dial *et al.*, 1989). Others have found that within academic departments' women and men advanced at the same rate or that sex differences in promotion rates disappear in younger cohorts (Tesch *et al.*, 1995; Nickerson *et al.*, 1990; Wilkinson and Lindae, 1986). However, the status of women in the Science and Technology careers in Sri Lanka has not been well reported despite the fact that the general economic and education indicators available in the country reveal a gradual increase in the participation of women in scientific careers.

With the introduction of free primary, secondary and tertiary education including university education in 1945 and the change in medium of instruction from English to the local languages Sinhala and Tamil, there was a rapid expansion of educational opportunities in Sri Lanka (ADB, 1999). With the above changes, the educational attainment of Sri Lankan women showed a marked improvement within past 40 years. The data collected by the Department of Census and Statistics (DCS) within the period of 1963-2000 revealed that the number of females not attending school decreased from 44.7 per cent in 1963 to 12.2 per cent in the year 2000 and the number of girls who attended secondary level education increased from 14.7 per cent in 1963 to 35.3 per cent in year 2000. The number of females passing G.C.E ordinary level examination also showed marked improvement from 2.6 per cent in year 1963 to 16.9 per cent in year 2000. Similarly, women achieving a degree or higher level of educational qualifications increased from 0.1 per cent in year 1963 to 2 per cent in year 2000. The number of female students enrolled to the universities in the areas of science and engineering showed gradual increase from 39 per cent in year 2002 to 46 per cent in year 2006.

Though the number of females achieving degrees in the area of science and engineering showed a marked improvement, their contribution to the professions in various fields of Science and Technology remain comparatively low. According to the national R&D survey 2006, women represented 41 per cent of the total scientists' population in the country. Of this number, 54 per cent of women scientists are employed in the Research Institutes and other Science and Technology related Institutions in the country followed by 41 per cent in Higher Education sector. Only 4 per cent are employed in Business Enterprises and Private Non Profit organizations.

According to the same survey, when investigated on the distribution of women scientists across different disciplines it showed that female participation was higher in the areas of Natural sciences (34 per cent) followed by Medical Sciences (29 per cent), Agriculture sciences (15 per cent), Engineering and Technology (13 per cent) and other S&T related fields 0.6 per cent respectively.

Since the attainment of higher educational qualifications of women in the country is increasing gradually in the country, it is important to investigate the existing status of women scientists, their career achievements and the constraints faced by them. Such an approach will help in taking necessary steps to eliminate or reduce the barriers and encourage more women participation in the S&T careers. Since this area has not been investigated in Sri Lanka so far, the Science and Technology Policy Research Division (STPRD) of the NSF identified this subject as an important issue that need to be investigated. The main objectives of the study are to find out the issues pertaining to career prospects of women scientists and to highlight policy recommendations to reduce the constraints and problems associated with their careers and, then, develop necessary plans to improve women participation in science and technology sector employment in the country.

2.0 Materials and Methods

The primary source of data for the study was collected from a survey conducted in 2007 on women scientists in the country. The postal questionnaire survey method and the direct interview method were employed to collect the information of the women scientists who are working in the Higher Education Institutes, Research Institutes, S&T Service sector organizations and Private sector organizations. The secondary data were obtained from records and reports on the higher education sector, Department of Census and Statistics, newspapers and many other reports on gender issues pertaining to the country.

3.0 Results

The survey sample consisted of total of 1,628 of women scientists as recorded in the national R&D Survey 2004. Out of the sample, 65 per cent responded to the questionnaire (1,058). The responded sample consisted of 27 per cent of scientists working in the fields of agricultural sciences followed by 27.8 per cent in natural sciences, 6 per cent in engineering and technology field and 26.1 per cent in other scientific fields including the social sciences and humanities.

The survey sample comprised 13 per cent women scientists of age below 30 years, 46 per cent between 31-40 years, 31 per cent 41-50 years and 10 per cent above

51 years (Table 8.1). 26 per cent of the sample investigated had PhDs as their highest academic qualification and this represented the 87 per cent of the women scientist recorded for having the PhDs in the country. The survey covered a 53 per cent of the women Professors working in the higher education sector in the country.

Table 8.1: The Distribution of Sample According to Age and Academic Qualifications

Highest Degree Qualifications	*Age*				
	<30	*31-40*	*41-50*	*above 50*	*Total*
Bachelor Degree or related degree	64	221	93	29	407
M.Sc./ M.Phill or related	76	192	76	29	373
Ph.D. or upper or related degree	0	70	156	52	278
Total	140	483	325	110	1,058

The data collected through the survey was analyzed from various angles to understand the current status of women scientists and the progress made in their careers at academic levels as well as professional levels.

The analysis conducted to investigate the association between age and academic achievements of the women scientists showed significant ($p<0.001$) association between age and attainment of the highest academic qualification. Accordingly, the most productive age for achieving Ph.D. was within age of 41-50 years. However, most women scientists were able to achieve postgraduate qualification at Master level within 31-40 years of age. The outcome of the interviews conducted with the women families. scientists revealed different perspectives on the subject. The marital status of the women scientists has direct impact on the attainment of postgraduate degree.

The most common reason given for the slow progress in the career development was that they need to devote more time for the wellbeing of their children and spouse. Hence, the women scientists are compelled to delay their academic achievements until such time that the children start schooling and they do not need constant attention of the mother. The other reason given for the said delay was the lack of opportunity for women to do a Ph.D. degree compared to men as most superiors prefer to have male research students than female research students due to extensive field investigations involved in most S&T subject areas. The women scientists working in S&T institutions other than the higher education sector institutes had mentioned that when they were young recruits, it was very difficult for them to obtain duty leave or funding sources to seek postgraduate education. However, when they are senior and are promoted to higher grades, they are allowed to do post-graduate degrees to fulfill the requirement compulsory for the next promotion. Nevertheless, this barrier could be a common issue applicable to male scientists as well.

Another factor investigated for measuring the productivity of women scientists was the number of publications produced by them. The number of publications depends on the amount of research activities undertaken and their findings. The ability to do direct research depends on the independent ability to designed and plan a research project, to obtain facilities needed to carry out the same and the wherewithal to execute the research programme. To obtain the necessary funding and infrastructure facilities, the scientists have to devote time and energy to compete successfully with the peers. Most women scientists at the onset of their careers find it very difficult to achieve these targets as they have to divide their time between their family and their career and, also, they have limited experience. However, the study revealed that some young women scientists who were fortunate to get the advice and constant guidance from their senior counterpart had been able to secure necessary funds and successfully execute their research projects. Once the funding is secured the next step is to conduct research and publish the outcome in peer reviewed journals or filing for patents. The investigation done on the rate of publications by women scientists in different age groups showed that the number of papers published in the local and international journals was significantly ($p<0.001$) high for the scientists of age group 31-40 years followed by age group 41-50 years. The study revealed that women scientists who are fully involved in academic or research activities produce more publications than the ones involved in engineering and technology activities or administration and management activities. The designation level of women scientists also showed a significant association ($p<0.01$) with the number of publications they produced. The reason could be that the number of publications is also used as a criterion for promotions in the scientists working in the academic and research sectors. Therefore, most scientists in 31-40 years pay more attention to producing more publications to meet their promotional requirements. However, a comparison with their male counterparts in the same age group showed that there is a marked difference in the number of publications produced. The female scientists produced less number of publications than their male counterparts. The main reason for this difference would be the time devoted by both groups for research activities.

The interviews conducted with different female scientists on the issue of publications revealed that it takes more time for female scientists to complete the tasks as they are more duty conscious compared to male scientists. Hence, they seems to devote more time than males to fulfill the other duties and obligations assigned to them such as academic activities, administration etc. as well as looking after their families.

The other constraint mentioned by the female scientists was the social barriers. Accordingly, their male counterparts have more freedom to move in the scientific community and, therefore, they can easily form links with other parties that assist them to climb up their social and career ladder, while a large proportion of women scientists lack this opportunity. Women scientists happen to do it in their own time and the hard way, taking a longer time to reach the same level. Some traditional and cultural ethics also hinder them in taking part in some research projects that involve extensive field work and laboratory work in late hours. However, women scientists stated their desire to work in these projects if certain facilities such as proper transport or accommodation or flexible working hours etc. are provided to them. The less involvement in research activities as well as giving low profile in other external activities (*e.g.* contribution to social and professional bodies etc.) lessen the opportunities for young women scientists to receive awards or other academic achievements that are secured by their male counterparts of the same age. However, most female scientists in late 40s mentioned that they were able to achieve these targets only in late 40s when they have less family responsibilities and more experience in their life.

Another criterion that was used to measure the productivity of the women scientists was the participation in local or international seminars and workshops at the capacity or resource person and participants. The rate of participation was observed in different categories as age groups, marital status and number of children. The analysis showed that women scientists with two or more children participate more at seminars/workshops and work after hours compared to those with only one child. This interesting outcome may be due to the fact that women scientists with a number of children may be confident that the children have company in the absence of the mother. Most of these women scientists mentioned that their spouse and family were very helpful and assist them in taking part in these activities. A significant association ($p>0.01$) was seen in the age of the scientist and the participation in seminars and workshops.

Table 8.2: Participation at Seminars and Workshops (During past 3 years)

Age Categories (Years)	*Participation (Per cent)*		
	Yes (Per cent)	*No (Per cent)*	*Total*
below 30	38	62	100
31-40	33	67	100
41-50	42	58	100
51 above	80	20	100
Total	40	60	1001

The scientists above age 50 years had participated more frequently in the workshops and seminars than the women scientists below 40 years of age. The designation level of the women scientists and getting opportunities to participate in the seminars or workshops also has a significant association ($p<0.005$). This is because the scientists in the senior category with more responsible roles and authority receive more chances to participate in workshops and seminars than the junior or middle level scientists. However, such opportunities are still less for women scientists in comparison with the males in the same category. Some women scientists having children mentioned that, there are occasions when selecting for foreign training and participation in seminars etc., that their heads do not bother to consult them but had decided that married female scientists would not possibly participate or undertake the task because of children or spouse. Therefore, the superiors tend to give the opportunity to someone else in the institution. Due to this reason, the married women scientists usually miss the chance of participating in workshops or seminars very often. However, they could have participated at such functions, if they were given the opportunity to choose. It has also been expressed that it is not possible to protest due to the fear of being misunderstood by the seniors even though the protest is justifiable.

According to the survey, this type of interference seems to be rare but still exists in the sector and may consider as indirect sex discrimination.

Women scientists above 50 years showed a tendency to get involved in academic activities more than research activities ($p<0.001$). There was also a positive correlation between age and involvement in administration activities. The reason for these phenomena is that when scientists get mature and reach the highest grades in their designation, they tend to get involved in administration work and other activities more than research activities.

The productivity of women scientists in relation to the gender of their superior was analyzed. Accordingly, 68 per cent of the respondents work under a male as their immediate superior while 81 per cent of the S&T sector institutions are headed by male CEOs. This clearly shows that the high rank job categories in the S&T sector are still dominated by male scientists and technologists in the country. The analysis showed that time devoted for research activities and extension field work was significantly higher for the women scientists working under female heads or superiors while the time devoted for meetings or attending seminars etc. is significantly higher for women scientists working under male superiors.

When looking at the recruitment procedure of the S&T institutions in the country it showed that there is no gender bias in the officially stated recruitment policies and norms in Sri Lanka. However, like in many other countries, there is some degree of indirect discrimination in recruiting women for science careers ($p<0.001$), especially in the private sector organizations and also for some job categories such as IT, engineering and technology etc. This is because many employers still believe that women are not capable of handling technical work and/or field work.

However, no significant pay discrimination has been reported in the survey but some women scientists in the private sector institutions reported that the male counterparts in their organizations receive pay hikes more often than women scientists due to the reason that such benefits are decided on the merit base.

The analysis also showed that there is no gender bias or discrimination when selecting for rewards or recognition or when selecting for higher administration positions in the country. The qualifications, experience and the achievements requested are equal for both male and female scientists. The women scientists were of the view that their achievements are affected by family responsibilities, cultural and social values etc., and such hindrance reduces their chances of being successful when competing with men who are not very much affected by said issues.

4.0 Discussion

The information gathered by the questionnaire survey and interviews conducted using women scientists at various levels of careers indicated that a number of factors affect the career prospects of women scientist in Sri Lanka. This can be investigated by looking at the problem in different levels. According to Sonnet and Holton (1996), the status of women in Science and Technology career depends on various factors that can be explained by two models in the Social-Science literature.

- The one called *'deficit model'* is based on structural explanation of scientific careers. According to this model, women scientists as a group receive fewer opportunities along their career path due to structural obstacles like legal, political and social that are in current practice or existed in the past that affect all women in general in the area of science leading to less career prospects.
- The second model called *'difference model'* describes the deeply ingrained differences in behavior, outlook and goals between women and men. In this model the root cause for the gender disparities in career achievements is internal to the individual.

The survey results indicated that the status of women scientists in the country too can be explained by using these two models. The importance that women scientists give to their career prospects and to sustainability of their family life varies and this is also influenced by external factors such as social values, family background, institutional culture etc. as well as individual goals and views. These factors, in different magnitude, affect the productivity of women scientists and the selection of scientific professions by females in the country.

The studies conducted by the National Science Foundation of America revealed that women scientists, particularly at the doctoral level, regardless of the employment sector, fare less than men in terms of status or rank, salary and promotion, although there is some lessening of all these differences among younger scientists and engineers today (Vetter,1981). Equal opportunities for male and female citizens are recognized by the Sri Lankan government in its constitution and therefore there is no direct discrimination in recruitment for careers or selection for training, education, etc. like in some other countries. As there is no gender-based positive discrimination for employment, women would be expected to compete with men for employment, funds for projects, positions in scientific organizations on equal terms. Since women who are qualified to work in the area of S&T are nearly equal in proportion to male, one should expect that the number of women scientists in the country is more or less equal to male scientist. However, the situation in the country differs by having a comparatively low number of women in these positions. The numbers of women representing and chairing professional scientific bodies, Institutional Management boards etc. are still low. For example, in the university sector, of the total number of professors, 22 per cent are women. Similarly, the majority of the higher administration level positions in research institutions and other S&T organizations are chaired by men. Therefore it is evident that there is some cause for this gender disparity in the S&T sector in the country.

When analyzing the responses of the women scientists of Sri Lanka, it is clear that there is no salary disparity on the basis of gender in the country. However, the interviews conducted with some selected groups of scientists revealed that there are many other money earning avenues for financial benefits such as consultants, tutors, researchers that are open to men but not to women. Further, male scientists get more opportunities to work outside the country compared to women scientists. Since men are not very much concerned about the lack of proper transport or accommodation facilities, and therefore, they have more opportunities to work in the projects that need extensive travel or need working in extra hours etc. that widen their career path. When considering the current working setup in the local S&T institutions, it is difficult for women scientists to work after hours. Transport problems at night, distance to their working place from their residence, personal safety, family responsibilities including caring children and family etc., make it more difficult for women to get fully involved in their career than men. Other than the above physical difficulties, the existing traditional beliefs and practices also somewhat restrain women to get involved in some career related activities to their full capacity.

In the society of Sri Lanka today, most women marry in their late twenties and the decisions about child bearing and child rearing mostly fall into the hands of the women. In this context, ambitions of women to pursue a career take secondary importance with women taking leave to attend family needs and other concerns, at least until children start primary schooling. As a result, the most productive period of women's career will get delayed and this delay conflicts with traditional tenure clocks (Grosz *et al.*, 2005). In subsequent years, this barrier gets translated into having poorer professional achievements than men at the same age and basic qualifications. This also affects the academic development of women with most of the women scientists obtaining their postgraduate qualification after the age of 41-50 years, while most of the men of the same age complete their post graduate degree during their thirties. Therefore, fewer women may be able to compete successfully with men for the same post (BAL, 2005). If steps are taken to make allowance for the time gap and making awards for career development independent of the number of years, it will allow women scientists to have equal participation in career activities and make it easier for them to get into scientific activities after the break in their career. Some women scientists expressed their concerns that they have gone backward several years from their career compared to the men recruited in the same year with them, due to maternity leave etc.

Apart from having post-graduate degrees or other academic qualifications, there are other certain specific steps and goals which are essential for career advancement in Science and Technology sector. A proven record for achievement in the form of publications in reputed journals and patents are some of the major criteria for selection when competing for promotion, receiving foreign training, selecting for rewards, having administrative positions, nomination for various reviewing bodies or election to prestigious institution or professional bodies (Sonnet and Holton, 1996). Several studies were conducted to explore the basis for the gender gap in scientists working in different disciplines (Yedida and Bickel, 2001; Carr *et al.*, 1992). Accordingly, constraints in the traditional sex roles, manifestation of sexism in the institutions, and lack of effective mentors affect the achievements of women scientists. Carr *et al.* (1992) reported that female academics who had children published less and received institutional support less than male colleagues who had children. Some studies also showed inequity relative to the allocation of resources, space, salary, outside professional activities, and positions of influence (http//web.mit.edu/fnl/women.html). In the present survey a significant number of women scientists working at junior level positions complained about the lack of opportunities, space and allocation of resources which make it difficult for them to work at their full capacity. Due to their social status in the country, women do not get very much involved in political activities and, therefore, when competing for higher management positions in the government institutions, they find that they are at a disadvantage for the selection.

Certain aspects of institutional culture and policy too, may pose challenges for women. Success as a scientist needs long working hours and time commitment that for many is incompatible with raising children (Benze *et al.*, 1998). Many surveys done in different countries documented that woman in scientific fields work on average fewer hours than men (Bal, 2005; Carr *et al.*, 1998; Tesch *et al.*, 1995). As mentioned earlier, the analysis done in the present survey showed that time devoted for research activities and extension field work was significantly higher for women scientists working under female heads or superiors while the time devoted for meetings or attending seminars etc. is significantly higher for women scientists working under male superiors. The main reason for this difference, as explained by many, was due to the fact that women as heads are more concerned about the safety and family issues of the other women scientists and make provisions/facilities to overcome such issues. That allows the subordinates to work after hours and get involved in their research and extension activities (personal interviews).

It is also obvious that the assistance of the spouse and sharing family responsibilities equally will assist wives to get more involved in their career and achieve its goals. This has been well acknowledged by the Sri Lankan women scientists who have participated more than others in training programmes, seminar etc. and who have done more publications.

5.0 Conclusions

Though regulations indicate equal opportunities, the data collected and the interviews held with a number of women scientists revealed that, in general, most of the women scientists are not very happy with the current situation prevailing in the various scientific organizations. Most of them are of the view that there is indirect discrimination for women when recruiting for employment or selecting for higher positions in the institutions. They feel that most of the administrators prefer to work with men rather than women because they are of the view that women are not capable of handling administrative positions compared to men. However, it was noted when looking at the opportunities for research activities, the university academics are generally happy with their positions in the universities, while women scientists working in research and other scientific institutions and private sector organizations are dissatisfied with the opportunities they receive. It is mostly at the level of administration positions in universities and other institutions that women scientists have a grievance in relation to gender discrimination. Therefore, policy initiatives should be taken to investigate this phenomenon and eliminate the barriers that prevent women scientists entering into more administration positions. Also, it is necessary to improve the research culture in Research Institutions and other S&T organizations.

Since women have to balance their careers with their family life, the necessary policy initiatives should be taken to compensate women in this aspect and make some provisions to assist them to cope with the situation. To help them with small children, it is necessary to establish child caring units and nursery schools in the Universities, Research Institutes etc. Creating Science Park and establishing various interlinked scientific organizations in one place would also help to create such facilities more effectively.

Assurance of priority for women scientists when offering transport facilities or residencies will also help them to work after hours or in their free time. Giving flexible working hours also would help women scientists to adjust their working time with their family activities. Special attention should be made when selecting for awards or positions in the institutions. The study revealed that the women's participation was very low in the professional bodies and institutional management bodies. Therefore some places should be created in these government bodies for women.

Finally it should be recognized that the number of women who enter to the labour market every year with qualifications in the area of science and technology is almost equal to men and, therefore, planning authorities in the country should take immediate initiatives to workout suitable policy planning measures to encourage active participation of women in scientific careers. Such approaches will increase the contribution of women through science and technology towards the socio-economic development of Sri Lanka.

References

ADB, 1999. Women in Sri Lanka: Country Briefing Paper, Asian Development Bank, Programme Department West.

Bal, V., 2005. Women Scientists in India: Nowhere near the Glass Ceiling, Current Science, 88 (6), pp. 872-78.

Benze, E.J, Clayton, C.P., Costa, S.T., 1998. Increasing Academic Internal Medicine Investments in Female Faculty. American Journal of Medicine, 105, pp. 459-63.

Carr, P.L., Ash, A.S., Friedman, R.H. 1998. Relation of Family Responsibilities and Gender to the Productivity and Career Satisfaction of Medical Faculty. Ann. Intern. Med 129, pp. 532-38.

Cole, J.R., Cole, S., 1973. Social Stratification in Science, University Chicago Press Chicago.

Dial, T.H., Bickel, J., Lewicki, L.M., 1989. Sex Difference in Ranking Attainment among Radiology and Internal Medicine Faculty. Acad. Medicine. 64, pp. 198-202.

Grosz, B., Dulca, C., Dymecki, S., 2005. Report from the Task Force on Women in Science and Engineering, Cambridge Mass: Harvard University.

http//web.mit.edu/fnl/women.html (accessed November 27, 2008).A study on the status of women faculty in science at MIT, the MIT faculty Newsletter, March 1999 Vol. XI (4).

Kaplan, S.H., Sullivan, L.M., Dukes, K.A., Phillips, C.F., Kelch, R.P., Schaller, J.G., 1996. Sex Difference in Academic Advancement: Results of National Study of Pediatricians. New England Journal of Medicine, 335:3, pp. 282-9.

Long, J.S., Fox, M.F., 1995. Scientific Career and Particularism. Annual Review of Sociology (21), pp.45-71.

National Research and Development Survey 2004, National Science Foundation, Maitland Place, Sri Lanka.

Nickerson, K.G., Bennet, N.M., Estes, D., Shea, S., 1990. The Status of Women at one Academic Medical Centre: Breaking through the Glass Ceiling. Journal of American Medical Association, 264, pp. 1813-7.

Sonnet, G., Holton, G., 1996. Career Pattern of Women and Men in Sciences. American Scientist, 84, pp. 63-71.

Tesch, BJ, Wood, H.M., Helwig, A.L., 1995. Promotion of Women Physician in Academic Medicine: Glass Ceiling or Sticky floor. Journal of American Medical Association, 273, pp. 1022-25.

Vetter, B.M 1981, Women Scientists and Engineers: trends in Participation. Science, Vol. 214 (18), pp 1313-1321.

Wallis, L.A., Gider, H., Thaler, H., 1981. Advancement of Men and Women in Medical Academia: A pilot study. Journal of American Medical Association, 246, pp. 2350-3.

Wilkinson, C.J., Lindae, H.W., 1986. Status of Women Academic Anesthesiology. Anesthesiology, Vol. 64, pp. 496-500.

Yedida, M.J., Bickel, J., 2001. Why aren't there more Women Leaders in Academic Medicine? The views of the Clinical Department Chairs. Academic Medicine, Vol. 76, pp. 453-65.

Chapter 9

Gender, Science and Technology in Tanzania: A Situation Analysis

Rose Rita Kingamkono
Tanzania, Commission for Science and Technology (COSTECH)
P.O. Box 4302, Dar es Salaam, Tanzania
E-mail: rkingamkono@costech.or.tz

ABSTRACT

The study on 'Situation Analysis in Science and Technology (S&T) in Tanzania with Gender Perspective was conducted in August 2004. The overall objective was to determine the situation regarding S&T and the available opportunities for acquisition and application of S&T with gender perspective. It analyzed gender sensitivity in education and professional training; employment; and decision and managerial positions in almost all economic sectors. Field surveys and interviews were conducted using semi-structured questionnaires and by consulting relevant published and unpublished documents. Eleven government ministries and nine non-governmental organizations were purposively selected for the study. The study also involved reviewing 13 national policies, 2 constitutions, 12 Acts of Parliament, and 3 strategies. Reviewed also were staff profiles of public universities, technical education institutes and research and development institutions as well as 16 reports and papers relevant to gender. Content analysis was used as a tool for analyzing the qualitative data. The paper points out disparities between women and men with regard to the enrollment in the education systems particularly so in science and technological areas as one goes up the ladder from primary to tertiary level of education. Consequently few women compete effectively with their male counterparts for better-paid jobs resulting in women being grossly under represented in formal employment, decision and managerial positions.

Keywords: Gender, Science, Technology.

1.0 Background

In developing countries, including Tanzania, women are central to development as a result of their gender-defined roles; making a major contribution to the production of food and the provision of energy, water, health-care and family economy. Yet women are, generally, under represented in education, policy and decision-making levels. High illiteracy rate of women leads to unwise choices and decisions about use of the resources. The importance of women literacy can be appreciated by the fact that a national or private income in the hands of the poor or male literacy explains only a small part of community health improvements whereas increased female literacy shows a greater effect. In a cross-country regression analysis (Smith and Haddad 2000) increase in female secondary school enrollment was the highest explainer of the reduction in malnutrition rates in pre-school children, explaining 43 per cent of the reduction followed by food availability (26.1 per cent); healthy environment (19.3 per cent); and finally women's status (11.6 per cent).

Tanzanians today agree that the gender differentiated roles between men and women in many communities put women at a disadvantage. In all ethnic groups in Tanzania, the culture dictates that girls should be brought up in preparation for motherhood and the male spouse take all the other domains in a household. The outcome of this practice has been limited access of girls to education being considered as a liability to the family if she does not get married. Limited access to education particularly higher learning for girls adversely affects not only their access to employment but also their abilities, capacities and capabilities to be healthy themselves as well as to improve the health of their families. This situation thus fails to contribute effectively to the development of the communities. This study was an attempt to analyze the Gender, Science and Technology (G, S&T) situation in Tanzania with a view to discerning opportunities and constraints for acquisition and application of S&T with gender perspective. This paper specifically analyzes gender sensitivity in education and professional training; employment; and decision and managerial positions in almost all economic sectors.

2.0 Methodology

Data were collected through field surveys using semi-structured questionnaires interviewing key informants and by consulting relevant published and unpublished documents to form an overview and identify gaps in information. Eleven government ministries and nine Non-Governmental Organizations (NGO) were purposively selected for the study. Reviewed also were staff profiles of public universities, technical education institutes and research and development institutions as well as sixteen reports and papers relevant to gender. The list of ministries, agencies and NGOs visited and documents reviewed are presented in Table 9.1. Content analysis method was used to analyze the qualitative information. In this case the recorded dialogue with respondents was broken down into smallest meaningful units of information.

Table 9.1: Division of Labour between Male and Female Members in a Household in Sub-Saharan Africa

Activities	*Adult*		*Adolescents*	
	Men	*Women*	*Boys*	*Girls*
Agriculture				
Farm preparation	x	x	x	x
Tiling/cultivation	x	x		x
Sawing		x		x
Weeding		x	x	x
Carrying and manure the field		x		x
Harvesting and carrying the harvest for storage	x	x	x	x
Preserving harvested food		x		x
Processing food		x		x
Home gardening		X		x
Household chore				
Fetching water, firewood and cooking		x		x
Cleaning household compound		x		x
Washing utensils and cloth		X		X

Source: Kavishe, 1993.

3.0 Results and Discussion

3.1 Education and Professional Training

In Tanzania, fewer females than males enroll especially in secondary schools and in higher learning institutions mainly because, traditionally women were not expected to go to school. The introduction of a school system during colonial period allowed few girls to enroll in schools. The little and biased education they received was to confine them to the kitchen. This tendency over time has created a gender gap between men and women in education and employment as shall be seen in subsequent section. After independence, a number of measures were undertaken to improve the education system.

The introduction of Education Ordinance of 1961 was the starting point for equal opportunities for both females and males. Major changes came with the Arusha Declaration of 1968, the Musoma Resolution on Universal Primary Education (UPE) in 1973 and the Education Act No.25 of 1978 that emphasized the right of every citizen to get education to the best of his/her ability. The Educational and Training Policy itself is also gender sensitive in the sense that it states as one of its major goals to promote education for girls in general. More recently, Tanzania introduced and implemented a 5-year Primary Education Development Programme (PEDP) since 2001, and Secondary Education Development Programme (SEDP) since 2004. PEDP seeks to establish high quality education for all while SEDP aims at preparing future national workforce and leaders. Both of these programmes are planned to specifically be gender sensitive. The strategies above have started bearing fruit, since, currently, primary education enrollment between boys and girls in terms of numbers is equal (Figure 9.1).

However, the number of female students who enroll at secondary education still lags behind enrollment of boys and the situation worsens up the educational ladder with a sharp decline at the joining of the Advanced Secondary Education (Figure 9.1). For those who succeed to join advanced secondary education, they normally join tertiary education successfully.

3.2 Enrollment in Tertiary Education

Since Advanced Secondary Education prepares students to join tertiary education institutions, the lower girl enrollment in Advanced Secondary Education (Figure 9.1) means that fewer girls/women than men will also enroll in tertiary educational institutions as evidenced in Figure 9.2.

The deep-rooted patriarchal mindset in Tanzania's population is responsible for these gender disparities in student enrollment particularly in higher learning institutions. The society still looks at the girls and women as inferior and expects them to act accordingly, even where performance shows that they are equal or better. Brought up to become mothers, girls are subjected to multiple domestic-based roles as they grow, compared to boys. These responsibilities of girls to the family negatively influence their performance in primary and secondary school examinations to enable them join tertiary learning institutions.

Since the gender division of labour does not expect boys to participate in day to day domestic chores, they get more time than girls to prepare for their examinations and are hence able to be selected for tertiary education opportunities. For example, a study in Tanzania on time allocation and energy expenditure of women and men in two villages during weeding and harvest seasons indicated that the total number of activities for women and girls was significantly higher than that of men and boys (Table 9.1). Further, societies in general do not expect girls to equitably compete academically with boys. Under scarce resources when a family has to decide on which child to invest in education, the tendency is to give preference to boys because 'boys are considered as asset to families and girls are liabilities, since they will ultimately get married and money will be wasted.

At age of completing Primary Education, both boys and girls normally would have entered puberty. For girls, who are considered as a liability to the family if they do not get married, this is the stage when they are to get ready for marital life, and hence willingly or unwillingly arranged marriages of some kind, particularly in rural areas, is not uncommon. These arrangements and attitudes deny girl's access to secondary and higher learning education, particularly S&T disciplines, which feature highly up the education ladder. They also predispose them to becoming in-the-family-way sooner than their boy-age mates at the expense of their health and social welfare. The overall effect of the above account are adverse effects on their employment an

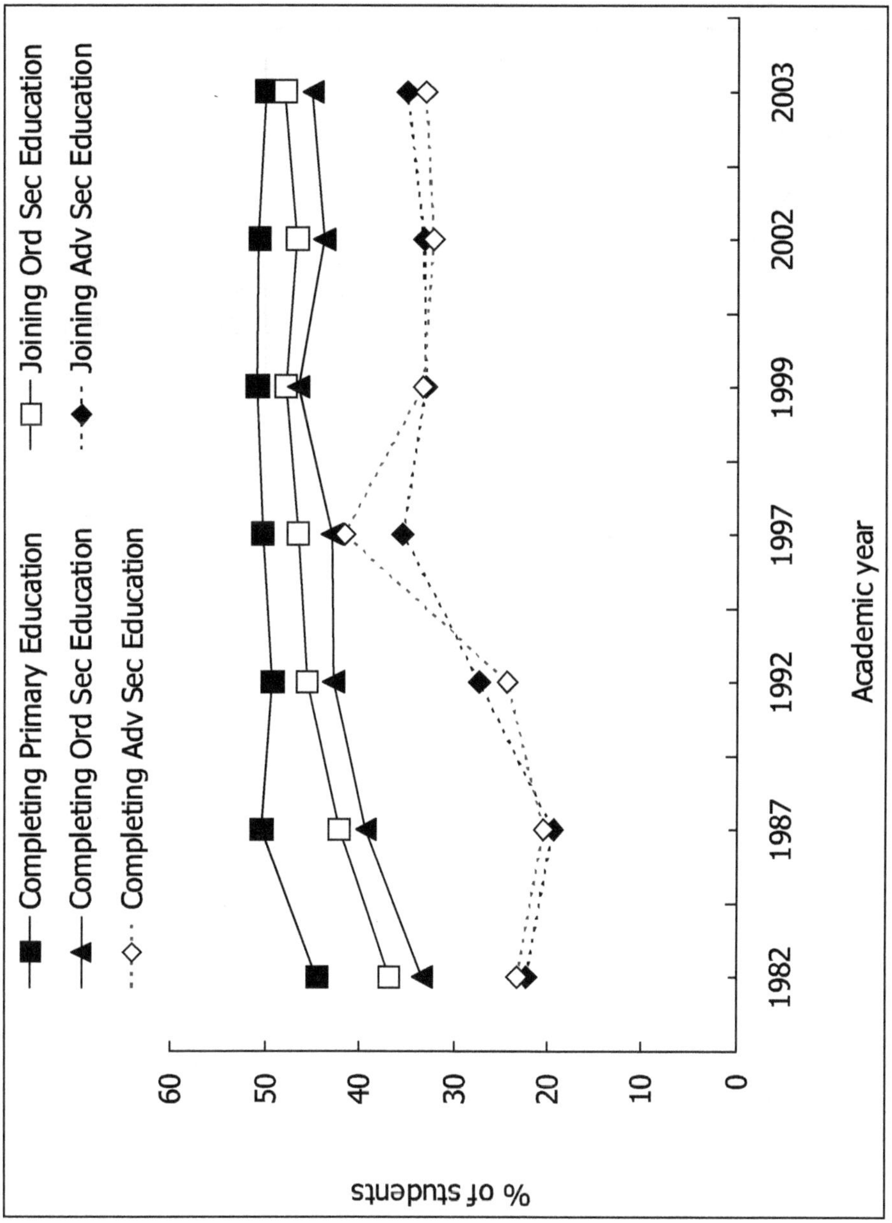

Figure 9.1: Girls Enrollment in Primary, Ordinary and Advanced Secondary Education (Ministry of Education and Culture, 2003)

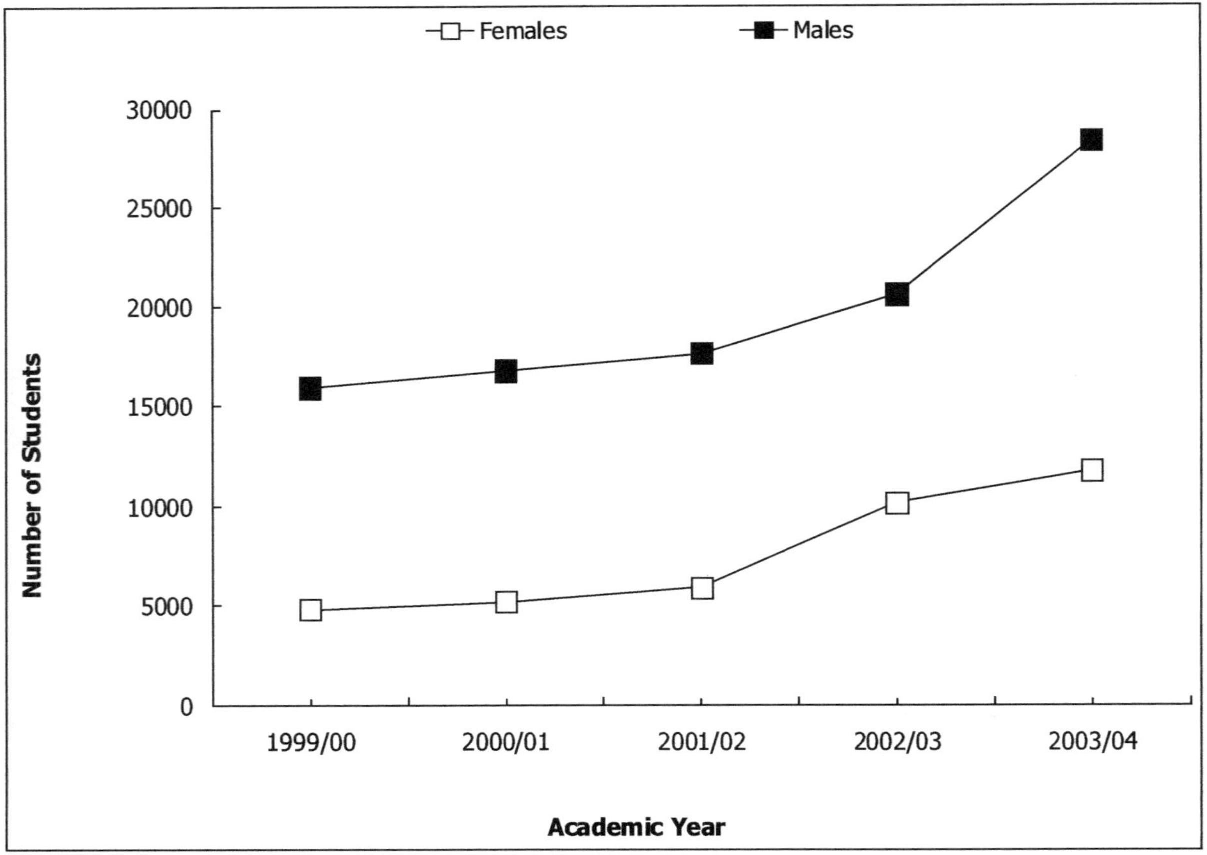

Figure 9.2: Total Undergraduate Student Enrollment by Gender–(MSTHE[1], 2004)[2]

1 Ministry of Science, Technology and Higher Education.

2 Total of 8 Public and 13 Private Universities, 3 Technical Institutions and 14 Other Institutions in Other Ministries

important barriers to nutrition and health progress and eventually to socio-economic development not only of the women themselves but also their families. Additionally, in lower school levels, the girls may be given jobs at school by teachers such as making tea for the teachers and entertaining visitors, which may keep them away from their class work. Sometimes, girls performing well in school get harassed by their fellow students, being alleged as getting assistance or bribed with good points from male teachers for in-kind services, which is a discouraging attitude.

3.3 Science and Technology

The general objectives of the Science and Technology (S&T) policy (1996) are to develop and manage S&T in a manner consistent with physical and human endowment in the country. Policy statements that address gender, S&T categorically recognize the need to raise the participation of girls in science by reviewing the school curriculum in order to strengthen and encourage participation and achievement of girls in mathematics and science subjects. Among the strategies in place is the encouraging of girls to study science subjects in secondary schools. The Ministry responsible for S&T and Higher Education then had introduced a system whereby it provides awards to best and talented girl students including payment of tuition fees to motivate more girls to undertake science subjects. In order to encourage others (men and women) to enter and be innovative in science and technology, the ministry through the Commission for Science and Technology (COSTECH) provides Tanzania Science and Technology Awards (TASTA) to inventors and eminent scientists. Also, all the 58 Folk Development Colleges in the country offer training with gender perspective.

3.4 Enrollment of Girls in Science Subjects

With the S&T policy statements in mind, the Education and Training Policy (ETP, 1995) in Tanzania stipulates that S&T shall be essential components of education and training in the whole education and training system. That policy statement not withstanding, as a whole, arts subjects attract more students than science subjects in tertiary education regardless of gender (Figure 9.3).

However, the situation is much more pronounced in girls or women enrollment for technical education (Figures 9.4 and 9.5) and science subjects (Figures 9.6 and 9.7). Even for those who opt for science subjects, they normally undertake the so-called soft sciences (biology, medicine and agricultural sciences, Figure 9.6).

This is expected for there are more failures in science subjects in secondary level education (Figure 9.8 and Table 9.2) and even in teacher's education certifying examination more failures are among girls than it is among boys/men (Figure 9.9).

3.5 Strategies Addressing Gender Disparities

Indeed, policy makers as well as other stakeholders are concerned about these disparities and some efforts are in place in Tanzania to improve the situation. The plan by the Ministry of Education and Culture and that of Ministry of Science, Technology and Higher Education is to increase the number of students taking science and improving the quality of education. This will be achieved through using the cost

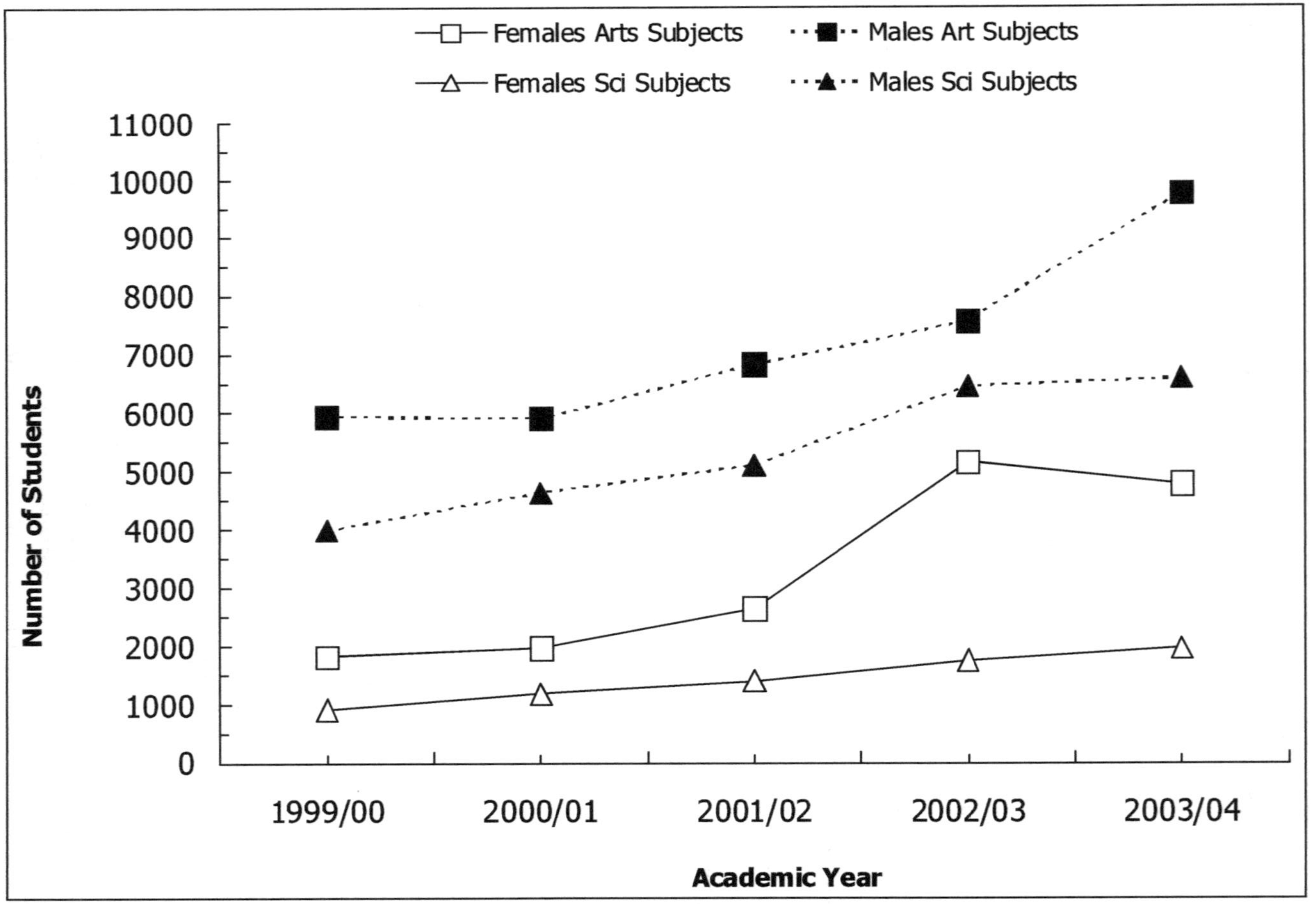

Figure 9.3: Undergraduate Students Enrollment in Arts and Science Subjects in Public and Private Universities by Gender (MSTHE, 2004)

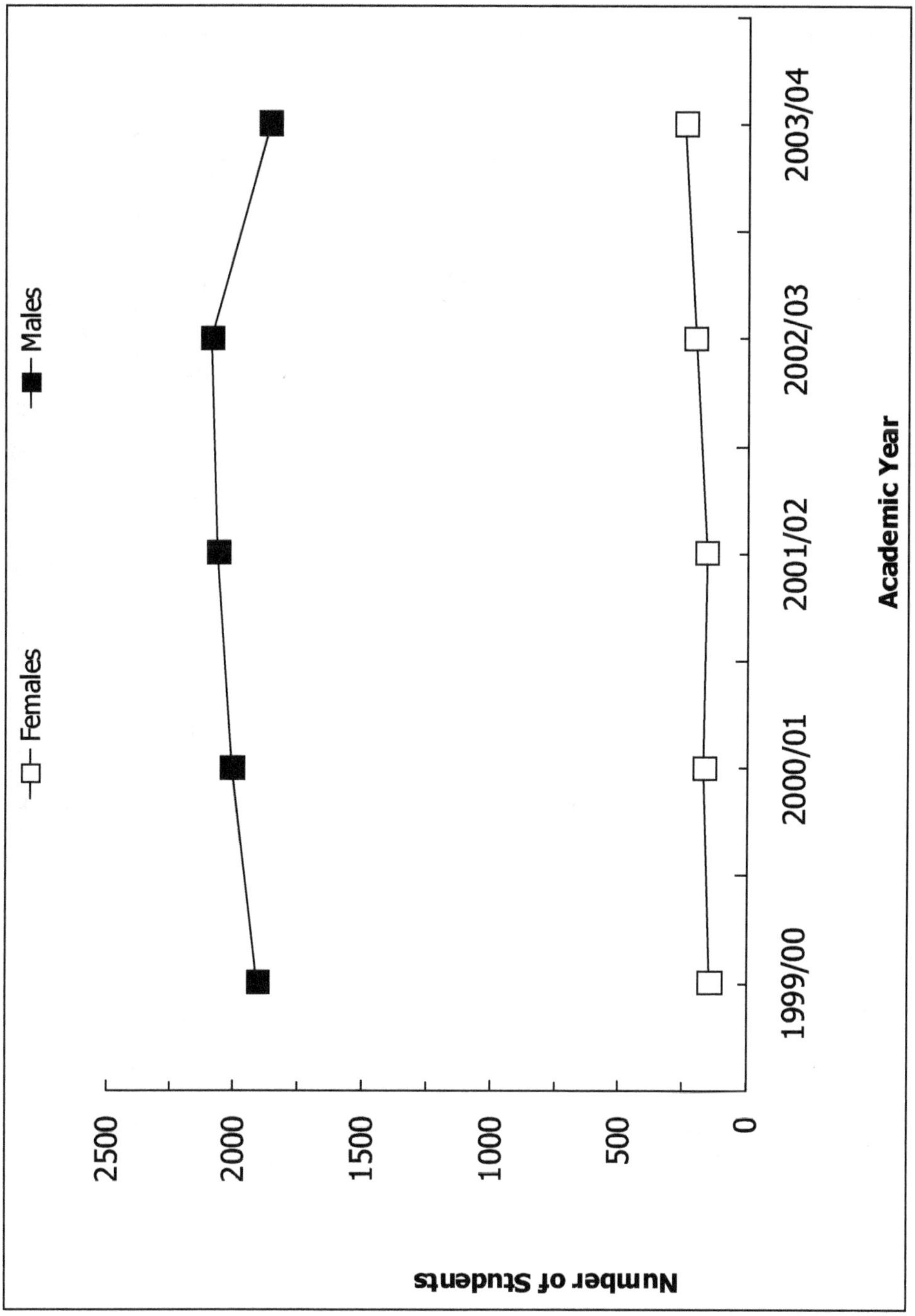

Figure 9.4: Student Enrollment in Technical Institutions by Gender (MSTHE, 2004)

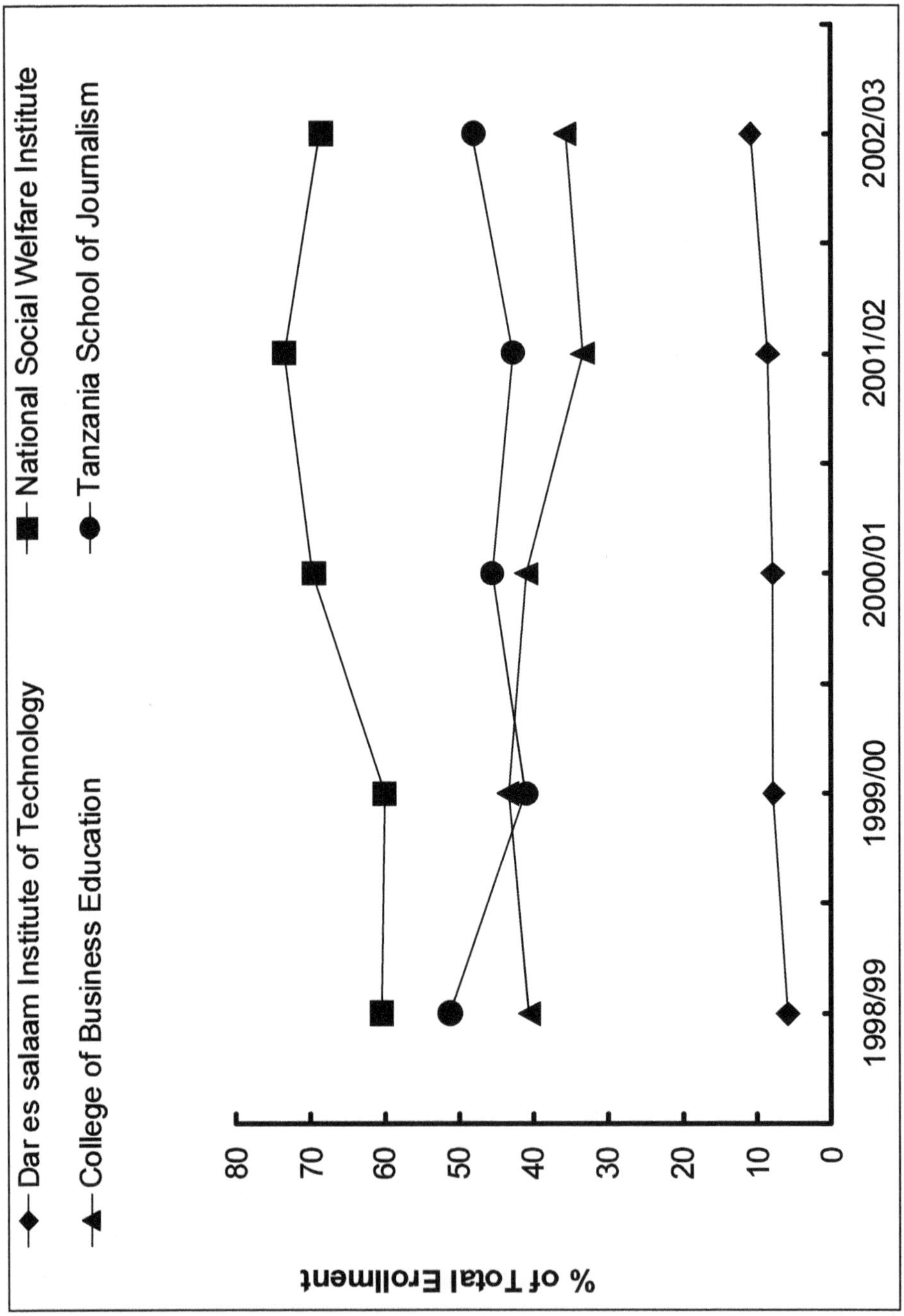

Figure 9.5: Female Students Enrollment (Per cent of total) in Various Non-Higher Tertiary Education Institutions (MSTHE 2004)

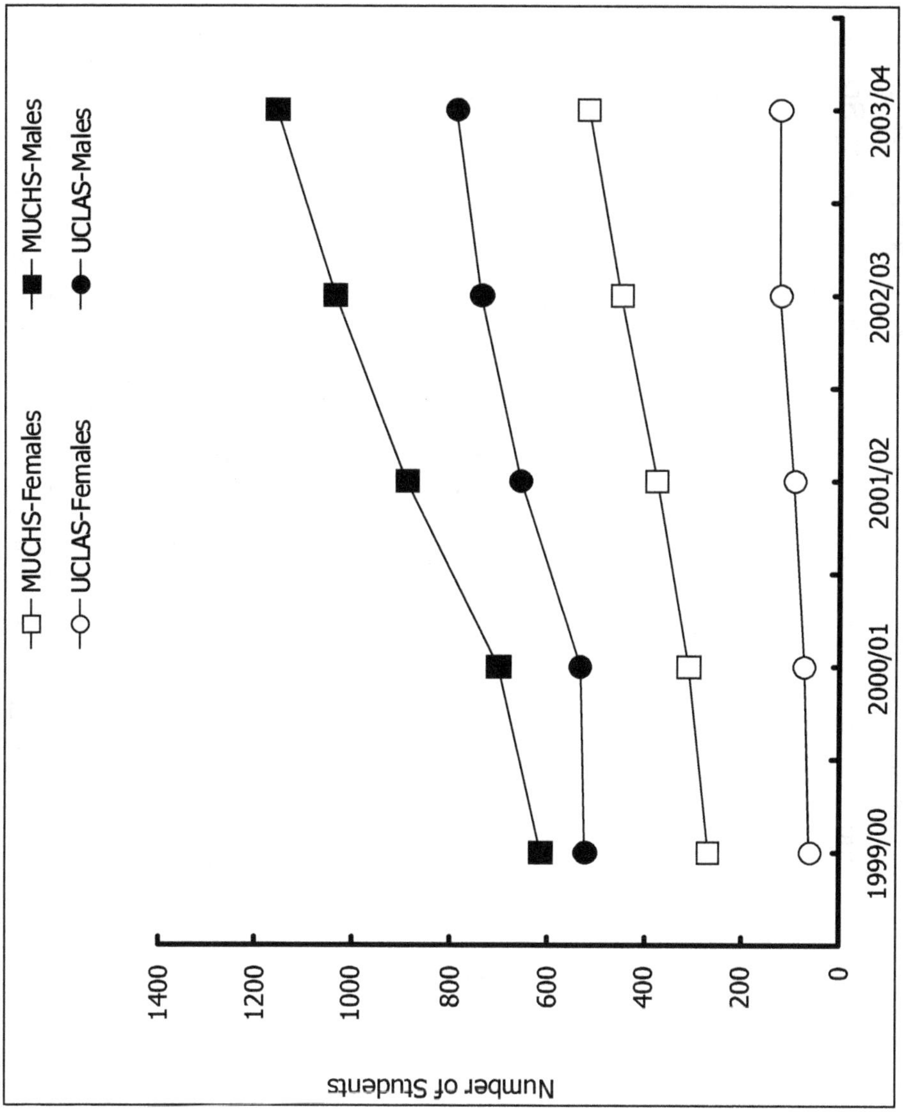

Figure 9.6: Students Enrollment at MUCHS and UCLAS by Gender (Source: UDSM 2004)

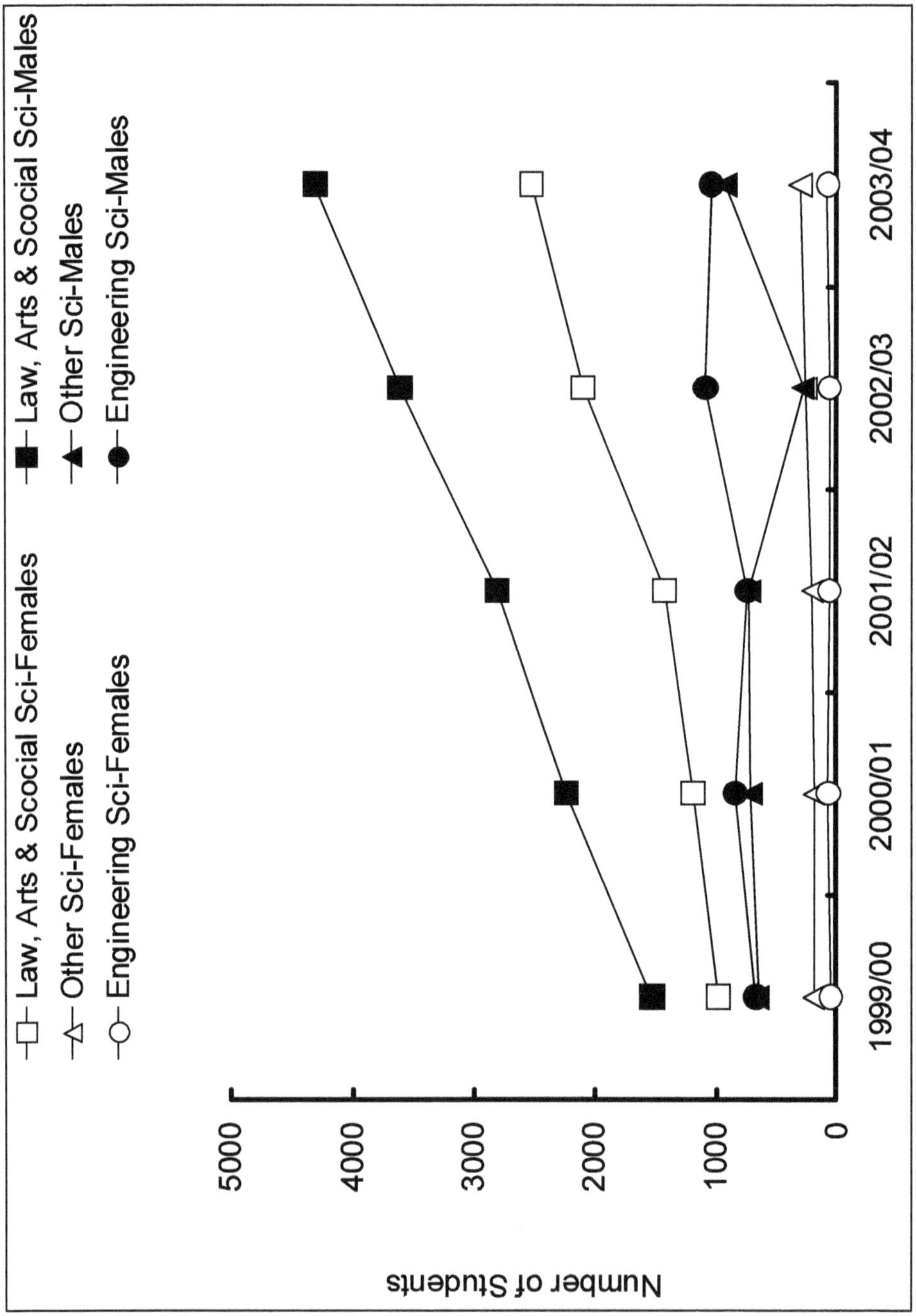

Figure 9.7: Undergraduate Students Enrollment at the University of Dar es Salaam by Gender (UDSM 2004)

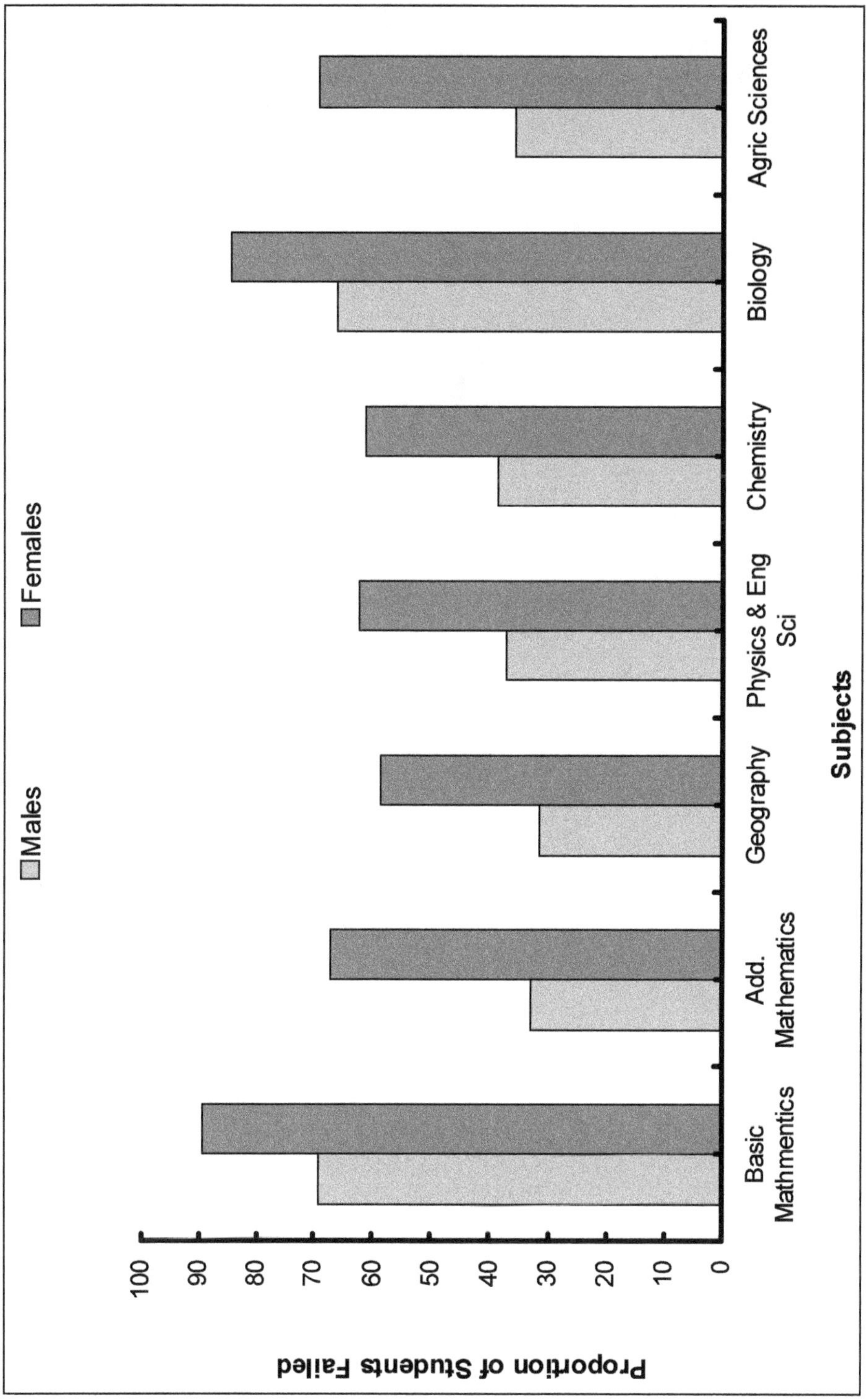

Figure 9.8: Students Failures in Science Subjects in Ordinary Level Secondary Education in 1998 by Gender (Source: MSTHE 2002)

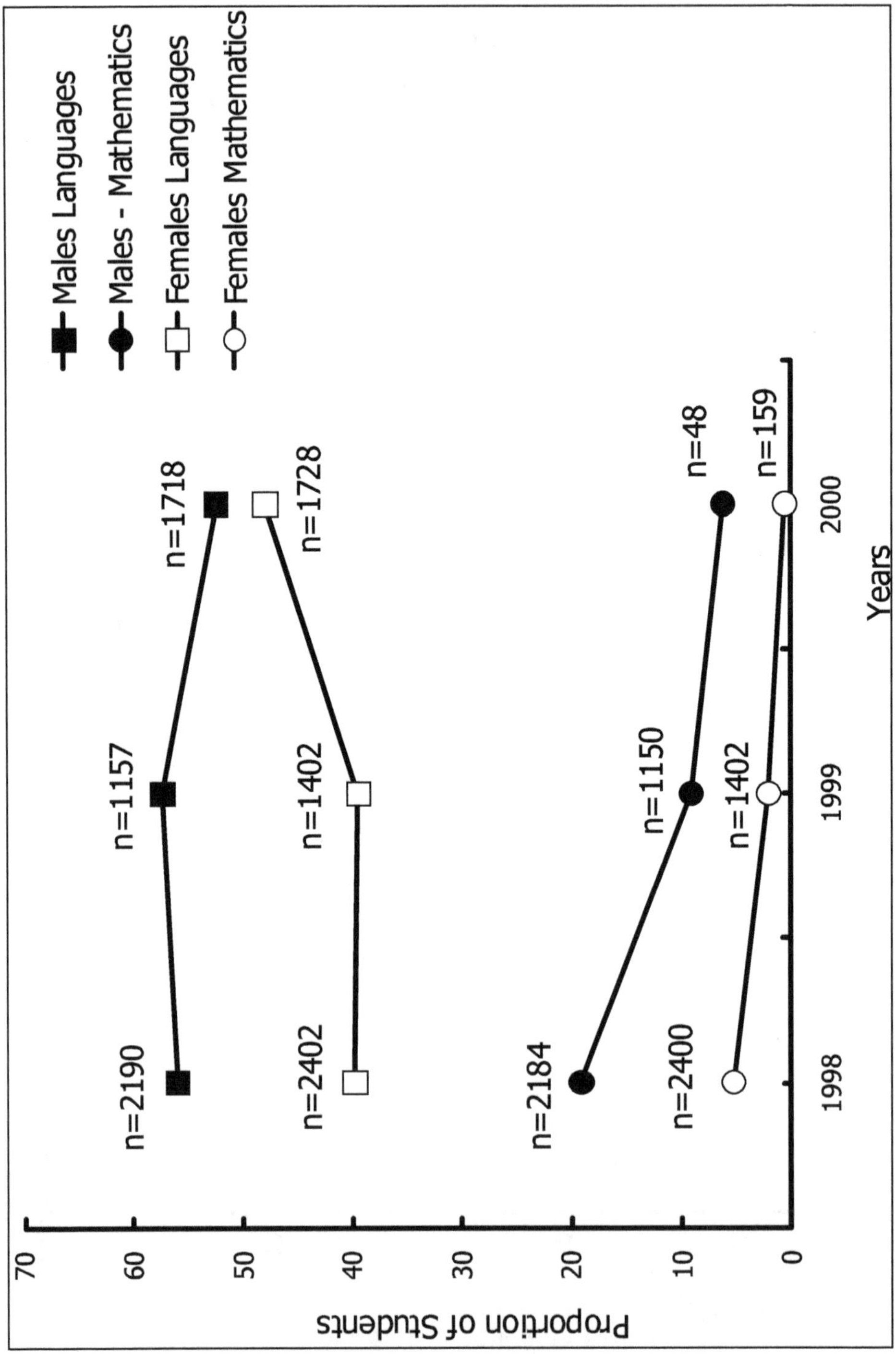

Figure 9.9: Students who Passed at Grade A-C in Teacher's Education Certificate Examination

sharing policy. These efforts have brought some positive results whereby girls' enrollment into science subjects and examination performance has improved albeit to a small extent (Table 9.2).

Table 9.2: Percentage of Total Enrollment in Mathematics and Science Subjects and Student Who Passed Examinations at Advance Secondary Education

Year	*Subject*	*Enrollment-Girls*	*Student who Passed*	
			Girls	*Boys*
1996	Mathematics	12.6	9.5	–
	Biology	33.8	31.8	–
	Chemistry	23.8	18.3	–
	Physics	18.8	14.9	–
2001	Mathematics	17.4	13.1	66.6
	Biology	34.3	32.8	63.9
	Chemistry	26.5	21.1	61.6
	Physics	21.8	17.4	66.6

3.6 Vocational Education

The vocational education training policy, which is demand driven, has allowed for the designing of programmes required in the market. These programmes provide women with improved skills, which have enhanced the performance, as well as improved status or added value to the jobs they undertake particularly in the Small and Medium Enterprises (SMEs) and informal sector.

However, since most Vocational Training Institutions were established to cater for males, females have been left out with only a few of them joining colleges offering hard science subjects/skills such as plumbing, mechanics and masonry. The challenge is to encourage females to enroll in such training and also establish more vocational schools to cater for the female dominated skills.

3.7 Sponsorships

It is interesting to note that, whereas women may want to advance themselves through the education ladder and acquire professionalism, the government/loans/ schemes sponsorship for students enrolling in private universities, technical education institutions and professional education institutions under other ministries seems rather less accessible to women compared to men. For example the biggest proportion of women who joined in such institution in 2003/04 academic year was sponsored privately or through NGOs (Figure 9.10). This is a challenge that needs to be looked into closely with a view to lifting out the barriers that block women in accessing the government and loan sponsorship schemes.

3.8 Employment Opportunities

Article 22 and 23 of the Constitution of the United Republic of Tanzania provides for all Tanzanian citizens the right to work and be remunerated. Tanzania has ratified

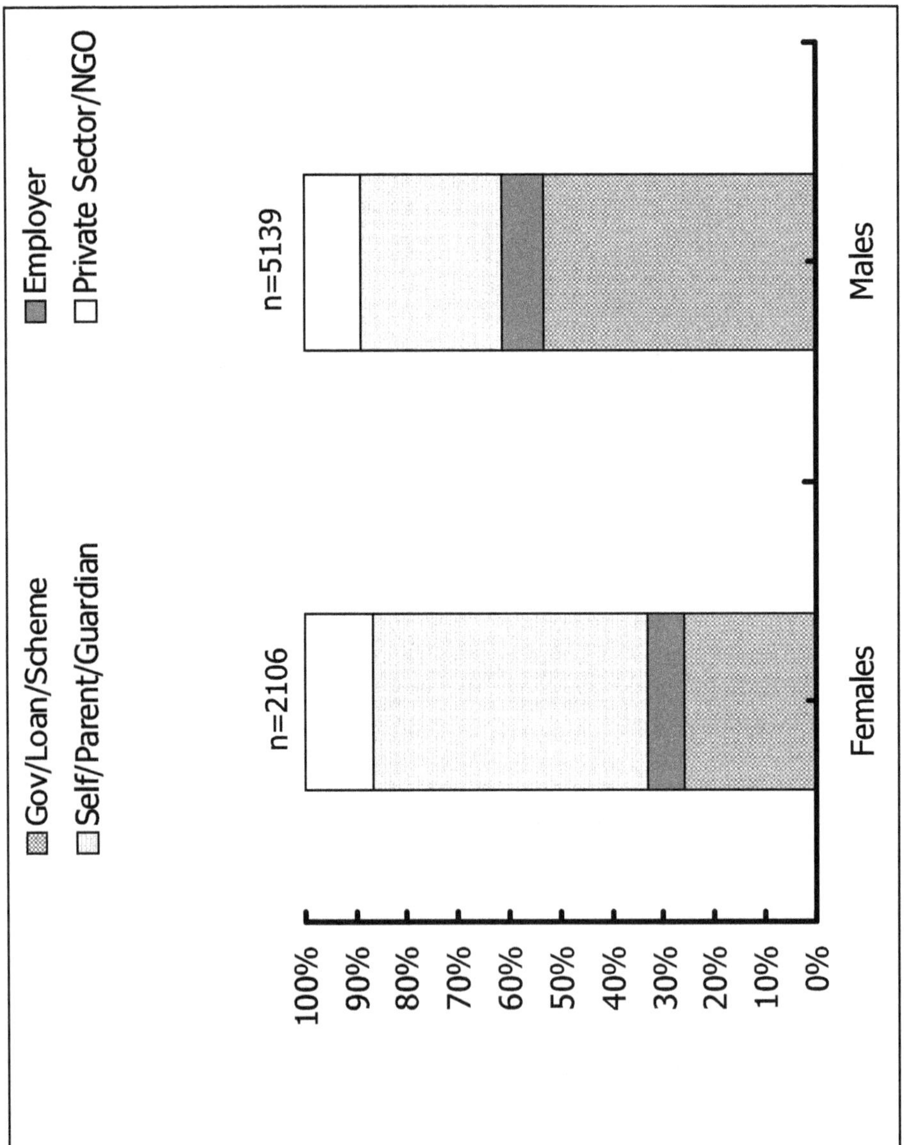

Figure 9.10: Student Sponsorship in Private Universities, Technical Education Institutions and Professional Education Institutions Under Other Ministries in Years 2003/04 by Gender (MSTHE, 2004)

International Labour Organization (ILO) conventions, including convention 100 and 101, which are specifically against gender discrimination in employment. The Small and Medium Enterprises policy together with the public service regulations have provided an environment for promoting equal opportunities and eliminating discrimination and bias in employment based on gender.

The government has domesticated these international labour laws/instruments through the enactment of the National Employment Service (NES) Act of 1999 and passing of the Employment and Labour Regulations (ELR) Act by the Parliament in 2003. These Acts provide for equal opportunities to men and women in access to employment services; prohibits discrimination at the workplace on the basis of gender, sex, marital status, disability and pregnancy among others; specifically for women, the Act protects pregnant and lactating women employees by providing social security hence shifting the cost from the employer to the social security fund; and the Act requires employers to report to the Labour Commissioner on their plans to promote equal opportunities. Other strategies, which are in place that may increase chances of women to access job positions in the nation, include the establishment by the government of the Public Labour Exchange Centre (PLEC) in 2002 to link job seekers with employers. Job seekers are provided with vocational guidance and counseling as well as with information about employment perspectives.

In 2003, 30 per cent of job seekers were women; the establishment of similar services by the private sector with minimal charges; establishment of a Women's Information Window, within the National Gender Machinery which provides women with information and advisory services on job availability, training, scholarships and internship opportunities; labour laws are being reviewed to become more gender sensitive and to enforce safety regulations in the workplaces for both women and men, whereby thirteen labour laws are under review in Zanzibar. Concerted efforts have also been taken to institutionalize gender equality in employment. For example, the Civil Service Management Employment Policy that has been established by the government under the civil service reform programme has formulated an environment that does not allow any form of discrimination against gender in work places. The employment policy, established in 1999, also assures both men and women equal opportunities in employment in all sectors. On the same line, the centre for job assistance for both men and women was established in 2002. Statistics reveal that about 42 per cent of those who visited the Centre in 2004 were women. The Centre also assists clients in jobs availability, how to access them, training opportunities and sponsorship for women.

3.9 Challenges in Women Employment

Despite the strategies above and good will of the government, low academic qualifications by most women as discussed in section 5.1.4 and 5.2 and low working experience, particularly of young women limits their ability to effectively compete for the better-paid jobs in the labour market with their male counterparts. Women are thus employed in lowly paid jobs. Figure 9.12 shows the distribution of employees in Non-Higher Education Research and Development (R&D) institutions by gender and qualification (Science and Arts combined) and Figure 9.13 shows distribution by

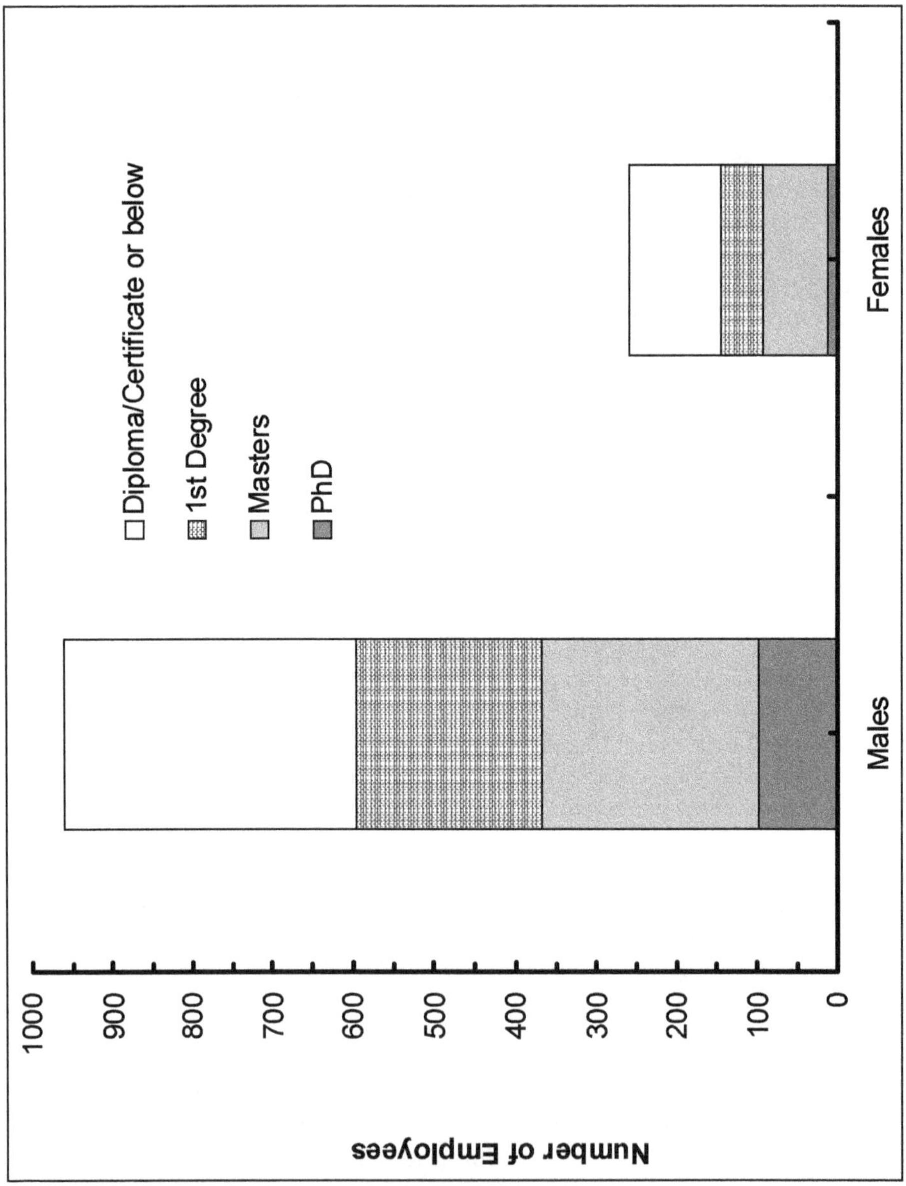

Figure 9.11: Number of Employees in Non-Higher Education R&D Institutions by Gender and Qualification (Science and Arts Combined)

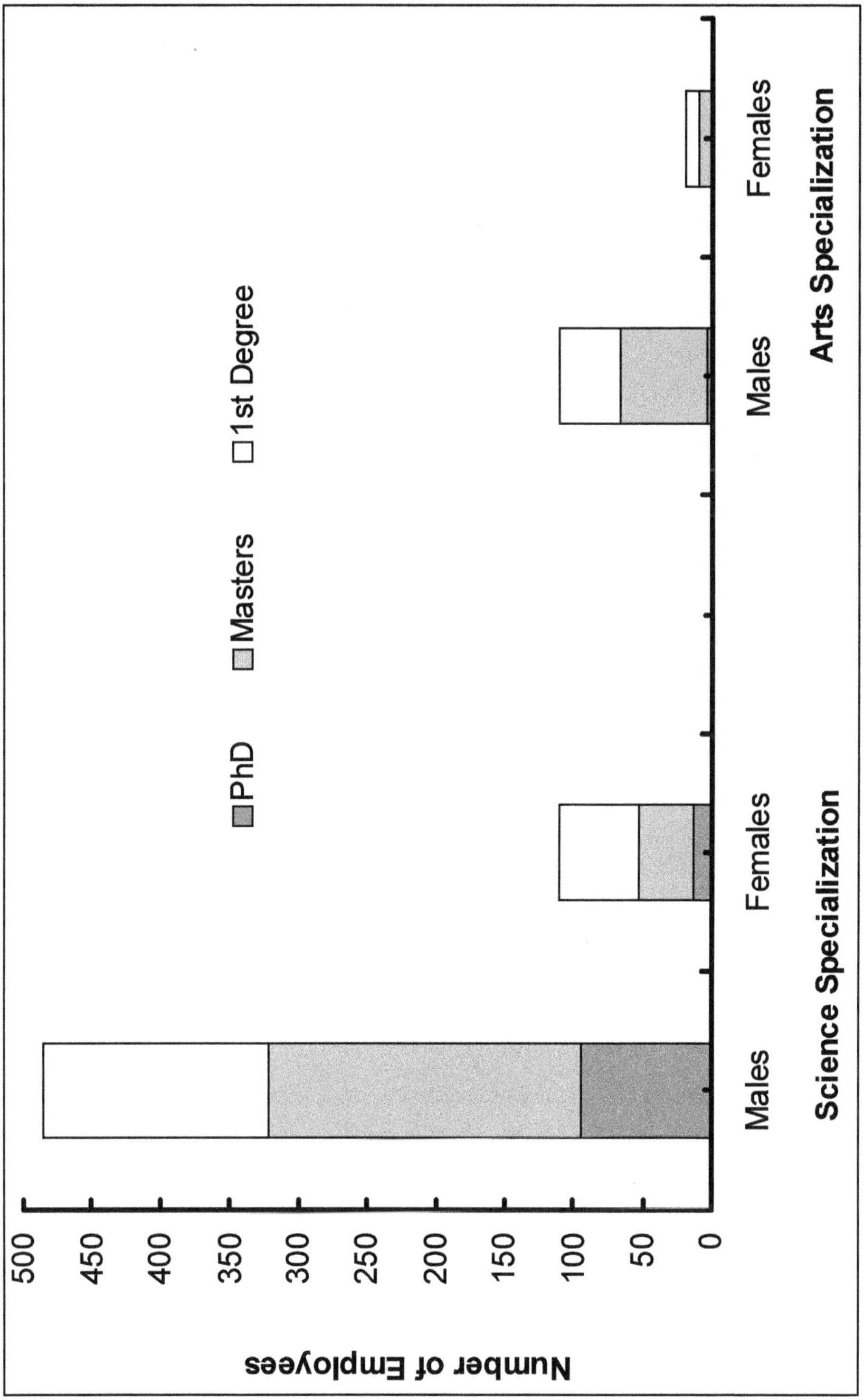

Figure 9.12: Number of Researchers in Non-Higher Education R&D Institutions By Gender, Qualification and Field

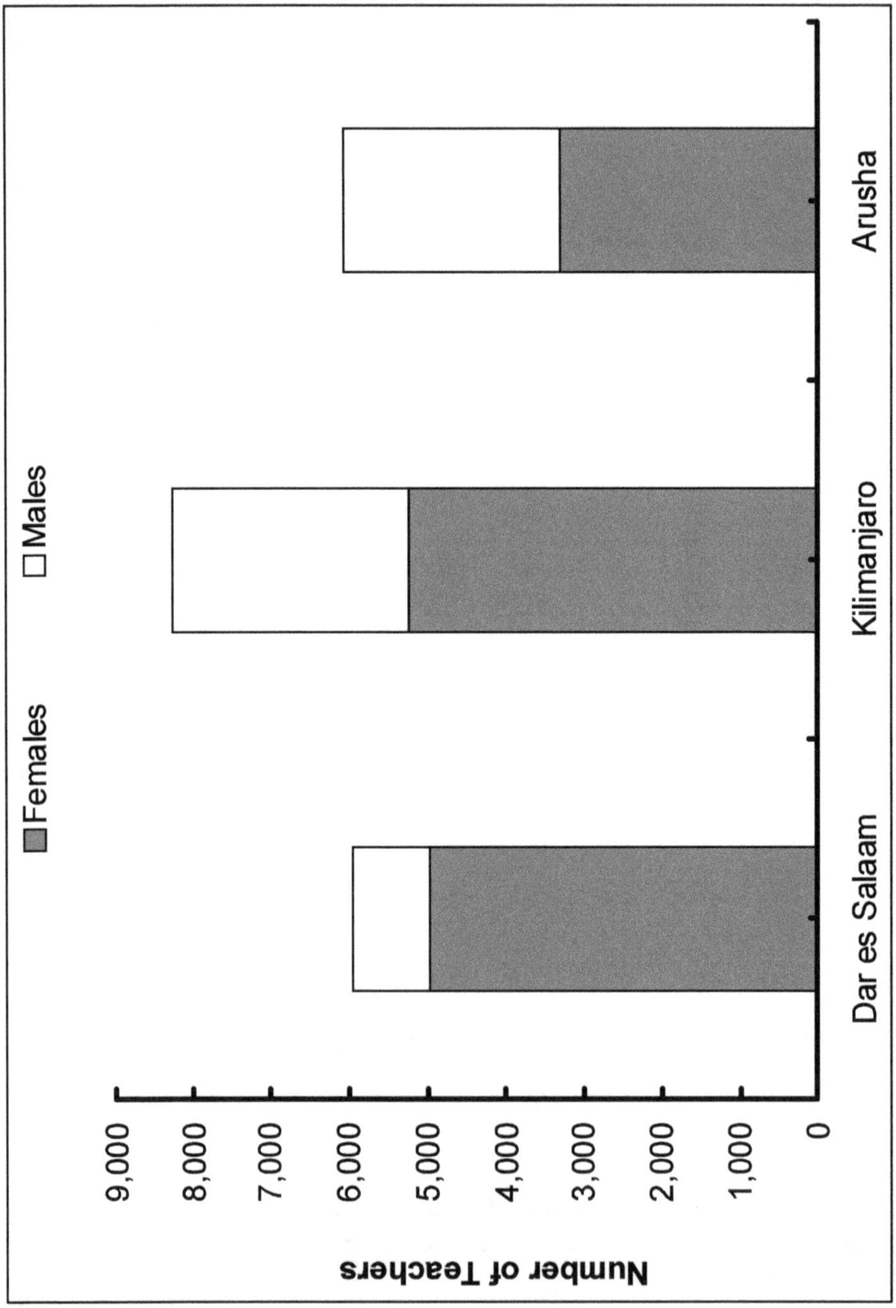

Figure 9.13: Number of Teachers at lower Primary and Secondary Schools in Dar-es Salaam, Kilimanjaro and Arusha Regions by Gender

science or arts specialization. In both these figures, women employees constitute only about a quarter of the male employees and distribution of their qualification and specialization is also proportionately distributed.

The recent year's retrenchment of workers in the government and parastatal organizations resulted in a decrease in the number of females in formal employment even more since both low academic qualification and experience were among the criteria used for retrenchment. Additionally, the household productive and reproductive role of women, making them vulnerable to malnutrition, had put them in a persistently poor health, predisposing them to poorer performance and frequent work absenteeism (WED, 1996), which was another criteria for retrenchment. Most women are employed in the private and informal sector, and for girls who have no, or only primary, education are employed as house girls or barmaids. In many cases this group is not only overworked and underpaid, they are also sexually harassed in the course or effort to retain a job at a risk of catching STDs including the deadly HIV/AIDS.

3.10 Employment by Gender in the Teaching Profession

Except for Dar es Salaam, Kilimanjaro and Arusha regions particularly in urban areas of these regions where female teachers outnumber male teachers at the lower levels of education (Figure 9.14), in all the rest of the regions, male teachers outnumber female teachers by far. The discrepancy is particularly pronounced in academic staff at tertiary level institutions especially in science subjects as can be seen in Figures 9.15 and 9.16. It is also important to note in Figure 9.17 that, much fewer females are qualified for becoming academic staff at tertiary institutions and hence there are just a handful of women who hold professorial positions and senior lecturership (Figure 9.18).

3.11 Decision-Making Positions

The power relations that prevent women from leading fulfilling lives operate at many levels of society, from the individual and private to the high public level. Achieving the goal of equal participation of women and men in decision making will provide a balance that more accurately reflect the composition of society and is needed in order to strengthen democracy and promote its proper functioning. Women equal participation in politics and decision-making is not only a demand for simple justice or democracy but also a necessary condition to ensure that women's interests are taken into account. Without the active participation of women and the incorporation of women's perspective at all level of decision-making, the goals of equal development and peace cannot be achieved. Women in decision making positions in governments contribute to redefining policy priorities, placing new items on the agenda that reflect and address women's gender specific concerns, value and experiences, and providing new perspectives on mainstreaming policy issues.

3.12 Gender Sensitivity in Decision-Making Positions

The constitution of the United Republic of Tanzania (URT) and most of the sectoral policies provide for equal opportunities for both male and female citizens to hold decision-making positions at any public office. Despite these provisions and

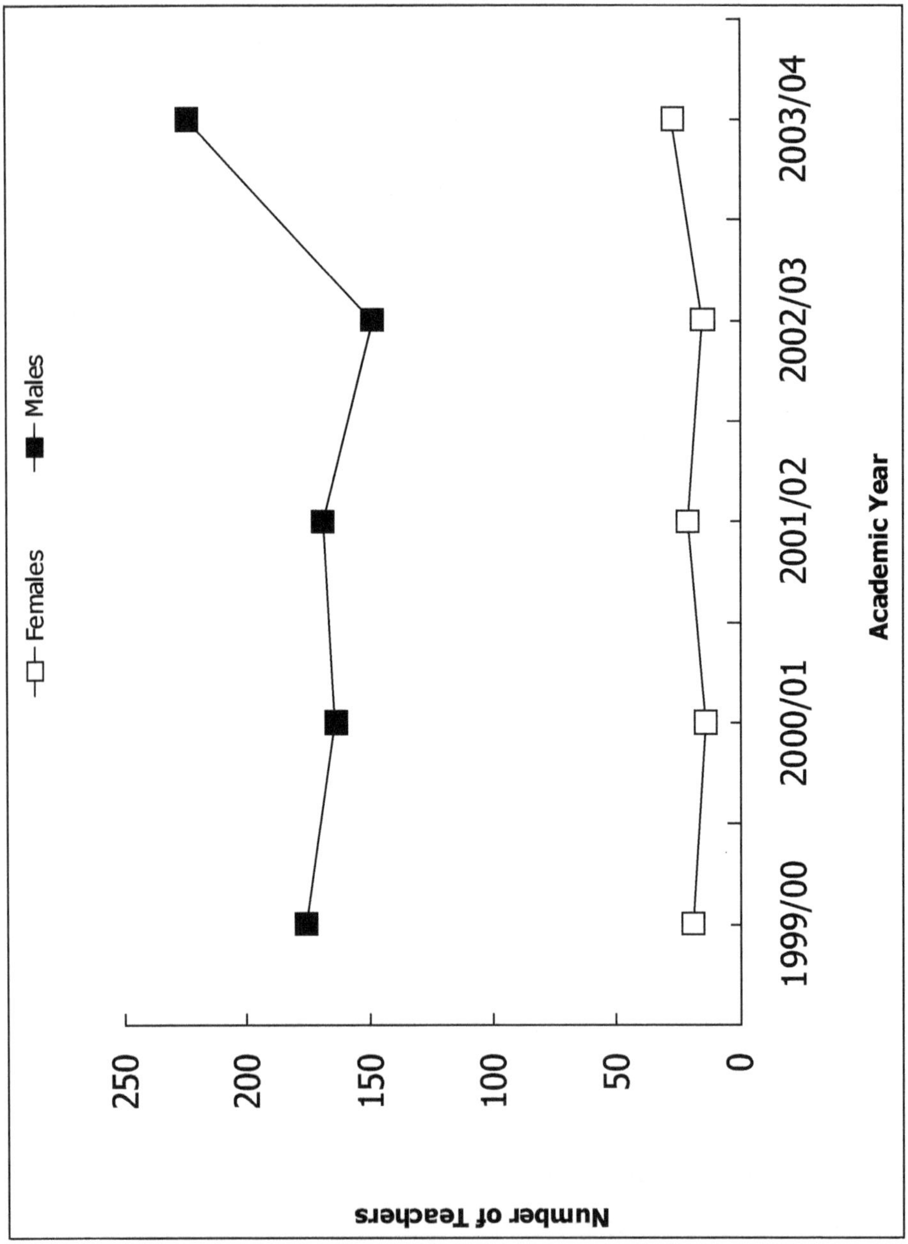

Figure 9.14: Number of Teaching Staff in Technical Education Institutions by Gender (Source: MSTHE 2004, UDSM 2004)

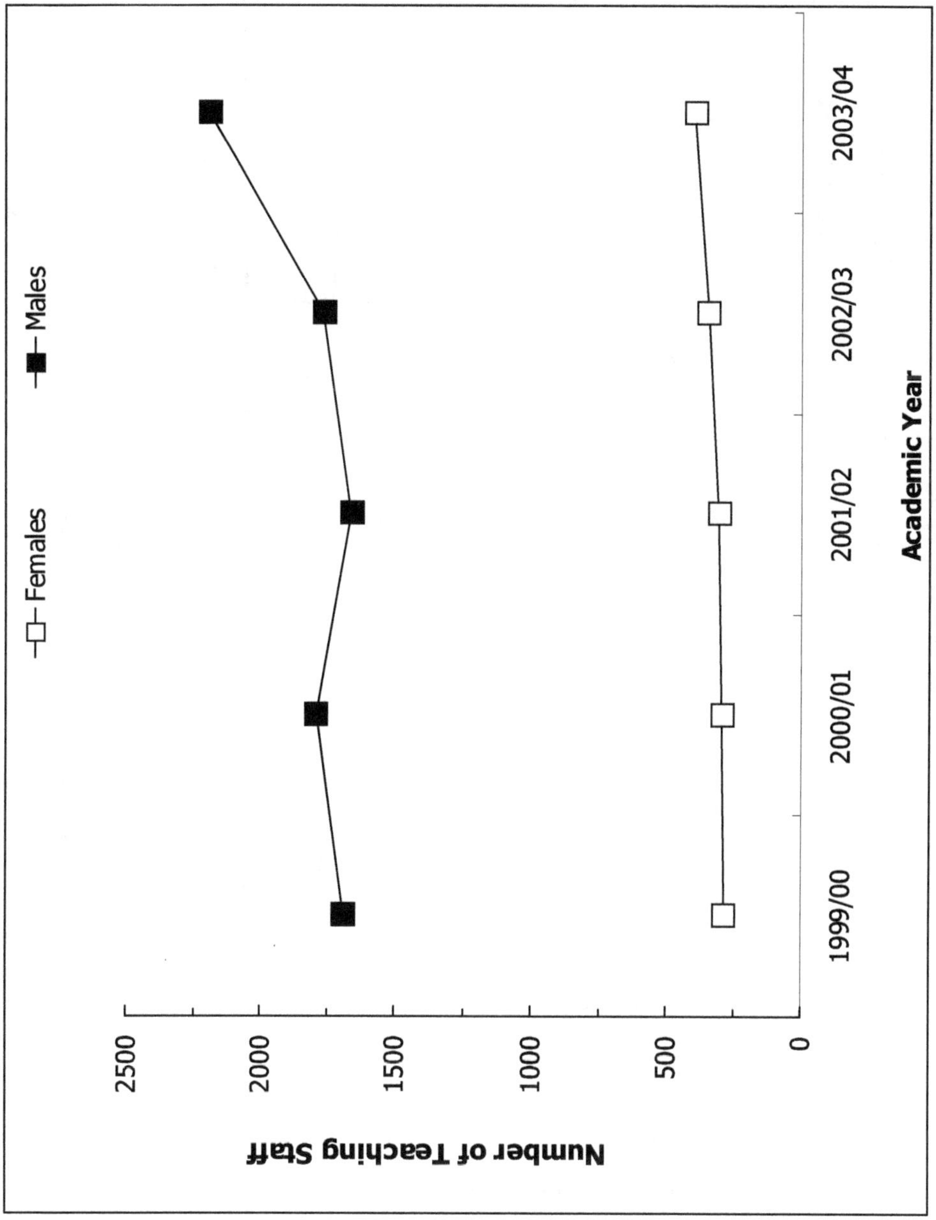

Figure 9.15: Number of Teaching Staff in Public and Private Universities by Gender (MSTHE, 2004)

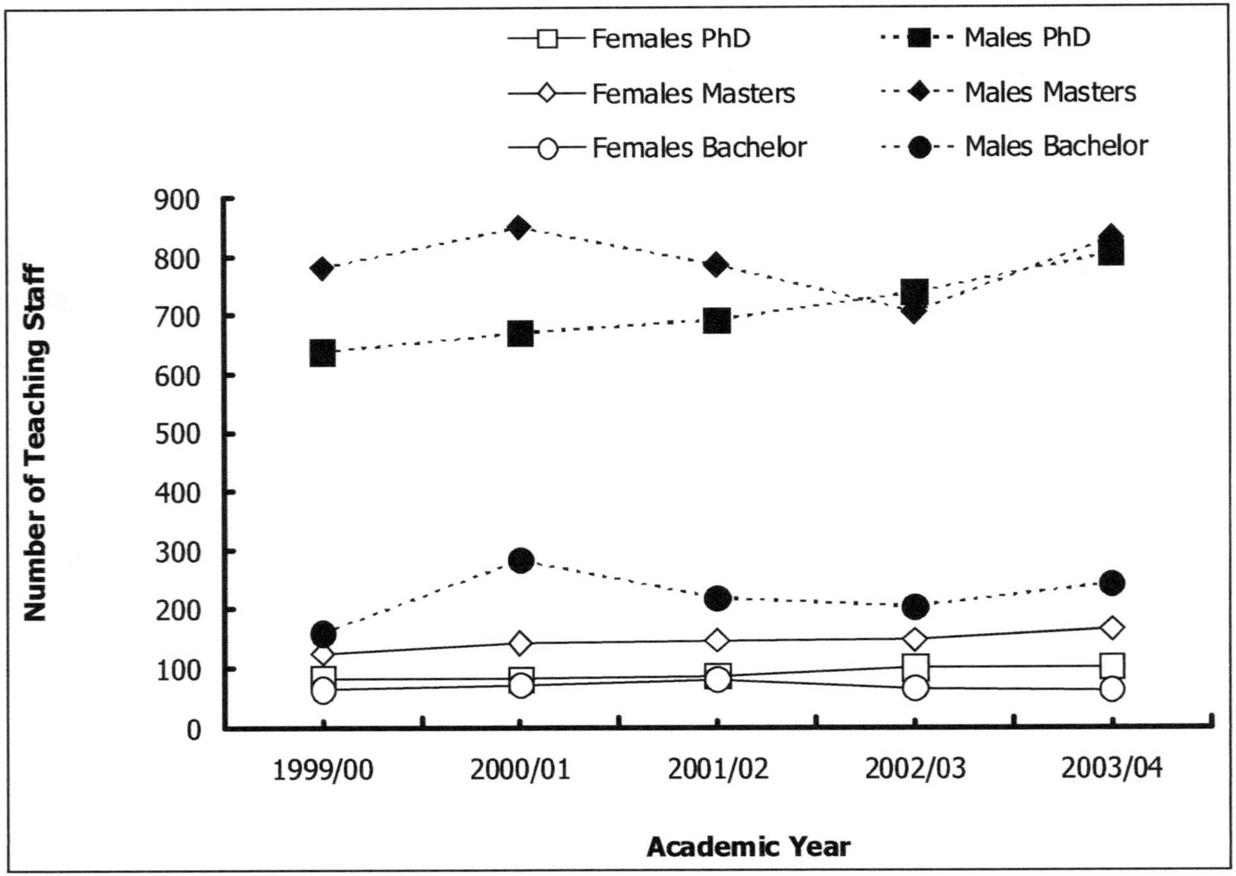

Figure 9.16: Teaching Staff by Qualification[1] (MSTHE, 2004)

1 In Public and Private Universities, Technical Institutions and Other Learning Institutions Under Other Ministries by Gender.

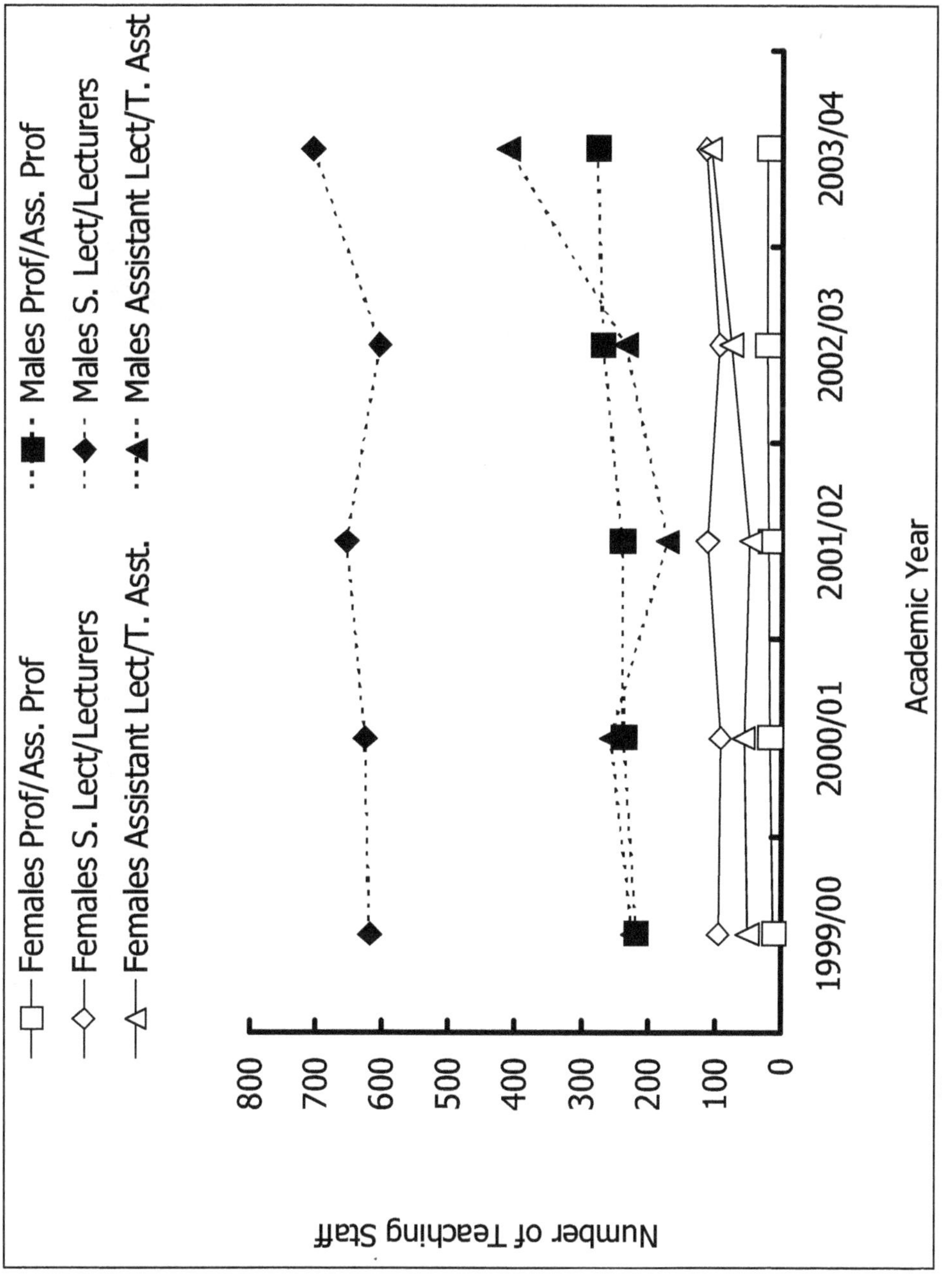

Figure 9.17: Teaching Staff by Rank in Private and Public Universities by Gender

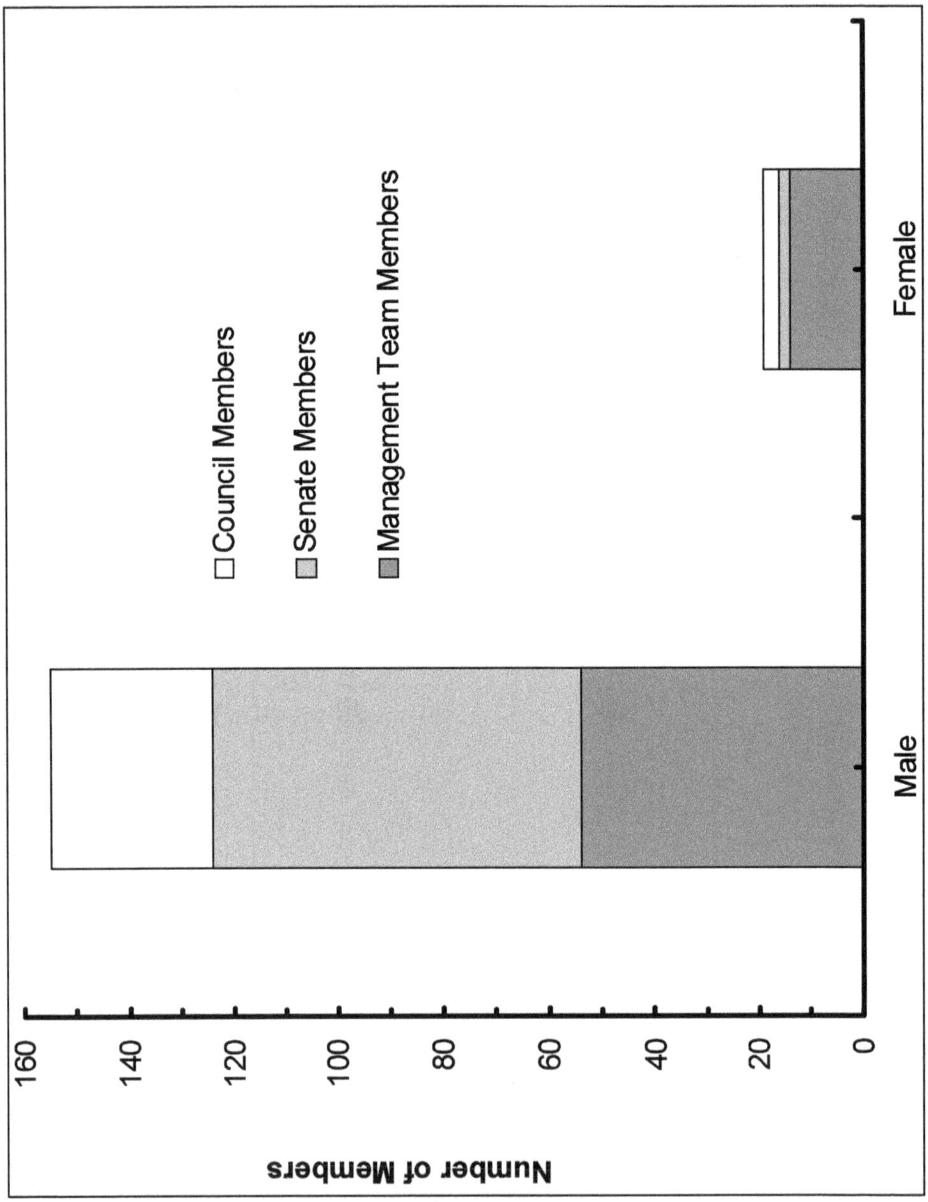

Figure 9.18: Management Team, Senate and Councilor Members at SUA by Gender in Year 2004

extensive gender sensitization in this regard, women in Tanzania continue to be under represented at all levels of government decision making positions. This is evidenced in Figures 9.19–9.22. The challenge that remains is the promotion of equal opportunities for men and women, as well as the recognition of the value of the multiple roles of women and therefore the mainstreaming of gender concerns in employment practices. The problem lies mainly in changing people's mindset. For instance the private sector employers do not often abide by all standards that promote equity at places of work.

4.0 Conclusions and Recommendations

Tanzania has adopted the Beijing Declaration and Platform for Action on Access of women to education, training and employment as among its priority areas for implementation. Further, except for few policies, most of them have gender sensitive policy statements. There are also a number of strategies and on going programmes aimed at addressing the gender disparities in almost all socio-economic sectors. However, despite all of these, one still sees significant disparities between women and men particularly with regard to the enrollment in science and technological studies as one goes up the ladder from primary to tertiary level of education with women occupying the losing end.

Consequently fewer women compete effectively with their male counterparts for the better-paid jobs. Women are thus employed in poorer paid jobs and are conspicuously less represented in political and decision-making positions. Additionally, although women are a significant part of the Tanzanian agricultural labour force, due to cultural barriers and socio-economic disempowerment, women have normally limited access or control of the necessary productive resources such as land, credit facilities and even information.

The observation above may be explained by the fact that, although most Tanzanians today acknowledge that gender-differentiated roles compromise women's social welfare, culture still dictates that women should be brought up in a manner that prepares them for motherhood rather than outdoor economic activities. This attitude and practice unfortunately leads to limited access of girls to education, which in turn adversely affects their employment as well as their access to educational facilities, particularly to higher learning and in S&T disciplines, which feature highly up the education ladder. The female socio-economic dis-empowerment and denial of fundamental human rights are important barriers to nutrition and health progress and eventually to socio-economic development not only of the women themselves but also their families.

The challenges above call for integrated efforts aimed at articulation of the different sectoral inputs with a gender perspective. Specifically, it calls for operationalization of the gender sensitive policy statements and implementation of the strategies already considered in the different policies with diligence aimed at making a significant desired shift. It also calls for reviewing of policies that are gender blunt so as to engender them but also crafting new gender strategies where there are none in place. Capacity building interventions at the grassroots level to address the various gaps identified above as well as addressing various policy issues on poverty reduction

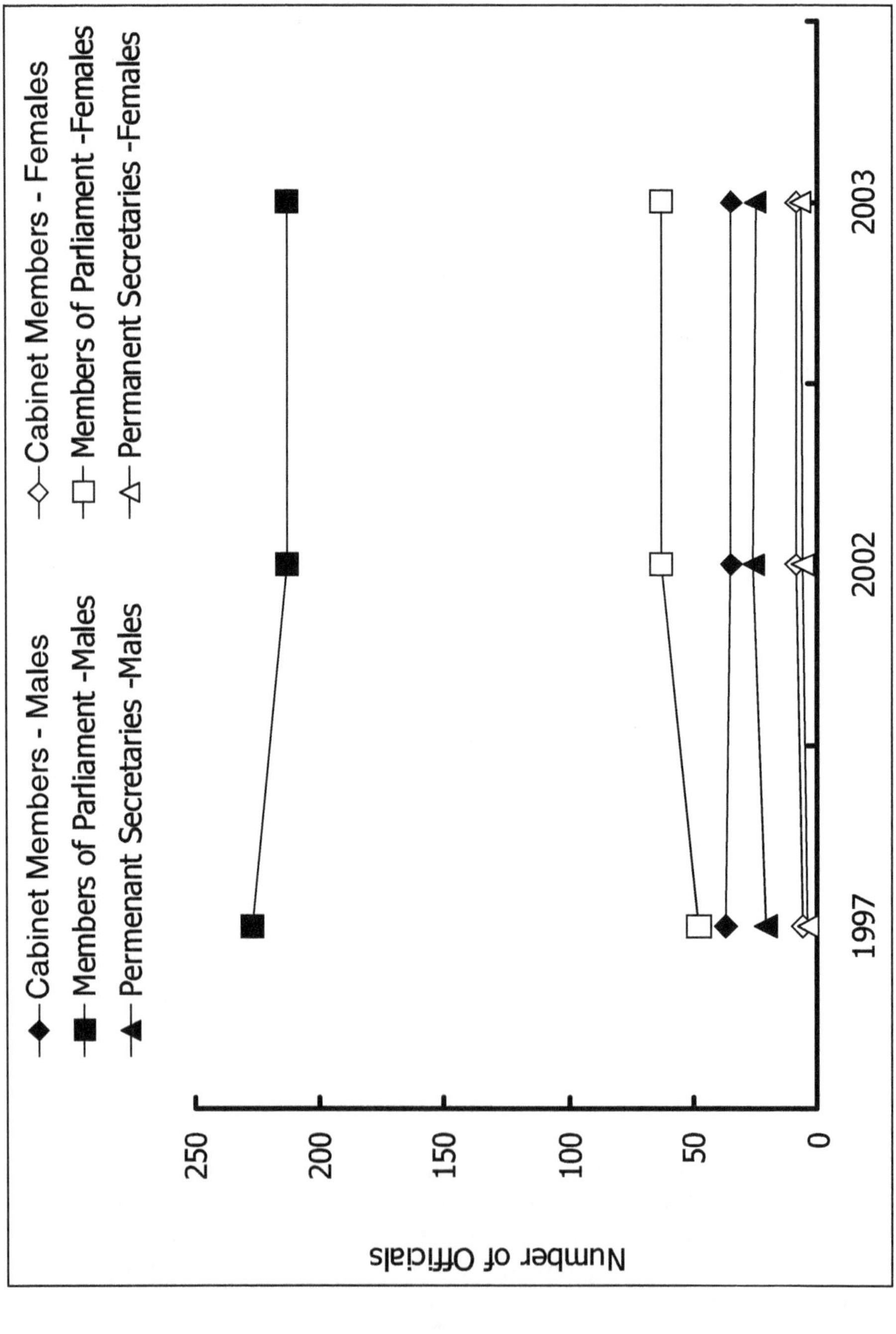

Figure 9.19: Number of Cabinet and Parliamentary Members and Permanent Secretaries in URT by Gender (*Source*: Civil Service Department)

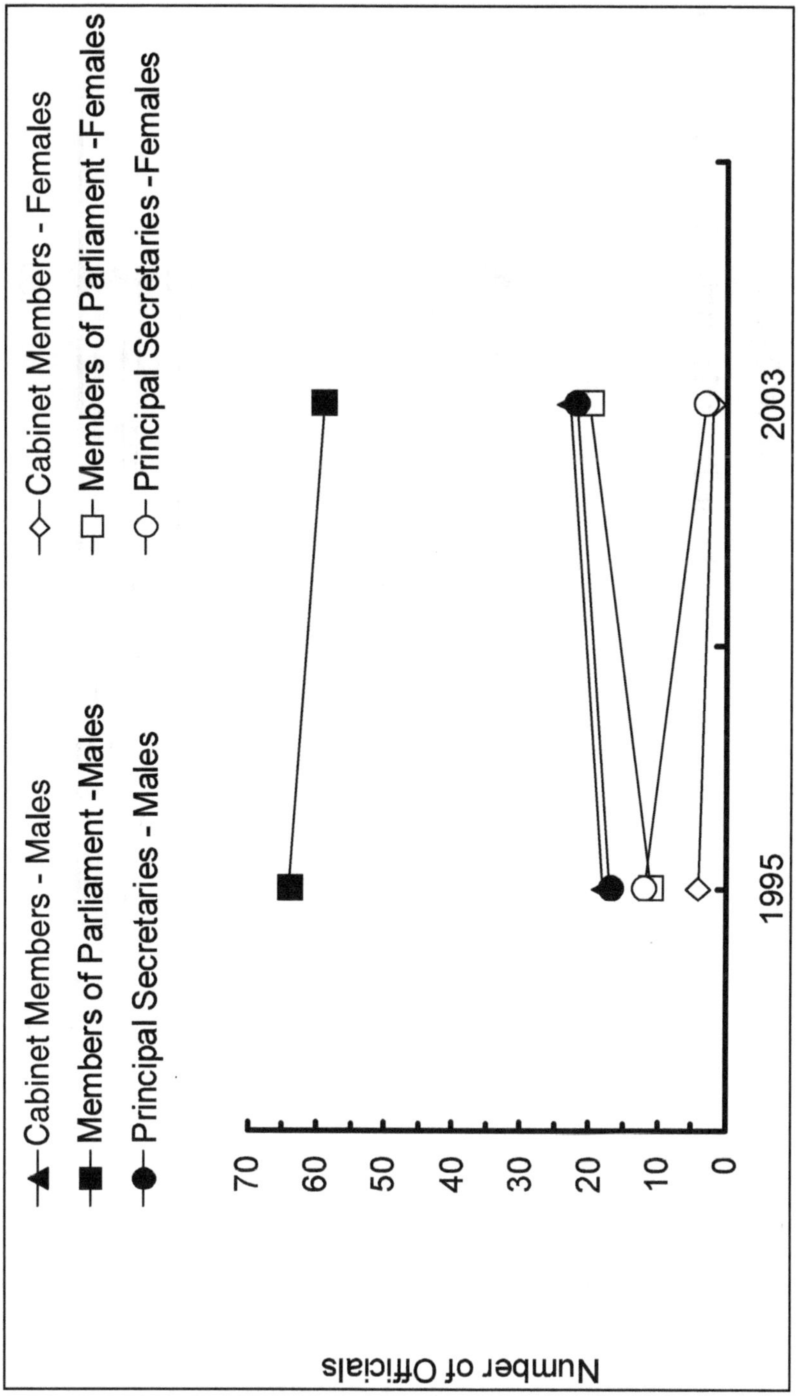

Figure 9.20: Number of Cabinet and Parliamentary Members and Permanent Secretaries in Zanzibar by Gender
(*Source*: Civil Service Department)

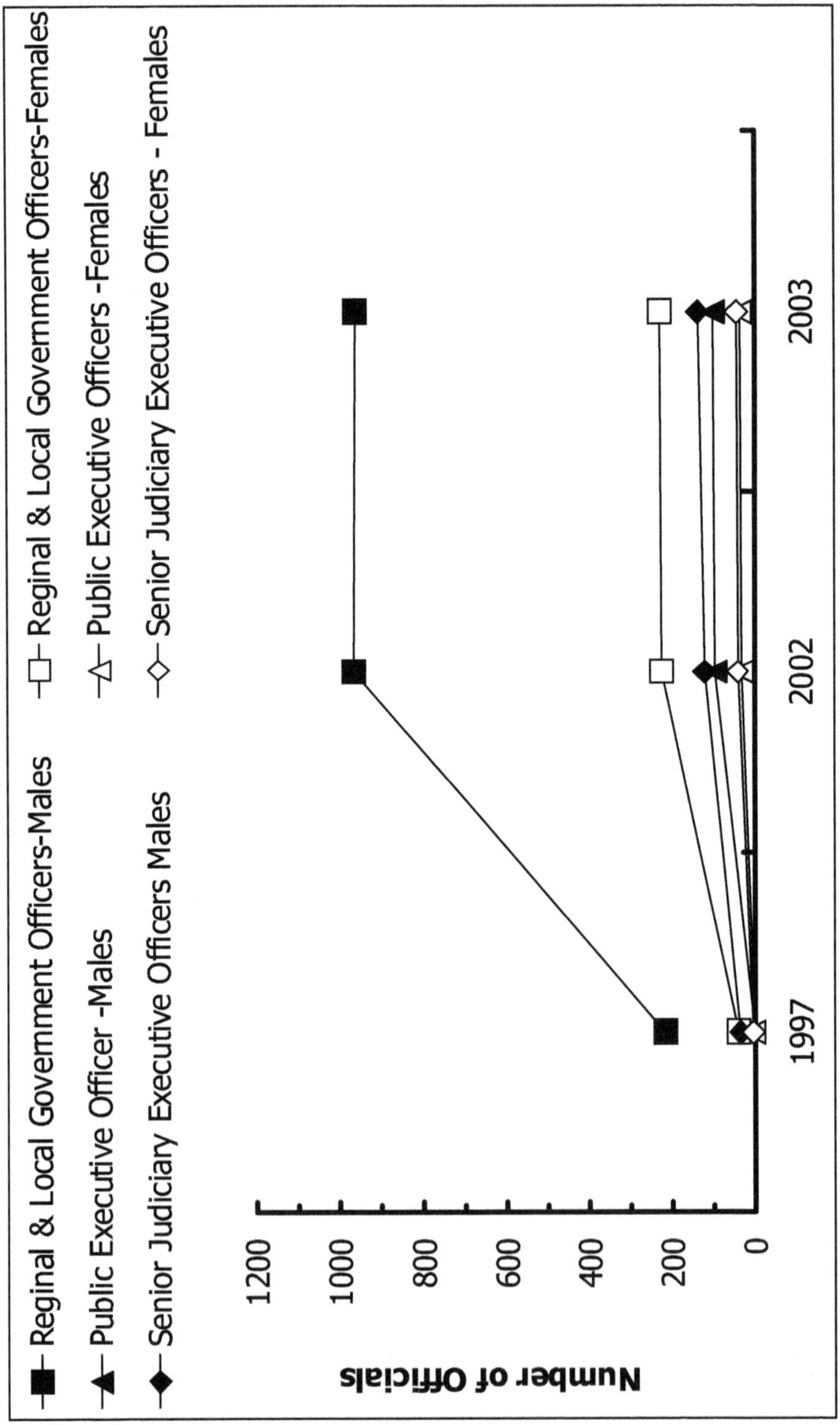

Figure 9.21: Number of Senior Executive Officers at Government, Regional and Local Government Levels in URT by Gender (*Source:* Civil Service Department)

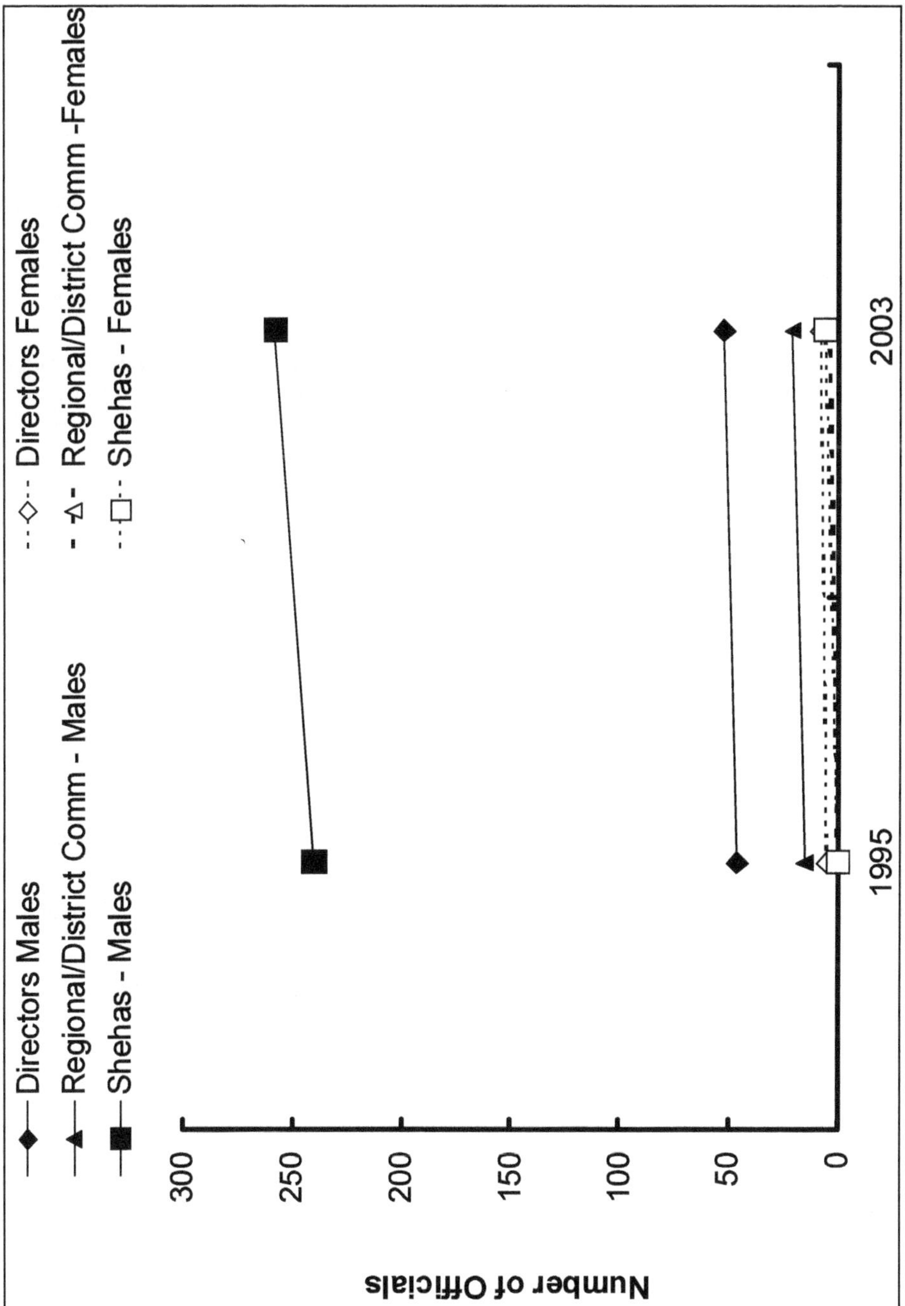

Figure 9.22: Number of Senior Executive Officers at Government, Regional and Local Government Levels in Zanzibar by Gender (*Source:* Situation Analysis of Women and Children in Zanzibar)

and involvement of men and women in influencing policy and programme formulation processes to promote ownership and sustainability of the programmes is of paramount importance.

References

Agriculture Sector Development Strategy (ASDS), 2001.

Agriculture and Livestock Policy, 1997.

Arusha Declaration and Musoma Resolution 1976 "Universal Primary Education (UPE)" Constitution of United Republic of Tanzania 1977.

Education Act 1978.

Education Ordinance 1961.

Gender Dimension Programme Committee, University of Dar es Salaam Report. October 2004. The Tanzania *Gender* Networking *Programm*e (TGNP), February 1999.... Mbilinyi, M. (Ed) (1990), Education with a *Gender* Perspective. A *report* prepared for SIDA. *Dar es Salaam.... Gender Dimension Programme Committee, University of Dar es Salaam. October 2004*; Situation Analysis of *University of Dar es Salaam*

Gender and Education in Tanzanian Schools, Report 1998.

Household Budget Survey 2000/01Report.

Household Budget Survey 2000-01 report, National Bureau of Statistics.

Kavishe, FP., 1993. Nutrition-relevant Action in Tanzania. UN ACC/SCN Country Case Study supported by UNICEF. A Case Study for the XV Congress of International Union of Nutritional Sciences. September 26 to October 1, 1993, Adelaide. Monograph series No. 1.

Mbilinyi, M. (Ed), 1990. Education with a Gender Perspective. A report prepared for SIDA. Dar es Salaam.

Ministry of Education and Culture, Dar es Salaam. Basic Statistics in Education 1995-1999; 1998-2002; 2003.

Ministry of Science, Technology and Higher Education. Basic Statistics on Higher Learning Institutions in Tanzania 1997/98-2001/2002.

National Census 2002 Report www.tanzania.go.tz/census/regions.htm–23k.

National Education Policy Paper National Website (http://www.tanzania.go.tz/) Tanzania Education Sector Policy Overview Paper, September 2006. http://www.enable.nu/publication/D_1_7_Tanzania_Education_Policy_Overview.pdf.

Policy on Women in Development 1992.

Political Handbook and NGO Calendar 2001, 2002, 2003.

Situation Analysis of University of Dar es Salaam Information Pertaining: to Employment, Retention and Enrollment by Subject and by Gender for the period of July 1997 to December 2002.

S&T Policy for Tanzania 1996.

Towards Gender Equality in Tanzania. The Tanzania Gender Networking Programme (TGNP), February 1999.

UNICEF, 1995. "The Girl Child in Tanzania". Today's Girl, Tomorrow's Woman". A Research Report Compiled by Mabala, R. and Kamazima, S.R. UNICEF: Dar es Salaam.

United Republic of Tanzania (URT). Policy on Women in Development, 1992

Women Education Development Group (WED) 1996. http://www.google.co.tz/search?q=Tanzania+Women+Education+Development+Group+(WED)+1996&hl=sw&start=40&sa=N.

Women, Gender, and Science Report, 1997 http://www1.umn.edu/humanrts/cedaw/tanzania1998.html.

Abbreviations

COSTECH	:	Commission for Science and Technology
ETP	:	Education and Training Policy
G,S&T	:	Gender, Science and Technology
ILO	:	International LabourOrganization
NGO	:	Non-Governmental Organizations
PEDP	:	Primary Education Development Programme
R&D	:	Research and Development
SEDP	:	Secondary Education Development Programme
S& T	:	Science and Technology
TASTA	:	Tanzania Science and Technology Award
UPE	:	Universal Primary Education
URT	:	United Republic of Tanzania

Chapter 10

Gender Challenges in Science and Technology in Democracies and Transitional Democracies

Barbara Wejnert
Professor, Department of Global Gender Studies, University at Buffalo, New York, United States
E-mail: bwejnert@buffalo.edu

ABSTRACT

This paper discussed the position and role of women in science and technology in contemporary democratic countries in comparison to countries that had experienced recent transitions to democracy. The analysis is based on empirical research conducted in the United States and democratizing countries of East and Central Europe and the former Soviet Union. I argue that although democracies, as well as transitional democracies promote equal civil rights and gender equality, unequal representation of women in science, industrial labourforce and technology is observed across all those states. Consequently, initiation of policies promoting equal participation of women in relation to men in technical and scientific professions remains to be one of the essential paths in achievement of gender parity and women empowerment regardless of existing political system.

Keywords: *Women, Science and technology, Research, Education, United States democracy, Transitional post-communist democracy.*

1.0 Introduction

Contrary to popular belief, throughout the modern era of technological and scientific discoveries and increasing societal education level and paid labourforce participation of women, women have experienced far more discrimination with regard to their employment status, earning abilities and achievement in science and technology. The unequal position of women in terms of their involvement in technological and scientific professions is observed in developed and developing countries, and across states with democratic and not-democratic political system. As Michael Binyon (1983:36) so cogently noted it is:"*True that in Russia women keep the wheels of industry turning. Virtually no factory could keep going without the female labor, no hospitals could function in a country where women constitute about 70 per cent of all doctors, and service industry would collapse without women hairdressers and waitresses, ticket-sellers and shop assistants...* ". At the same time, despite prioritization of gender equality, the earning capacity of women lagged behind that of men (Binyon1983). Moreover, access of women to technologically based work is limited. Therefore, the presence of women in blue-collar jobs; engineering/mathematics and finance in well democratized and democratizing states is persistently stereotyped as representation of women " non-traditional jobs", and classified as women involvement in "typical male fields".

2.0 Women in Science and Technology in American Democracy

2.1 Employment

Since 1950, regardless of the increase of women participation in the labourforce in the United States across all age categories (Figure 10.1), women are continuously combating the invisible barrier that is strong enough to hold women back from top-level jobs simply because they are women rather than because they lack job-relevant skills, education, or experience. Such barrier is called the glass ceiling, and it depicts an exclusion of women from top management positions.

The Glass Ceiling has prevented women from getting ahead in the work force. Thus, although women's participation in the U.S. labour force has increased, and women occupy 44 per cent of management jobs in American companies, top

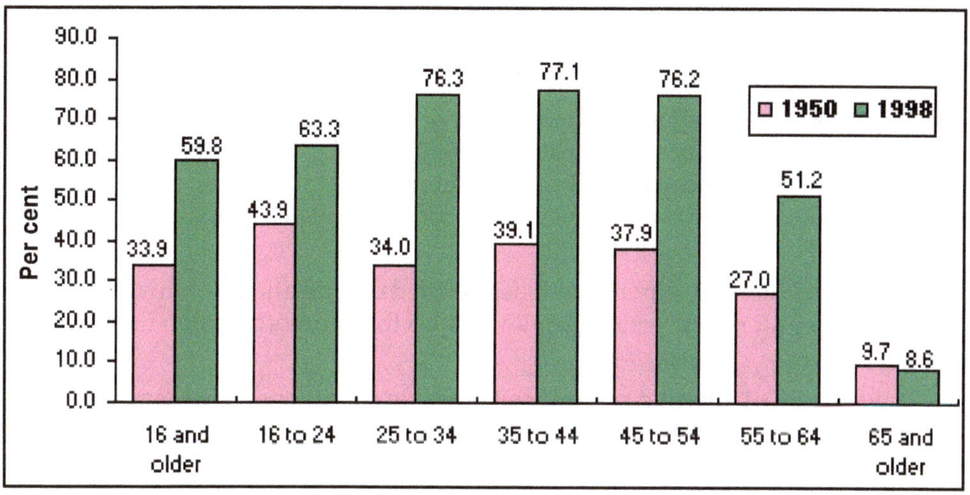

Figure 10.1: Labour Force Participation Rates of Women by Age, 1950 and 1958 (*Source*: Figure compiled base on Statistical Abstract of the United States, U.S. Census Bureau, versions, 1995-2000)

management ranks remain dominated by men. Consequently, by 2008 only 3 per cent to 5 per cent of the top management jobs in private sector companies are filled by women.

2.2 Managerial Positions

One of the most powerful ways to shatter the glass ceiling is to increase women presence at the top managerial positions. The higher the percentage of lower level management jobs filled by women, the more likely the establishment will have women in top management positions. As research show, the greater the emphasis on promotion and development, the more likely the establishment will have women in top management positions.

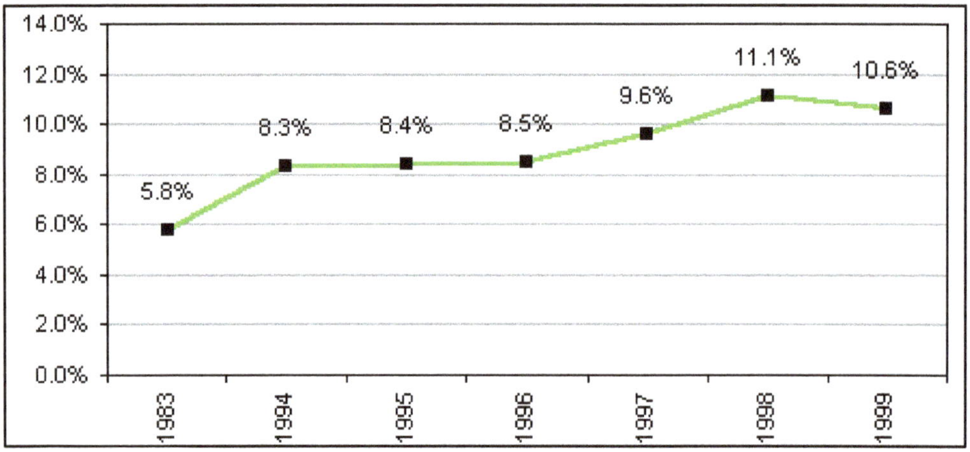

Figure 10.2: Employed Engineers–Per cent Female
(*Source*: Statistical Abstract of the United States, U.S. Census Bureau, versions, 1995-2000)

Nonetheless as of today, more women are among labors in blue collar job than managerial positions. Blue-collar are the job performed by those who work for wages, especially manual or industrial laborers. Among those employees are construction jobs, ironworkers, electricians, carpenters, stone masons, surveyors, and other union members.

Being a substantial minority, women formed organizations to provide support to their issues and rights. Among organizations that support women in blue-collar jobs are

Tradeswomen, Inc. that provide fair and safe conditions for women who work in the building trades; and an Office for Women's Rights at the City of Seattle that develop special recruitment programmes and workshops to provide information to women.

Not only women in blue collar job are underrepresented. Similar situation exists among women in engineering. Since the last three decades the number of women engineers is steady increasing, nonetheless still only about one out of ten engineers are women. To stimulate women to achieve full potential in careers as engineers and leaders, to expand the image of the engineering profession as a positive force in improving the quality of life and to demonstrate the value of diversity, women engineers formed Society for Women Engineers.

Nonetheless, across the broad spectrum of the U.S. engineering jobs, salaries for women engineers remains much smaller than for male (Figure 10.3).

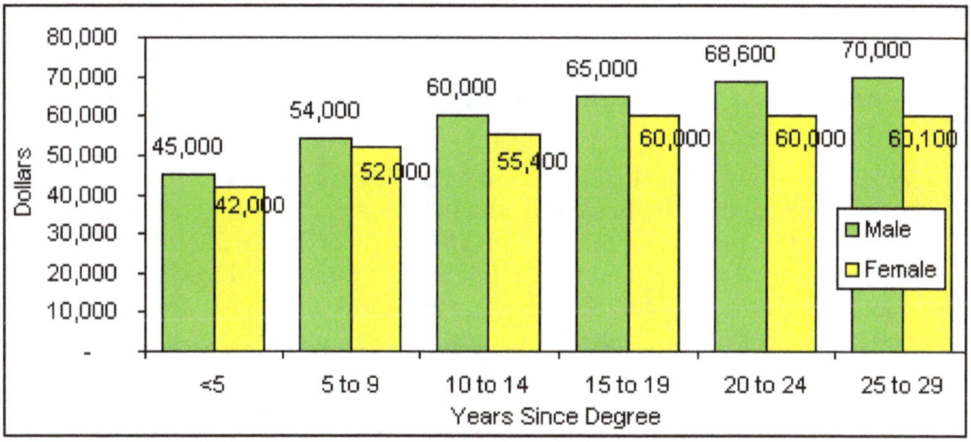

Figure 10.3: Median Annual Salaries by US Engineers– All Degree Levels by Years since Degree

(*Source*: Figure Compiled based on National Science Foundation, 2000. *National Science Board Science and Engineers Indicators-2000.* Arlington, VA: Appendix 3-8)

Not only in science and technology but in finances women are visibly underrepresented at the top position. These positions, commonly known as job on Wall Street (the financial centre of the United States in New York) are characterized by salaries at the top income brackets in the country. Women being a minority on Wall Street face frequent gender discrimination in promotion to the highest positions and in increase of salaries in comparison to man. The known discrimination cases brought by women to courts against their companies include Salomon Smith Barney, Goldman Sachs, Morgan Stanley, and Merrill Lynch. In few cases after winning the discrimination cases, finally women could break through the glass ceiling and land in the executive suites of these companies. The Equal Employment Opportunity Commission (EEOC) has been designing rules to ensure women are well-represented at the top levels of financial firms.

Regardless of an increasing representation of Women on Wall Street, women are still vastly underrepresented in the top ranks and climbing in the ranks of managing directors relatively slowly – 17 per cent in 2003 compared with 13 per cent in 2001. In addition, the economic downturn negatively affects overall percentage of women working on Wall Street, *i.e.*, such employment declined to 37 per cent in 2003 from 41 per cent in 2001 and 43 per cent in 1999.

2.3 Women's Movement

To organize themselves and to have the women voices heard, women organized annual "Women on Wall Street" conference – a meeting addressing issues faced by women on Wall Street. These conferences were initially attended by only 350 people—participation increased to 2,000 men and women by 2003. There some positive outcomes of such meetings, For instance, after the meeting in 2003, the Deutsche Bank offered mentoring circles and an additional 12 weeks' paid leave after maternity leave.

3.0 Women in Science and Technology in Transitional Democracies of Eastern Europe

Women representation in science and technology is also severely challenged in democratizing countries. In those countries the number of working women steadily increased over the decades following World War II, nonetheless the concentration of female workforce was in agriculture, unskilled industrial, and low skilled service sector jobs, such as secretarial, accounting, sales (Drakulic 1993, Titov 1984). These jobs typically were ranked as not difficult and consequently resulted in lower wages. In Czechoslovakia and Yugoslavia, the average wage for women was only 68 per cent and 60-80 per cent respectively, of the average wage of men (Drakulic 1993). In Hungary, where the earning opportunities for women improved since the 1980's but reached only the level of 70 per cent of men's wages and the fundamental gender-based division of labour was never eradicated (Lampland 1990).

With the post-communist transition to democracy the situation in the labour market worsened affecting the opportunities for women employment and achievement, especially in typically science and engineering professions (Amott and Matthaei 1996).

3.1 Employment

In early years of transition, rising unemployment has been a particular threat to citizen's sense of security but women have suffered most from the declines in employment (Wejnert and Spencer 1996). Institution of democracy and associated with that market economy led to decrease of the need for laborers, and women were first laid off from jobs and were able to hold on the very low paid and least prestige jobs. Such trends caused, on one hand, the rapid increase of women among registered unemployed, on the other, an increase of gender gap in earning between female and male workers. Moreover, although more women than men were registered as unemployed, due to the gender gap in earning in communist time and, since unemployment benefits were structured according to wage rates, women received lower unemployment benefits than men in all East European and Russian countries. For example, in Hungary in 1991, 79 per cent of unemployed women, as compare to 55 per cent of unemployed men, received benefits lower than the official minimum wage. At the same time, one-seventh of all families were headed a single female, accounting for the rearing of almost one-sixth of the children below fifteen years of age (Wejnert 1996).

Similarly, in Poland in the 1991, the unemployment rate was 13 per cent (in 1999 was about 10 per cent) but there were three times more women than men registered as unemployed, while the number of jobs offered to men was seven times higher than to women. By the end of October 1990, women constituted 51 per cent of the one million unemployed, and there were 37.3 per cent unemployed women for every vacancy compared with 9.5 per cent unemployed men. By the 1994 male constituted 58.8 per cent of the overall labour while women 44 per cent but out of overall 13.9 per cent unemployed, women constituted more than half—52.2 per cent while man 47.8 per cent of the total number (Polish Statistical Office [Glowny Urzad Statystyczny] 1995). Similar situation was faced by women in other countries. In Bulgaria, in mid-1990s, 43.1 per cent of employed women expressed fear that they will lose their jobs, as compared with 38.5 per cent of employed men (Petrova 1993).

Access to the better-paying, professional jobs, such as in engineering, law, top business administration, and managerial, was at best limited for women. In Bulgaria, only 1.6 per cent of women worked in management, decision making, and higher administrative spheres in 1988. In Yugoslavia, although the number of working women steadily increased from 26 per cent in 1961 to 36 per cent in 1981, the number of women in managerial positions steadily decreased from 15 per cent in 1962 to 12 per cent 1981 (Ojulic 1991). Contrary to popular belief, during the period of democratic transition and consolidation of democracy greater gender asymmetry was created, particularly in professional field of science and technology. Lower earning and lower position were in sharp contrast to the significant strides that had been made by women in educational attainment at the university level.

3.2 Education

In former Soviet-bloc states, the number of female university students grew from 23 per cent of the total number of students in 1948-49, to over 50 per cent in 1988-89, reaching ratio of 24 per cent of female students to the total number of women aged 20-24 years. In comparison, the ratio for men was 23 per cent (United Nations Development Programme 1992, table 34).

By the beginning of the transition to democracy in 1989-1990, women represented 56 per cent of all specialists with a secondary education or with two years of collegiate education, and 38 per cent of specialists with at least full university degrees (Siklova, 1993). At the start of the 1990's, when the number of school-age girls was equal to the number of school-age boys, female students in secondary schools (high schools) outnumbered male students in Poland, Hungary, Czechoslovakia and the Soviet Union. At the same time, female students composed more than half of the university student population in almost all former Soviet-bloc countries, *e.g.*, in Poland 52 per cent, in Hungary 53 per cent (GUS 1993:430; KSH 1993:67-71, United Nations Development Programme 1992:191) (Figure 10.4).

In the Soviet Union there were an equal percentage of female and male students majoring in the previously predominantly "male" specialties of science at universities (Figure 10.4). Thus, it is clear that the differential wages and occupational positions between men and women were not due to an inferior educational background for women.

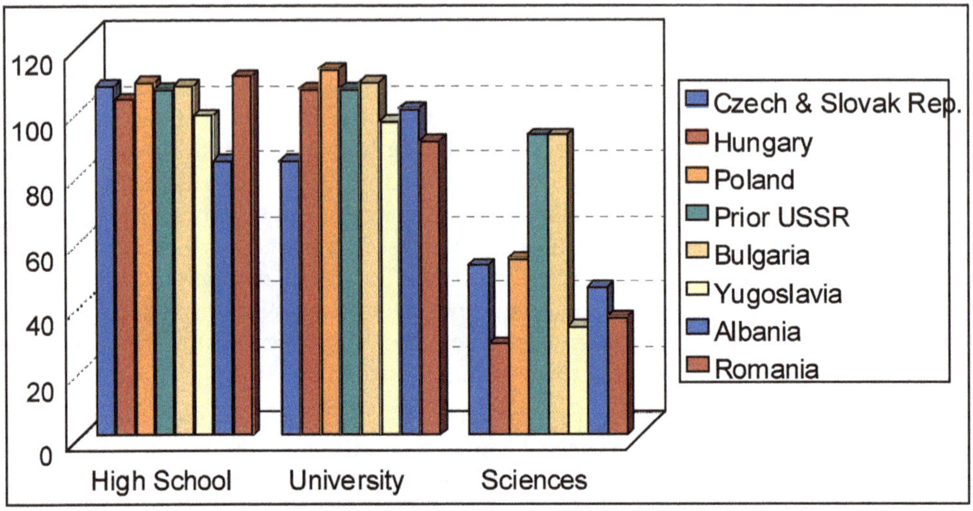

Figure 10.4: Female Students in Relation to Male Students in High Schools, Universities and Majoring in Natural Sciences in Former Soviet-Bloc States

(*Source*: Adapted from "Human Development Report 1992" p. 191. All figures are expressed in relation to the male average, which is indexed to equal 100. The smaller the figure the bigger the gap, the closer the figure to 100 the smaller the gap, and a figure above 100 indicates that the female average is higher than the male.)

3.3 Managerial and Administrative Positions

Gender asymmetry was broaden due to limited representation of women on the managerial or public domain positions (Olujic 1991). For instance, In 1982 in East Germany, women constituted 33.6 per cent of the People Chamber (parliament), but no women held positions in the Politburo (a governing body), and only one women served as a minister (in the ministry of education). In 1983 in Hungary, women (see Figure 10.5) held only 27 per cent of the National Assembly members, and only three out of twenty one minister positions. In transitional democracy of Poland women represented only 10.45 per cent of members in *Sejm* (Polish Parliament) and 1 per cent in *Senate* of post-communist parliament in 1994 (GUS-Polish Statistical Office 1995) and there was only one woman minister and only one woman as committee chair in the Polish government, while deputy positions held by women in Poland increased from 12.4 per cent in 1964 to 23 per cent in 1983, they still constituted the lowest female representation among the Eastern European countries. Similarly, only 10 out of 102 judges in the Polish Supreme Court were women.

Although women's under representation in governing bodies could be a result of women's reduced interest in politics, an analysis of women political dissidents of the 1970's and 1980's do not support such hypothesis. Over the period of political struggle for democracy in East Central Europe, women were represented equally to men in democratic, oppositional movements.

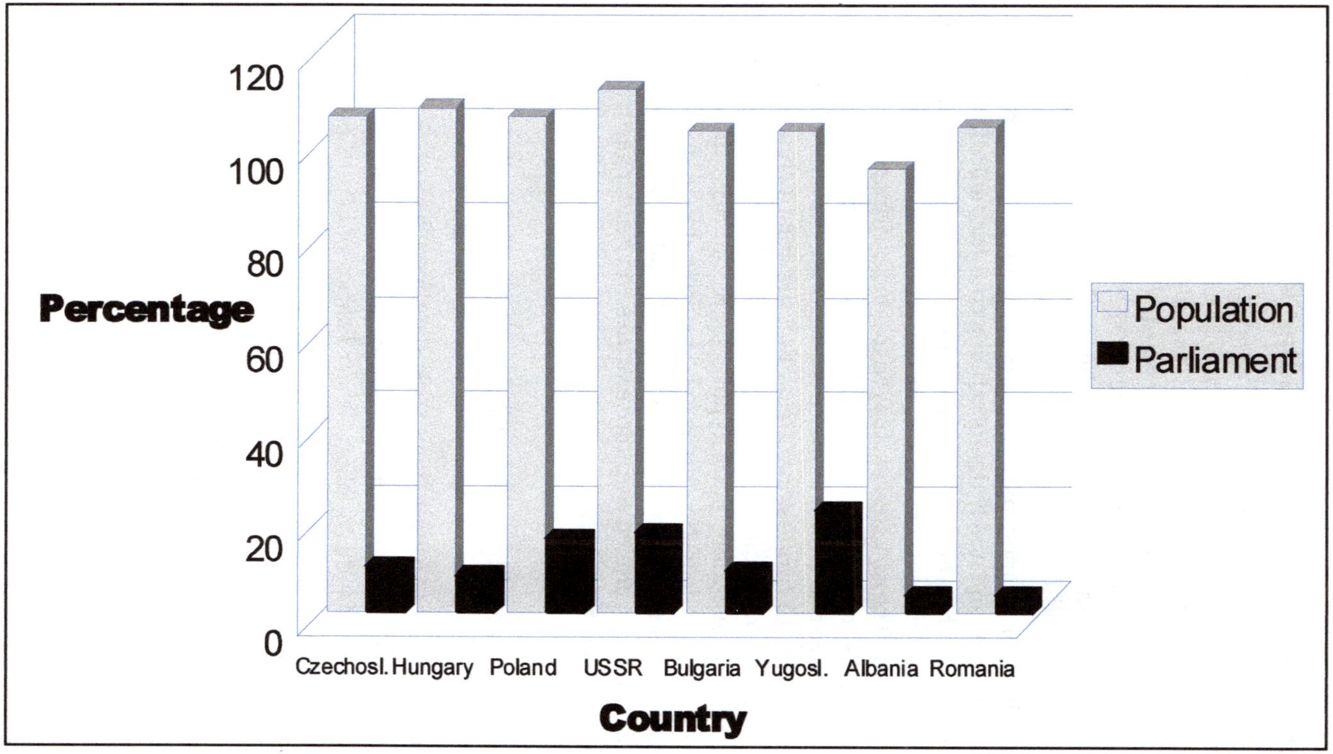

Figure 10.5: Females in Relation to Males in Population and in Parliaments in Former Soviet-Bloc States, in 1990

(*Source:* All Figures are expressed in relation to the male average, which is indexes to equal 100. The smaller the figure the bigger the gap, the closer the figure to 100 the smaller the gap and a figure above 100 indicates that the female average is higher than male. Source: United Nations, 1992. Human Development Report, p. 191.)

In contrast to their broad participation in the network of political opposition, but following the pattern of women under representation in governments, leaders of dissident movements excluded women from major political activity (Wejnert 2002).

4. Summary of Results and Concluding Remarks

The gender gap in academic science is a topic of ongoing policy and scholarly debates. "Studies in fields as diverse as engineering and biology have found that women scientists suffer from an attainment gap along at least three important dimensions: productivity, recognition, and reward" (Waverly, Murray and Stuart, 2006: 665). The limited evidence that exist about the 'academic entrepreneurship' such as patenting, consulting, and scientific advisory boards, also suggests a gender gap of considerable magnitude

The comparison of women participation in science and technology in the democratic United States and newly democratizing countries adds to this debate. As presented study indicates, an introduction of civil freedom, equal rights and democratic principles is not sufficient measure of the possibilities and opportunities that are open to women. The case studies showed to gender disparity in representation of women in science and technology, the establishment of policies promoting gender equality in high-prestige and well-paid positions becomes only the initial step on the path of opening access to professional fields that are relatively close to women. Stereotype perception of women's low potential to achieve in science seems to constitute additional constraints. Such hypothesis would need to be tested in the future. However, in case the testing turns to be positive it would shed light on the problem of interdependency of democratization and limited women representation in engineering and science, limited earning capacities and gender gap in participation in scientific entrepreneurship..

References

Amott, Teresa and Julie Matthaei. 1996. *Race, Gender and Work.* Boston, MA: South End Press.

Binyon Michael. 1983. *Life in Russia.* New York: Pantheon Books.

Drakulic, Slavenka. 1993. Women and New Democracy in the Former Yugoslavia, in: Funk, Nanette and Magda Mueller, *Gender Politics and Post-Communism*, New York: Routledge: 123-130.

GUS, Polish Statistical Office [Glowny Urzad Statystyczny] 1995. *Rocznik Statystyczny*, p.81.

Lampland, Martha, 1989. Unthinkable Subjects: Women and Labour in Socialist Hungary. *East European Quarterly*, 4:389-398.

KSH (Hungarian Statistical Office). 1993. Mayar Statisztikai Zsebkonyv, Hungary: Budapest: 67-71.

National Science Foundation. 2000. National Science Board. Science and Engineering Indicators-2000.Arlington, VA: Appendix 3-8

Olujic, Maria, 1990. Economic and Demographic Change in Contemporary Yugoslavia: Persistence of Traditional Gender Ideology. *East Euroepan Quarterly*, XXIII, 4:477-485.

Petrova, Dimitrina., 1993..The Winding Road to Emancipation in Bulgaria, in Funk, Nanette and Magda Mueller, *Gender Politics and Post-Communism*, New York: Routledge: 22-30.

Siklova, Jirina, 1993. Are Women in Central and Eastern Europe Conservative? In: Funk, Nanette and Magda Mueller, *Gender Politics and Post-Communism*, New York: Routledge: 74-84.

Titkov, Anna, 1993. Political Change in Poland: Cause, Modifier, or Barrier to Gender Equality? In: Funk, Nanette and Magda Mueller, *Gender Politics and Post-Communism*, New York: Routledge: 85-99.

United Nations Development Programme. 1992. *Human Development Report*, Oxford University Press.

Waverly, Ding, Fiona Murray and Toby Stuart. 2006. Gender Differences in Patenting in the Academic Life Sciences. *Science*, vol 313, August 4: 665-667.

Wejnert, Barbara. 2002. The Contribution of Collective Protests to the Softening of Communist Regimes in East Central Europe. In B. Wejnert. (Ed.) *Transition to Democracy in Eastern Europe and Russia.* Westport, CT: Greenwood Press/ Praeger Press, p.65-91.

Wejnert, Barbara. 1996. Political Transition and Gender Transformation in the Communist and Post-Communist Periods, in: B. Wejnert and M. Spencer (Eds.), *Women in Post-Communism*.Greenwich, Connecticut, JAI Press: 3-19.

Wejnert, Barbara and Metta Spencer, 1996.*Women in Post-communism.* Greenwich, Connecticut: JAI Press

Chapter 11

Country Report on Mauritius

Shamima Abdul Carrim Delbar
Ministry of Industry, Science and Research,
Port Louis, Mauritius
E-mail: shayou@intnet.mu, scarrim@mail.gov.mu

1.0 Background

Mauritius is a small island situated in the Indian Ocean, east of Madagascar. It has a total surface area of some 2,040 sq km which also includes Agalega Islands, Saint Brandon and Rodrigues.

The island is of volcanic origin and is almost entirely surrounded by coral reefs. It was home to the DODO, a large flightless bird related to the pigeon which became extinct in the seventeenth century.

Mauritius has a population of 1,274,189 as at July 2008. It has the highest life expectancy (74 years) in Africa and also the highest level of literacy. It forms a mosaic of different races, cultures and religions since Mauritians are descendants of immigrants from the Indian sub-continent, Africa, Europe and China. The cultural diversity and racial harmony of the island make of Mauritius a unique place. Most Mauritians are multilingual, being fluent in Creole, French and English. English is the official language. Bhojpuri, Hindi, Urdu, Tamil, Marathi, Telugu and Mandarin are also spoken. The three major ethnic groups are Hindus making up the majority (48 per cent) followed by the Roman Catholics (23.6 per cent) and 16.6 per cent Muslims.

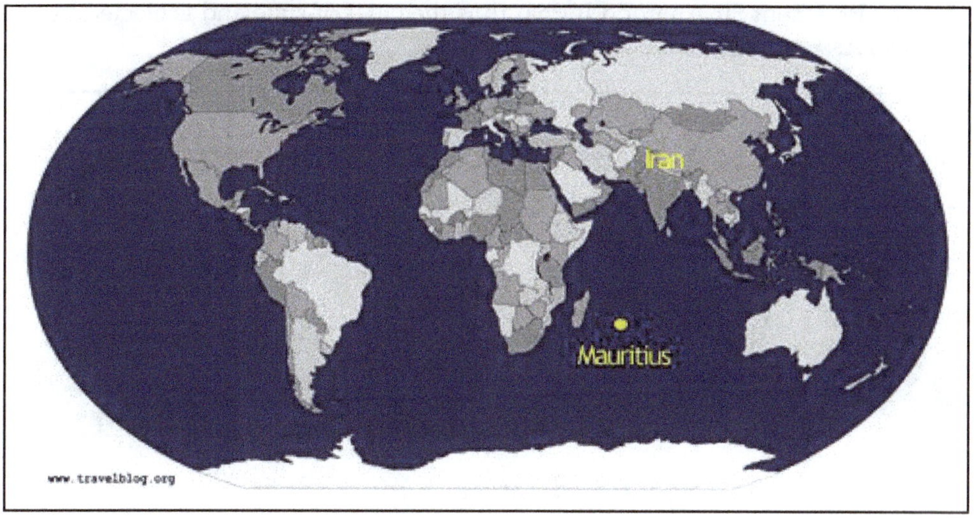

2.0 History

Arab sailors reportedly visited Mauritius during the Middle Ages, and on maps of about 1500, it is shown by an Arabic name 'Dina Arabi'. Although known to Arab and Malay sailors as early as the 10th century, Mauritius was first explored by the Portuguese in the 16th century and subsequently settled by the Dutch–who named it in honour of Prince Maurits van NASSAU–in the 17th century.

The French assumed control in 1715, developing the island into an important naval base overseeing Indian Ocean trade, and establishing a plantation economy of sugarcane. The British captured the island in 1810, during the Napoleonic Wars and gave it its independence in 1968. In 1992, Mauritius became a Republic with the President as Head of State but without constitutional powers.

3.0 Government

Mauritius is a parliamentary democracy with executive powers resting with the Prime Minister. National elections are held every five years and Dr Navinchandra Ramgoolam, our present Prime Minister was elected in office in 2005.

The President Sir Anerood Jugnauth is the chief of state and is elected by the National Assembly for a five year period.

4.0 Economic and Social Indicators

Main Indicators in 2008

Population	1,274,189
Growth rate of population	0.8 per cent
Labour force of which	548,900 353,600 Males and 195,300 Females
Unemployment rate	7.4 per cent
Inflation rate	8.8 per cent
Growth rate of economy	5.6 per cent
Life expectancy	74 yrs
Tourist arrivals (2007)	906,971
Literacy rate	84.4 per cent

Mauritius is a successful developing economy having recorded an average GDP growth of 5 per cent over the last twenty years. The economy is expected to grow at 5.6 per cent in 2008.

From a mono crop economy based on the cultivation of sugar cane, the island has metamorphosed into a robust and diversified economy, resting on agro industry; export oriented manufacturing, tourism, financial services and the ICT sector.

Our main exports are textile and clothing, sugar and cut flowers and we are a net food importer.Our main markets for export are Europe, U.S and Madagascar while our import markets are India, China, France and South Africa.

The manufacturing sector contributes about 20 per cent to the GDP. The sector has traditionally consisted of textiles mainly garment and apparel making which has dominated the economy for three decades. However, the erosion of trade preferences with the dismantling of the Multi Fibre Agreement coupled with the emergence of low cost producing countries has led to a redefinition of this pillar.

Mauritius has built its success on a free market economy, political and social stability, good governance and a modern infrastructure. The Government is committed to further expansion of the country's infrastructure across a range of sectors including water, electricity, ports and transportation.

In an effort to reduce cost of doing business in Mauritius and to promote the application of Information and Communication Technology (ICT), reforms to liberalise the industry have also been undertaken.

Mauritius is now shifting gears to move to a higher stage of economic development. New business opportunities in higher value added products and new emerging sectors have been identified and are now being promoted.

New strategies to attract investment in the industrial sector are now aligned with Government objective to move from trade preferences to global competitiveness with the aim to build an Attractive, Modern, Inclusive, Green, and Open Mauritius (AMIGO) as announced in the 2008-2009 budget.

Mauritius has an impressive track record on the International front where it is ranked 5th out of 181 countries for starting business in the World Bank Doing Business Survey 2009.

5.0 Investment

Foreign investment has always been welcomed in Mauritius.

In the 1970's, Mauritius started to diversify from a monocrop economy into an industrialized one. The Export Processing Zone concept was adopted whereby incentives were given to manufacturers who could import the raw materials duty free and export duty free to Europe amongst others. Investors flooded in from this moment especially from Hong Kong and set up mainly Textile and Clothing companies. Some of these investors also entered in joint ventures with the locals and this was the beginning of a new era for Mauritians especially women to join the labour force in droves. To date, the EPZ concept is no more as duties have come down considerably from a maximum of 80 per cent to 30 per cent and these are expected to be further reduced. All manufacturing enterprises whether exporting or producing for the local market can import their raw materials and production equipment duty free.

Since the complete suspension of exchange controls on balance of payments transactions in 1994, capital as well as profits can be repatriated freely, without prior approval from the Central Bank.

Government has developed several schemes aimed particularly at encouraging foreigners to invest in Mauritius; for instance, the Offshore sector, the Permanent Residence Scheme and the Regional Headquarters Scheme. In order to benefit from these schemes, investors must hold an investment certificate issued by the Board of Investment.

6.0 Live in Mauritius

Mauritius offers more than work and investment opportunities. Buying property in Mauritius has never been so easy. The integrated resorts scheme (IRS) or our small scale Real estate scheme allows you to own a world class villa.

Non citizens are also welcome to come and invest or work as professionals in our opened economy. The occupation permit allows an eligible non citizen to live and work in Mauritius for up to three years. An application can be made under any of the following categories:

- Investor. The business activity should generate an annual turnover exceeding $100,000

- Professional The basic salary of the professional should exceed $1,000 a month
- Self employed Income from the self employed business activity should exceed $20,000 annually

The retired non citizen can also enjoy a residence permit which allows him to reside in Mauritius for three years. All he has to do is to undertake to transfer his/her local bank account in Mauritius at least $ 40,000 US dollars annually or its equivalent in convertible foreign currency, for his/her living expenses in Mauritius.

Applicants under all categories may be eligible for a Permanent Residence Permit after three years. A Permanent Residence permit is valid for 10 years. To qualify, all the above criteria for investment must be satisfied for three years.

7.0 Tourism

Tourism, third pillar of the economy after the manufacturing sector and Agriculture, contributes significantly to economic growth and has been a key factor in the overall development of Mauritius.

Tourist arrivals have been expanding significantly, thus rising from 718,861 in 2004 to 906,971 in 2007, a 21 per cent increase within three years. Government has set a target of two million tourists by 2015.

The principal tourist markets were France (26.5 per cent), United Kingdom (11.8 per cent) and Reunion (10.6 per cent). Mauritius is predominantly a holiday destination for beach-resort tourists. It possesses a wide range of natural and man-made attractions, enjoys a sub-tropical climate with clear warm sea waters, attractive beaches, tropical fauna and flora complemented by a multi-ethnic and cultural population that is friendly and welcoming. These tourism assets are, its main strength,

especially since they are backed up by well-designed and run hotels, and reliable and operational services and infrastructures.

8.0 Women in the Labour Force

Mauritius has a labour force of some 548,900 persons out of which some 195,300 are women. Although the country has experienced an increase percentage of women in its labour force from 25 per cent in 1972 to 36.4 per cent in 2008, women continue to be concentrated in low skills and low paid jobs compared to their male counterpart.

Almost 50 per cent of working women are engaged in the 'Agriculture' and 'Manufacturing' sectors. Overall women are academically more qualified than men but still earn less in the labour market.

9.0 Empowerment of Women

As Mauritius is moving up on the ladder of industrialization, the shift from a labour intensive economy to a capital intensive one has affected women in the first place as they were the first ones who joined the labour force in large numbers when the export processing zone came into being.

Women in the EPZ lost their jobs in large numbers but the Government has designed strategies to assist these women so that they can be reintegrated in the labour market. The "empowerment program" was set up and women are able to take advantage of training in new areas where the scope for employment is better. Women are further encouraged to become self employed and acquire skills in new areas such hospitality, hairdressing, production of agro based products such as strawberries, pickles and spices and others.

10.0 Emerging Sectors

With the rise of globalization and the emergence of new economic players, Mauritius is compelled to restructure its economy to maintain its global competitiveness and relevance. New sectors have been identified and these evolve around the ocean, environment, renewable energy, water conservation and others.

The land based oceanic industry aims at exploiting commercially the position of Mauritius on the path of the deep sea current, through technologies and processes that are sustainable and environmentally friendly. The scientific properties of the water have been indisputably proven and confirmed that this water is pure, free from surface pollution and other contaminants, and has rich mineral and nutrient properties. Mauritius will be the first country to pump water from a depth of 1,000 metres and that too in the cleanest of oceans. Commercial applications are projected in the following areas: foods and beverages, health promotion and well being, medicinal and agrochemicals, renewable energy using the coldness properties of the water

11.0 Institutions to Promote Science and Research

There are two such institutions operating under the aegis of my Ministry *i.e.* The Ministry of Industry, Science and Research

11.1 Rajiv Gandhi Science Centre

It is a non-formal educational institution for the promotion of science and technology among students and members of the public. The centre is the first of its kind in our region. It has for objectives to:

- Supplement school education in a non-formal way by programming science demonstration lectures, science fairs, science seminars and other outreach programmes for children.
- Inculcate creativity among young people.
- Popularise science with an aim of enhancing the public understanding of science and technology.

11.2 The Mauritius Research Council (MRC)

It operates as an apex body with the objectives, among others, to foster, promote and co-ordinate research and development in all spheres of scientific, technological, social and economic activities and to advise the Government on all matters pertaining to scientific and technological policies. The activities of the MRC cover a broad spectrum of issues from identification of priority research fields to evaluation, dissemination and commercialisation of results of such research and applied research.

The MRC has already made significant contributions towards the setting up of the Land-Based Oceanic Industry. This new industry could generate more wealth from the ocean and enable the creation of a number of direct and indirect jobs. The products of the new oceanic industrial cluster are aimed mostly for export and also for the tourist industry and the local markets.

11.3 Regional Cooperation

Mauritius is committed to regional integration as the building block for global integration. As a small island developing state with a number of inherent constraints such as size, remoteness, exiguity of markets and prone to natural calamities, we are compelled to be actively present in all fora to draw attention to our concerns and to promote our interests. As such, Mauritius is an active member of several regional organisations such as the Southern African Development Community (SADC),the Common market for Southern and Eastern Africa (COMESA), the Indian Ocean Commission (IOC) and also the Indian Ocean Rim Association for Regional Cooperation (IOR-ARC). Iran is a member of the latter

The strategic location of Mauritius makes it an ideal gateway for business between Africa and the different continents-Asia, Europe Australia. Companies from these regions can manufacture or undertake value added activities in Mauritius and export duty fee to countries of COMESA and SADC.

12.0 Assistance

Following a cabinet reshuffle in September 2008, my Ministry has been assigned the portfolio of science and Research and is currently having wide consultation with different stakeholders in view of formulating a national policy framework for science and Research.

In this context my Ministry has written to the Executive Director of IOR-ARC who is based in Mauritius requesting an indication of the possible areas of cooperation that could be envisaged between IROST and our institutions.

We therefore look forward for further fruitful interaction in the near future.

Part II

Women Development: Technology as a Tool for Empowering Women

Chapter 12

ICT as a Tool for Empowering Women: A Case Study of Rural Rajasthan

Nirupama Prakash

Chief, Women Studies and Societal Development Unit,
Group Leader, Humanistic Studies,
Coordinator, Hospital and Health Systems Management
Birla Institute of Technology and Science, Pilani, Rajasthan, India
E-mail: rupa@bits-pilani.ac.in, rupa_prakash@hotmail.com

ABSTRACT

This paper highlights a community based project undertaken by Centre for Women Studies at Birla Institute of Technology and Science, Pilani, Rajasthan using ICT (Information, Communication Technologies) interventions to empower women and sensitize men towards gender based violence, female foeticide and health related Millennium Development Goals of making efforts to reduce infant mortality, maternal mortality and prevention from communicable and sexually-transmitted diseases. The main objective was empowering women and communities on health, social and development issues through mass media and promoting various ICT tools. Centre for Women Studies and One World South Asia (OWSA) used information and communication technologies as an empowering tool for rural development with prime focus on social equity. This included series of capacity building programmes for male and female village volunteers on specific issues such as gender violence, child marriage, skewed sex ratio and health related millennium development goals. This rural based project with ICT methodology which has been undertaken in the traditional state of Rajasthan has created a pool of gender sensitized community and empowered women who are now computer literate and techno savvy.

Keywords: *ICT, Project, Technology, Development, Gender.*

1.0 Gender and Society

The term gender is used to refer to behavioral differences between males and females that are culturally based and socially learned. Cultures differ widely in their norms and values regarding gender roles, the normative expectations concerning appropriately "masculine" or "feminine" behavior in a particular culture. The terms "male" and "female" are used for sex differences that are biological in origin. While the terms "masculine" and "feminine" refer to the corresponding culturally specific gender differences that are social in origin (Appelbaum and Chambliss, 1995).

Global comparisons show that, by and large, societies do not consistently define most tasks as either feminine or masculine. With industrialization the importance of muscle power declines, so people have even more options and gender differences are further reduced. Thus, gender is simply too variable across cultures to be considered a simple expression of biology. Instead, as with many other elements of culture, what it means to be female and male is mostly a creation of society.

Considerable research points to the fact that young children tend to form single–sex play groups.

Peer groups teach additional lessons about gender. Boys favour team sports with complex rules and clear objectives such as scoring a run or making a touchdown. Because such games nearly always have winners and losers, they reinforce masculine traits of aggression and control. Girls, too play team sports. But, girls also play hopscotch, jump rope, or simply talk, sing, or dance. These activities have few rules, and rarely is "victory" the ultimate goal. Instead of teaching girls to be competitive, female peer groups promote the interpersonal skills of communication and cooperation, presumably the basis for girls' future roles as wives and mothers. Boys, reason according to abstract principles. For them, "rightness" amounts to "playing by the rules." Girls, on the other hand, consider morality a matter of responsibility to others. Thus, the games we play offer important lessons for our later lives. (Macionis, J., 2006)

Sexuality may refer to Sexuality (biology)/sex or Sexuality (gender)/gender identity. Gender identity is a person's own sense of identification as male or female. The term is intended to distinguish this psychological association, from physiological and sociological aspects of gender.Culture is defined as way of life. Culture is divided into material culture and non-material culture. Material culture includes innovations, products, Tools / technologies, ICTs whereas non-material culture refers to beliefs, values, customs, language common to a group.

Norms are the agreed-upon expectations and rules by which a culture guides the behavior of its members in any given situation. Norms vary widely across cultural groups. Members of a culture must conform to its norms for the culture to exist and function. (http://www.cliffsnotes.com/WileyCDA/CliffsReviewTopic/topicArticleId-26957, article Id-26853.html). Cultural norms are behavior patterns that are typical of specific groups. Such behaviors are learned from parents, teachers, peers, and many others whose values, attitudes, beliefs, and behaviors take place in the context of their own organizational culture. (http://www.ncrel.org/sdrs/areas/issues/envrnmnt/drugfree/sa1lk2.htm)

2.0 Socio-cultural Dimensions of Gender-Indian Context

Indian culture is one of the most vibrant cultures in the world and rural culture is the backbone of Indian culture. Technology is rapidly changing the life styles and recent past has witnessed technological impact on rural India. However, urban-rural divide, rural poverty, education, gender disparity, maternal health and various cultural problems in rural India continue to be issues of concern. (Mondal, 2007)

In a recent study on "Maternal Health Situation in India-A Case Study", it has been pointed out that India has progressed rapidly on the socio-economic front but progress in improvement of maternal health has been slow. The review of safe motherhood efforts in India shows that in spite of major initiatives taken by the government in last 10 years, even today nearly half of all deliveries take place at home and antenatal care services coverage is poor and maternal mortality ratio remains around 300. The challenge is how to make safe motherhood strategies in the future more successful.

High priority needs to be given to safe motherhood and have policy and programmes designed to implement evidence-based strategies and detailed micro-level programme planning. The study highlighted that the main strategy for maternal mortality reduction of the government focuses on safe/institutional deliveries in the governmental and non-governmental sectors under National Rural Health Mission. Efforts to develop competencies needed for Skilled Birth Attendants (SBAs) in the entire cadre of staff nurses and ANMs (Auxillary Nurse Midwives) are being undertaken. Regular training of select medical officers to administer anesthesia has been taken up. Also multi skill training of medical officers, ANMs and Para-medics is being initiated to close specialist skill gaps. Intensified IEC is to be pursued to ensure behavioral changes that relate to better maternal survival and women's health *i.e.* spacing, age at marriage, education of the girl child (Vora *et al.*, 2008).

Many factors coalesce dramatically to make the position of women in Rajasthan a desperate one. Ancient traditions, religion, caste-based patriarchy, the economic effects of global capitalism combine with environmental stress factors make the way forward extremely complex. Patriarchy, discriminatory customs and values, caste based discrimination, high illiteracy and high rates of poverty seem pervasive. Despite all efforts towards social justice, women continue to be perceived as burdens.

Some of the factors contributing to this are: Sex ratio: there are 910 women for every 1000 males in the population of Rajasthan. (Regional differences show Dholpur and Jaisalmer districts to be the worst at 795/810 respectively). Female infanticide, anemia, poor nutrition, maternal mortality, (558 per 100,000), child deaths from poor nutrition and water-related diseases are factors. (80 per cent of all women of child-bearing age suffer from anemia; infant deaths are 79/1000; the maternal mortality rates are among the highest in the world. Women and girl-children are caught in a cycle of malnutrition. Sati and child marriage are age-old customs in Rajasthan. Incidences of sexual abuse and domestic violence are high. The state has one of the lowest rates for female literacy in India. Recognizing all this, the state has initiated various schemes for empowerment of women. Women in rural Rajasthan are now voicing their views to live in dignity which means meeting the basic needs; freedom from violence and access to justice. So the whole life-cycle of discrimination against rural women begins to be transformed (Grey, 2007).

3.0 Technology and Women Empowerment

Women empowerment is the key objective of many discourses on empowerment. It has become a contested concept. Strategies for empowerment, inclusion and 'capacity building' are assuming growing importance as community members, governments and specialists seek new community based solutions to ensure the long-term sustainability of rural communities.

The World Bank has identified empowerment as one of the key constituent elements of poverty reduction, and as a primary development assistance goal. Gender mainstreaming has been made a priority in development assistance. The promotion of women's empowerment as a development goal is based on a dual argument: that social justice is an important aspect of human welfare and is intrinsically worth pursuing; and that women's empowerment is a means to other ends. A recent policy research report by the World Bank,2001 for example, identifies gender equality both as a development objective in itself, and as a means to promote growth, reduce poverty and promote better governance. A similar dual rationale for supporting women's empowerment has been articulated in the policy statements put forth at several high level international conferences in the past decade (*e.g.* the Beijing Platform for Action, the Beijing+5 declaration and resolution, the Cairo Programme of Action, the Millennium Declaration, and the Convention on the Elimination of All Forms of Discrimination against Women [CEDAW]) (Malhotra *et al.*, 2002).

With its strong emphasis on social justice, recent literature on advancement of women, community participation, and community development advocates processes for participation that are more inclusive and empowering than have traditionally occurred.

Science and technology are decisively important tools for the advancement of women. Women's involvement in the applications of science and technology will lead to beneficial and productive uses. Science and technology are particularly useful in four critical areas: education and training, economy (also linked with poverty), health, and communications/media.

Women's ICT access in the developing world is generally confined to small elite of high income, urban-based citizens. Even in the industrialized countries, where overall connectivity is high, gender-specific constraints shape and define ICT use. Information and communication technologies could give a major boost to the economic, political and social empowerment of rural women, and the promotion of gender equality. However, this potential will only be realised if the gender dimensions of the Information Society in terms of users' needs, conditions of access, policies, applications and regulatory frameworks are properly understood and adequately addressed by all stakeholders. Poverty, illiteracy, lack of computer literacy and language barriers are among the factors impeding access to the ICT infrastructure, especially in developing countries, and these problems are particularly acute for rural women. But women's access to ICTs is constrained by factors that go beyond issues of technological infrastructure and socio-economic environment.

Low levels of education and illiteracy, reinforced by poverty, account in large measure for the problems women face in accessing and using ICTs. The low ratios of girls in science and technology courses also reinforce the negative dynamics that limit women's access to decision-making positions in the fields of science and technology.

The UN Fourth World Conference on Women (Beijing, 1995) was a watershed in realizing the power of information technology as a tool for women's mobilization, information exchange, and empowerment. Since 1995, the 'gender and ICT' agenda has steadily gained legitimacy as a serious area of concern, mainly through careful work by women's organizations and a few international agencies and donors.

Knowledge has become a key component in development discourses that traditionally focussed on capital formation and economic growth. Access to knowledge and information is instrumental in holistic development of women and marginalized communities. Women, especially in rural areas of developing countries remain underserved in maternity services such as institutional registration, prenatal care, routine check-ups, and preparations for safe delivery, post natal care and management of pregnancy related complications.

It has been pointed out that Information, Communication Technologies (ICT) have a potential role in spreading basic health and hygiene awareness, information and monitoring services on nutritional needs and maternal health, health surveillance and in creating a nationwide management and information service for delivery of health services (Malhotra, 2008).

4.0 Obstacles to Women's Access

For those women with the resources to access and use the new information and communication technologies, there are real benefits. For society as a whole, ICTs offer immense possibilities for reducing poverty, overcoming women's isolation, giving women a voice, improving governance and advancing gender equality. But this potential will only be realised if the factors that contribute to the current gender digital divide are recognised and addressed.

4.1 Literacy and Education

Illiteracy is a serious problem. ICTs require various kinds of literacy and the inability to read and write is a major barrier to women's access to ICTs. Technological and scientific literacy are also a pre-requisite to using new technologies successfully. Women face challenges in pursuing education at all ages due to lack of time to attend classes, family and domestic responsibilities, and socio-cultural practices that rate girls' education lower than boys.

4.2 Language and Content

Language excludes many from the Internet. Often women are not interested in using ICTs because the language and content does not speak to them, does not reflect their identities and does not teach, affirm or reinforce them. Women's viewpoints, knowledge, experiences and concerns are inadequately reflected on the internet, particularly perspectives from women in developing nations.

Gender stereotypes predominate and perpetuate those reflected in the print media. There is a need for women to develop, promote and publish their own perspectives and knowledge to ensure that they are represented on the Internet using their own voices.

4.3 Socio-cultural and Institutional Barriers

Generally, women have less access than men to ICT facilities where they exist. Numerous invisible barriers limit women's participation in the Information Society. One of the more pervasive but intractable problems is "techno phobia," or fear of technology. Women often have complex relationships with technology and machines as a result of being socialised over time to believe that machines and technology are a man's domain and not for women and girls, thus generating a gender bias in attitudes towards studying or using information technology. The social factors that produce these gender differences operate in both institutional and informal settings. An awareness of these socio-cultural and institutional barriers is quite essential.

4.4 Access, Control and Effective Use

Women's access to and control over ICTs is not equal to that of men. Infrastructure is largely concentrated in the urban areas, while the majority of women in the developing world are located in remote and rural areas. While ICTs can deliver potentially useful information, like market prices for women in small and micro-enterprises, for example, it is only one aspect of a longer chain of resources necessary to effect sustainable development. Where there is no access to roads, transport or credit, and other development inputs cannot be obtained, the access to and use of ICTs will be limited in its impact.

It is therefore equally important to support the provision of ICT facilities in these areas. (Primo, 2003)

4.5 Content Creation

To encourage women to use ICTs, the content available through ICTs should be relevant to their lives. Women's viewpoints, knowledge, experiences and concerns are inadequately reflected on the Internet, while gender stereotypes predominate. If women are to make use of the Internet for education, advocacy or income-generation, more relevant content, pertaining to both substance and language, must be made available through both new and traditional technologies. New technologies such as computers and the Internet should not deflect attention and resources from technologies that have been around for longer such as radio, television, video and print.

The relevance of ICT initiatives also falls short where they focus on "plugging in" women and other marginalized groups into existing global information flow, without any attention to local knowledge systems and content. There should therefore be greater attention paid to recognizing women and the poor as information producers, and providing relevant training in collecting, packaging and disseminating local knowledge, based on an understanding of local information needs, and that of women specifically. Such information may well be more useful for local communities in meeting their everyday challenges than "foreign" information available on the Internet.

4.6 Combining Access with Social Empowerment for Women

There are an increasing number of projects that specifically target women, and that combine access to information and training with efforts to empower marginalized women, often through the simple act of giving women a platform for voicing their concerns. There is a need to encourage information sharing about best practices that can be replicated, while attending to local issues and social dynamics in different places.

5.0 Case Study

The Centre for Women Studies at BITS, Pilani, Rajasthan forged collaboration with One World South Asia, New Delhi regarding an ICT enabled project on creation of Village Knowledge Centres with active participation of women in rural Rajasthan during February 2006 to November 2007.

The goals of the Project entitled "Socio-Economic Upliftment of Rural Women Through Technology Intervention: Realizing Millennium Development Goals Through ICTs" were empowering women and communities on health, social and development issues through mass media and promoting various ICT tools.

Two village knowledge centres (called Advocacy Centres) which empower rural women and the village communities with knowledge on Science and Technology, Health and Environment relevant to the local needs were opened near Pilani in Rajasthan – one at Jherli Village near Pilani and another at Dhandhar Village Panchayat on February 13th and 14th, 2006 respectively.

The Advocacy Centres provide knowledge relevant to the women and communities using computers, software and CDs; train the village communities in gathering and building the village knowledge bank. The Advocacy Centres were run by the Centre for Women Studies of Birla Institute of Technology and Science with support from One World South Asia, Delhi. The Advocacy Centres were managed by female and male village volunteers, who were trained by experts from One World South Asia (OWSA) New Delhi.

The project involved use of information and communication technologies as an empowering tool for rural development. Volunteers were trained in building the knowledge databases and also in knowledge sharing using software. It also builds up the use of community radio for disseminating information in the local language.

The main objectives of this ICT project were to address Millennium Development Goals directly related to health as given below:

- Efforts to reduce Infant Mortality.
- Efforts to reduce Child Mortality.
- Efforts to improve maternal mortality.
- Efforts towards prevention from communicable and sexually-transmitted diseases.

The project worked on the following methodology:

1. Introduction of gender issues.
2. Establishment of Knowledge (Advocacy) Centres in two Gram Panchayats namely Jherli and Dhandhar.
3. Formation of community groups for regular meetings in the community.
4. Promoting radio as a tool for narrowcasting in the community/Listener's Club.

6.0 Activities

1. Baseline survey
2. Capacity building
   - Digitising the family survey
   - Content collection
   - Content digitization
   - Content updation
   - Smooth running of knowledge centres through hub and spokes model
   - Conducting Focused Group Discussions on MDGs
   - Discussion, sensitization on communicable diseases and their prevention
3. Dissemination of information
4. Peoples' advocacy programmes from time to time
5. Local newsmagazine.

Centre for Women Studies at BITS, Pilani in collaboration with One World South Asia, New Delhi organized two days training programme during April, 2006 at BITS, Pilani for 26 village volunteers who had been selected to work as knowledge workers in the villages under the aegis of the project.

The objective of the training was to expose the volunteers to use audio and web-based tools for connecting communities and empowering people and to reaching and voicing the voiceless. The concept of life-cycle approach with focus on four MDGs (Millennium Development Goals) related to health was discussed as to how these four MDGs could be addressed and integrated with issues of education, poverty reduction and conducive environment. The training programme also focused on the following ICT tools:

- Audio as a powerful tool.
- Pre-recorded audio programmes addressing the issues in the communities especially women issues through listeners' clubs.

Five CD players and five sets of CDs with 29 pre-recorded programmes on women issues were given by the OWSA team to the village volunteers through the Sarpanch of Jherli and Dhandhar panchayats. The Village Volunteers were asked to listen to the CDs based on above mentioned issues; write the messages in the register provided to each of them, get them fully acquainted with the knowledge; play the programme on the CD player in the community and address the questions raised by the audience.

The volunteers were trained as to how the listeners' clubs could be formed in the villages and how regular scheduled meetings could be organized with men and women groups at the advocacy centres. Village volunteers were given training on how to collect local knowledge, add global knowledge provided by Centre for Women's Studies and One World South Asia, address the issues related to MDGs and bring out a local news magazine as a tool for dissemination of knowledge with the help of CWS (Centre for Women's Studies) and OWSA.

Capacity building on Hindi typing was given to the volunteers for sharing of knowledge-based content. The training Workshop also focused on capacity building on web-based software in connecting communities namely, open eNRICH, which supports local language *i.e.* Hindi for sharing of knowledge within two panchayats and also with the Hub at CWS and with OWSA.

Open-eNRICH is a web-based solution for community information, communications, and knowledge management which is meant to empower communities through enrichment with local content and connection to information and services that are tailored to their own needs. It acts as a platform for voicing the thoughts and feelings of the community.

The content generated through Open-eNRICH was disseminated through audio programmes. The main objective of the project was to connect communities and empower women and their families through ICT tools.

As a part of the methodology, baseline survey was conducted amongst 92 households covering 6 villages, three each from both the village panchayats. Dhandhar, Ghumansar, Kulharia ki Dhani in Dhandhar Village Panchayat and Jherli, Garinda, and Raila in Jherli Village Panchayat. These households were chosen on basis of simple random sampling method. Dhandhar covered 14.1 per cent, Garinda 5.4 per cent, Ghumansar 2.2 per cent, Kulharia ki Dhani 4.3 per cent, Jherli 43.5 per cent and Raila covered 30.4 per cent of the total households surveyed.

Regarding the source of drinking water, survey revealed that 13 per cent of the households have house taps, 70.7 per cent use public taps, 15.2 per cent own a well and 1.1 per cent use hand pump.

Regarding prevalence of diseases, the survey revealed that members of 35.9 per cent households had suffered from Malaria. 81.8 per cent of them were given anti malarial medicines. 64.1 per cent are aware how malaria spreads and 43.5 per cent of the households are aware that malaria can relapse if not treated properly.

Members of 5.4 per cent households suffered from Tuberculosis during past one year from the commencement of the study. 80 per cent respondents said that DOTS treatments were available to them and 18.7 per cent were aware about how long DOTS medication should be done.

1.1 per cent of the households had members with diagnosed HIV infection/ suspected case and 98.9 per cent of the respondents had no HIV infection case. 85.7

per cent of the households said that no counseling was provided regarding HIV/ AIDS. 58.7 per cent were aware and 41.3 per cent did not know about the modality of spreading of HIV/AIDS. 92.4 per cent said that the hospital does not provide medicine for ART (Anti Retro Viral Treatment). 68.5 per cent of the households did not know how to take care of HIV+ive person at home.

Regarding infant and mortality deaths it was found that amongst 85.9 per cent of the households there were no infant deaths during delivery or before 5 years of age.14.1 per cent reported about infant deaths. Of these 66.7 per cent died within few days, 16.7 per cent died within one year and 16.7 per cent deaths occurred during 1 to 5 years. In 20 per cent cases infant died because of complication at birth, in 13.3 per cent cases malnutrition was the cause of death and 66.7 per cent died of diseases. As for the health history of the children 48.9 per cent of the mothers did not have immunization cards for their children. 57.1 per cent had taken BCG, 66.7 per cent had DPT 1, DPT 2 and DPT 3 for their children. 69.1 per cent had taken the Polio drops. 62 per cent had taken the hepatitis B1, B2, B3 injection. 71.4 per cent had not been vaccinated against measles and 56.9 per cent had been vaccinated against Typhoid.

69.9 per cent of the children had not suffered from Diarrhea in past two months. 67.6 per cent were aware of the oral re-hydration solution. 50 per cent use ORS for children having diarrhea. 22 per cent of children suffered from Acute Respiratory Infection (fever, cough, rapid breathing) during two months prior to the baseline survey and 21.1 per cent were taken to the health facility. 10 per cent of children were reported to be suffering from malnutrition. The study revealed that 26 per cent of the respondent use contraceptive for family planning whereas 74 per cent do not use. 43 per cent of the wives had undergone sterilization surgery.

7 per cent of the husbands had undergone sterilization surgery. 31 per cent of the women consumed Folic Acid tablets during pregnancy. 34 per cent of the women received ante natal check up and rest 66 per cent did not receive any ante natal check. 39 per cent of women had taken tetanus injection during pregnancy.

62 per cent of the women have 1 to 3 children, 33 per cent of the women have 3 to 6 children, 4 per cent of the women do not have child and 1 per cent of the women have 6 to 9 children. 88 per cent of the women delivered at home, 10 per cent of the women delivered in hospital and 2 per cent of the women delivered in PHC. 55 per cent of the delivery was attended by untrained dais, 35 per cent of the delivery was attended by trained Dai and only 10 per cent of the deliveries were attended by trained doctors. 38 per cent of mothers started breastfeeding the infant in one to two hours, 33 per cent of the mothers started breastfeeding after a long time, 22 per cent of the mothers started breastfeeding in less than one hour. 7 per cent of the mothers started breastfeeding next day.72 per cent of the mothers gave prelatic to their children. 78 per cent of mothers started giving solid/mushy food to the infant in addition to breast milk between 6-9 months.13 per cent of them gave beyond this period. 8 per cent of the mothers gave between 3-6 months. 99 per cent of the mothers had breastfed their children for longer duration and 1 per cent of the mothers had breastfed her child for two months.

Village volunteers comprising of women and men were trained to digitize and analyze the data of the baseline survey.

Two days training programme was again organized at Jherli Advocacy Centre near Pilani on 15-16 November, 2007 for village volunteers working as knowledge workers. Using ICT technologies such as computers, audio tools and web based software; volunteers were trained to enhance their awareness level on health related millennium development goals focusing on maternal health, importance of girl child and health related issues. Listener's Club helped in dissemination of information in the village community to bring about positive changes regarding non-material culture addressing issues on gender, health, education and superstitious beliefs.

In addition, village volunteers were trained in gathering and building the village knowledge bank and to publish a local magazine. The name of the monthly local magazine was decided as *Gyan Deep Patrika.* They were trained in building the knowledge database and also in knowledge sharing using software. It also focused on the use of community radio for disseminating information in local language.

The training programme focused on capacity building on how to collect local knowledge, how to generate content for their local magazine keeping in view the diversified communities and literacy level of the village communities. Emphasis was also made on improving the communication skills and using simple language while developing contents for the local news magazine and involving the community people in this process. The Knowledge Workers were trained on designing and formatting the local magazine.

Editorial team of *Gyan Deep Patrika* was also formed. The idea was to address issues related to MDGs and disseminate the knowledge on socio-economic, gender, health and developmental issues with the involvement of community people at local level. The positive outcome of the project was evident in increased awareness on gender sensitization, making the rural community techno-savvy and computer literate. Some of the volunteers have been employed at BPO, Pilani.

Thus ICT intervention has been instrumental in reducing biases in local cultural norms and has helped in creating a pool of gender sensitized community with empowered women.

7.0 Conclusions

ICTs offer immense possibilities for reducing poverty; improving governance and advancing gender equality provided they are made more accessible and consciously applied towards the achievement of these objectives.

In order to support, broaden and include more women in these electronic webs, more needs to be done. We need to educate and train more women in order to challenge the often deeply entrenched techno phobia nurtured from an early age. ICT policy intervention is critical in order to include a gender perspective in all discussions related to ICT policy.

Despite the imbalance in access and provision for women, they are creating local content, creating space for their own voices, promoting their own knowledge and beginning to redirect the information flow. Participatory methodologies and women training programmes have created safe and nurturing environments wherein women can engage with the technologies rather than the prejudices and exclusionary environments often created in mixed learning spaces.

Acknowledgements

Due acknowledgement is made of University Grants Commission, New Delhi and OWSA, New Delhi for supporting the Case Study reported in this paper. Acknowledgement is also due to BITS, Pilani for providing necessary infrastructure for the study.

References

Appelbaum, R.P. and Chambliss, W.J., 1995.*Sociology*, Harper Collins College Publishers, NY, USA, pp. 271.

Grey, M, 2007. Women's Struggle in Rural Rajasthan: Seeking a Life of Dignity http://www.wellsforindia.org/pdf/empowerment.pdf accessed on November 11, 2008.

Macionis, J, 2006. Sociology, Pearson Education, Delhi, pp. 331.

Mondal, L, 2007. Rural India: Culture and Development, ABD Publications, Jaipur.

Malhotra, A; Schuler, S.R and Boender, C., 2002. Measuring Women's Empowerment as a Variable in International Development, World Bank Workshop on Poverty and Gender: New Perspectives.

Malhotra, G., 2008.The Role of Information and Communication Technologies in Addressing Safe Motherhood in South Asia in *Marriage and Family Review*, Haworth Press, Taylor and Francis Group, Philadelphia, U.S.A, Vol. 44 2-3, pp.357-363.

Primo, Natasha, 2003. Gender Issues in the Information Society, UNESCO Report on World Summit on the Information Society, Paris.

Vora, K. S, Mavalankar, D.V, Ramani, K.V, Upadhyaya, M and Sharma, B., 2008. Maternal Health Situation in India: A Case Study, Working Paper Series, Research and Publications, Indian Institute of Ahmedabad.

http://www.iimahd.ernet.in/publications/data/2008-03-02Mavalankar.pdf Accessed on November 25, 2008.

Chapter 13

Reaching the Un-reached: Technology Model for Women's Empowerment and Sustainable Development

Vinita Sharma

Science and Society Division, Department of Science and Technology, Ministry of Science and Technology, Delhi – 110 016, India

E-mail: vinita.sharma@gmail.com

ABSTRACT

It is well-known that technology can contribute to a higher standard of living, with enhanced opportunities for health care, education and employment for women. Recognizing this Government of India initiated schemes for empowering women through S&T as early as in 1984. The focus of these interventions was to demonstrate the application of science and technology to provide newer opportunities for income generation, reduce drudgery, improve health and environment, genomic conservation of natural resources and scientific validation of traditional knowledge thereby leading to a better quality of life for women. Partners in this exercise have been select voluntary organizations having an S&T base and key R&D laboratories in the country. Initiatives have been taken both in the farm and non-farm sector. The role of the Ministry of Science and Technology in these initiatives has been catalytic where technology development and demonstration aspects have been focused; however the experience gained in technology management at the grassroots level through success stories, setbacks and issues related to gender have been rather vast. The emphasis in these initiatives has been to improve upon existing skills, provide managerial capabilities and to understand the science behind the processes/products. This has made them more open to improved/emerging technologies for improving production efficiency and reducing drudgery in their day-to-day work. Since rural women have special understanding of natural resource management they also play a crucial role in re-nurturing and re-greening rural India. Need is to use S&T to ease women's work-load inside and outside the house involving them as equal partners, recognizing their knowledge, experience and the significant role they can play in sustainable development. This paper describes the "systems approach" adopted with proper networking to involve rural women as Para technologists in processing and preservation of horticulture produce for value addition and income generation at the village level.

Keywords: *Technology model, Women empowerment, Science and technology, Sustainable development, Skill development, Rural women, Food processing, Horticulture produce, Employment.*

1.0 Introduction

In India, rural women are extensively involved in arduous farm operations and agricultural activities from planting to harvesting and post-harvest operations. Their knowledge as seed selectors, seed storage and seedling production, livestock management, post harvest processing, processing of minor forest produce (MFP) has contributed a lot to the viability of agro-diversity and production in the highly diversified Indian sub continent. In terms of grassroots realities, traditionally women have been using and managing natural resources, collecting food, fodder and fuel, MFP products, medicinal plants etc. for their livelihood. They possess vast knowledge of eco-system and properties and potential uses of forest products. But, women's control over such resources and right to land is negligible, being controlled by men. In the case of dairy production, wherein women account for 93 per cent of the total employment, the majority of the dairy co-operative membership is assumed by men, leaving 14 per cent to women. The same is true to the fishery sector, which supports 5 million fishing communities in coastal areas, where women perform all processing, drying/curing and other related activities.

In the village, overexploitation of the environment contributes to degradation of productive resources: the land, the soil, water and forests. Ecological destruction through deforestation, population pressure, urban consumption pattern and rapid industrialization, wrong land use practices and degradation of village commons have displaced rural women from productive activities–diminishing their means of livelihood. As a result women spent more time to cover long distances in search of food, fuel and water ranging from 45 minutes to five hours in addition to work in the field and at the household level. A study by Development Alternatives, a voluntary organization working in the Bundelkhand region of central India reveals that an average woman in that region spends up-to four hours a day fetching water. Another study, in Orissa, Eastern India, shows that women walk an average of 7 kilometers a day, to gather various forest products, up from 1.7 kilometers 20 years back. A recent study conducted by the Centre of Science and Environment, Delhi in a Himalayan village in Chamoli, Uttrakhand unfolds that women spend 59 per cent, children 36 per cent and men 15 per cent work hours every year on backbreaking tasks. Their work is also compounded by the migration of male population from rural to urban areas in search of employment. Extensive studies also show that women producers have poorer access than men to all resources from land to credit and technology. In such conditions and prevailing social constraints, rural women work in a labour-intensive and non-monetized economy leading to feminization of poverty, thus, in real sense excluding them from full participation in developmental efforts to eliminate poverty.

In view of above and ground realities (Box 1), there is a growing need in addressing development and environmental concerns through input of S&T keeping in view gender perspective. Therefore, the need of the hour is to innovate and deliver sustainable rural technologies for sustainable livelihood by which women do not get marginalized in their communities.

Since rural women have special understanding of natural resource management and more sensitive to environment, rooted deep in their roles and work tasks, they can play a crucial role in re-nurturing and re-greening rural India. Need is to use S&T to ease women's work-load inside and outside the house involving them as equal partners, recognizing their knowledge, experience and the significant role they can play in sustainable development. In this process, emphasis should be given to reduce the number of hours of work and adding value to each hour of work. Women will have to be diverted from full time domestic work to economic activity with gender sensitive approach at various levels. With organized orientation and training, there is a great potential for empowering women with appropriate and sustainable technologies in areas like agriculture, biotechnology, information technology and service sectors in the rural areas. Technology based entrepreneurship development in such areas for self employment will enable them to improve their income generation capability; provide their children good nutrition and health and access to education; enhance their quality of life and contribute meaningfully to national development.

Research findings have shown that with the aid of appropriate developmental programmes, women as 'mother' can become effective agents of change in eradicating poverty in the current generation and to preventing the transmission to the next generation.

In India the Planning Commission has been was charged with the responsibility of making assessment of all resources of the country, augmenting deficient resources, formulating plans for the most effective and balanced utilization of resources and determining priorities. Jawaharlal Nehru was the first Chairman of the Planning Commission. The Prime Minister Dr Manmohan Singh as Chairman of the Planning Commission has spearheaded the formulation of the 11th Five year plan (2007-2012). The plan document provides clear goals for gender empowerment requiring S&T interventions:

- Reducing the incidence of anaemia and malnutrition among adolescent girls to break the cycle of ill-health and maternal and infant mortality.
- Access to basic services such as clean drinking water and sanitation facilities without.
- Address the issue of declining child sex ratio.
- Initiate a second green revolution needed to raise the growth rate of agricultural GDP to around 4 per cent. Recognizing the demographic realities — particularly the increasing role of women.

- Take recourse to the complementarities between environmental sustainability and woman's wellbeing.
- Measures to ensure that society recognizes women's economic and social worth, and accounts for the worth of women's unpaid work.

Box 13.1: Women and Rural India: Key Facts and Figures

- Indian population comprise of 48.1 per cent women and 51.9 per cent men.
- Rural population: 71 per cent of India's population. Farm production: 55-66 per cent of total labour.
- Male-based sex ratio: Rural areas:9.4 per cent ; Urban areas: 8.9 per cent work 1064h, a man 1212h and a women
- 70 per cent of the population earn their livelihood from agriculture
- Work force participation rate of women: 22.7 (male 51.6 per cent) Rural areas:27.2 per cent, Urban areas: 9.7 per cent
- Agriculture and allied sectors employ 89.5 per cent of total female in rural India
- Female literacy: 38 per cent as compared to male literacy rate of 66 per cent Rural areas: 30.4 per cent ; Urban areas: 63.9 per cent
- Rural poverty: 39 per cent;
  Urban poverty: 30 per cent.
- Proportion of pregnant women with anemia:88 per cent
- Human Development Index rank and gender-related Index rank: same 105 out of 146 countries–indicating low life expectancy at birth, low educational attainment, low income and gender disparities
- Women have heavy work load with dual responsibility for farm and household production
- Women's average contribution in overall farm production: 55-66 per cent of total labour.
- In the Indian Himalayas a pair of bullocks work 1064h, a man 1212h and a woman 3485h in a year on a one-ha farm.
- Taking the economy as a whole, women perform two-thirds of the work, but, earn only one-tenth of the income.
- Inter-state and intra-state variations in gender roles in agriculture, environment and rural production
- Unequal access to land and property rights.
- 90 per cent women in forest and tribal belts earn their livelihood by carrying heavy loads on their heads.
- In dairying 75 million women are engaged as against 15 million men, while in animal husbandry the ration is 25 million to 1.5 million.
- Country is loosing 1.5mha of forest cover annually, whereas fuel-wood contributes to 84 per cent of the total household energy consumption in rural India – contributes to hardship for women
- A chulha-user inhales 40 mg of carbon monoxide(CO) per kg of woods as against 17 mg/kg and 43 mg/kg of CO inhaled by active and passive smokers respectively: injurious to health.

Source: Census of India, 1991; ESCAP, 1997; UNDP Human Development Report, 2002.

1.1 S&T Based Empowerment of Women

Gender Perspective

It has been widely experienced that though technology is gender neutral transfer of technology to women need careful planning for it to be successful. Technology based intervention for both income generation, drudgery reduction or health cannot be successful unless women can benefit from it within their own social situation. The technology needs to be based on local resources, local markets and conducive with the social ethos. Only gender sensitive genuine people-cantered approaches to development can succeed on any appreciable and sustainable scale. The lesson learnt over last fifty years is that women, and particularly the most oppressed and marginalized, must own and manage programmes meant for their benefit. Women have, therefore, to play a key role in changing and altering existing power structures as equal partners and managers.

Empowerment and Health

The health of a nation depends upon the health and well being of its women. The 11th five year plan (2007-2012) of the country recognizes the pivotal importance of women's holistic health and recognizes that patterns of female mortality in India present compelling evidence that women enjoy less access than men to adequate health care and nutrition. The most critical periods for women in India are between the ages of one month to four years and the span between 15-34 (Peak child bearing age) and then again between 50-60 (postmenopausal). The greater morbidity among women is associated with lower consumption of food within the household and consequently lowers nutritional status.

It is these factors that have prompted a conscious effort by scientific departments in the country to evolve strategies to identify specific health problems of women including occupational health and to develop an intervention strategy.

The Department of Science and Technology has taken an initiative to address a felt need at the grass root and develop a hygienic good quality sanitary napkin, adhering to all norms being followed for high value products of multinationals.(Box 3). To address the health issues in women a multicentric programme is being implemented in the country to develop a model for intervention for non-communicable diseases in post reproductive age women. This pioneering programme focuses on nutrition related interventions based on location specific needs. This programme has attracted a lot of attention country wide, as issues of non communicable diseases in older women have not yet received adequate attention. Similarly a nutrition based intervention programme for prevention of anaemia in both women and adolescent girls has been initiated. It is expected that the nutrition based approach is likely to yield long term benefits as compared to conventional supplements with one or two micronutrients. Also along with women the entire family would be benefited if consumption of green leafy vegetables is increased.

Case Study 1: Community Based Approach for Prevention of Anaemia among Young Rural Women in India

According to the National Family Health Survey III (2005-06), important reasons for widespread anaemia in the country are inadequate intake and absorption of iron from cereal based diet and inadequate consumption of green leafy vegetables and citrus fruits. Apart from these factors, frequent pregnancies with shorter intervals, high prevalence of infections and infestations, faulty feeding practices and lack of dietary diversification are also some of the major reasons. To develop a strategy for management of anaemia a sample survey was conducted in rural areas of Maharastra which showed that young mothers in rural India have low dietary intakes (both macro and micronutrients), poor consumption of green leafy vegetables (GLV) and fruits, poor nutritional status (anthropometry) and a very low priority for self health care. The need for improving nutritional knowledge is beyond doubt as prevalence of anaemia in 16 districts of India shows that over 80 per cent in pregnant women and about 90 per cent in adolescent girls suffer from anaemia Though the National Anaemia Control Programme (NACP) which relied on a drug based approach has been in existence for over three decades anaemia still continues to be one of the main issues resulting in wide spread morbidity in women. It was evident that supplementing the drug approaches an integrated approach with elements of nutritional knowledge, socio-culturally acceptable and involving a participatory approach was the need of the hour. A pilot programme was initiated in rural Maharastra in partnership with a scientific institution, voluntary organizations, home science colleges and schools. A series of Informal meetings led to the developmental strategy to launch the programme including design of questionnaires and formats to collect data scientifically. The objectives of the programme was to ensure improvement in intake of micronutrient rich green leafy vegetables (GLVs), ensure all year round availability of GLV's, Kitchen garden activities, live demonstrations of GLV recipes, simple methods of preservation of GLVs, consecutive rounds of Hb estimation and preparation of booklet of recipes from GLVs. The results of the completed study showed a very good response for nutrition and health education through participatory research. The prevalence of anaemia was reduced significantly (almost by 50 per cent) in the first year and remarkable the reduction in prevalence was sustained in second year too. Primary data of anaemia prevalence in adolescents, non pregnant and pregnant young mothers

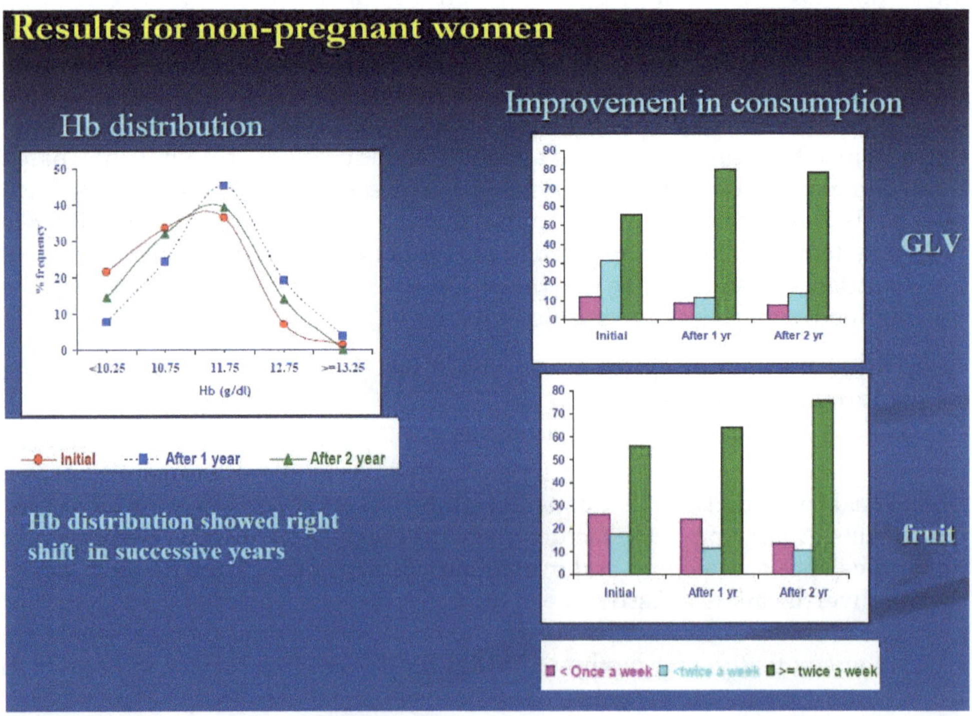

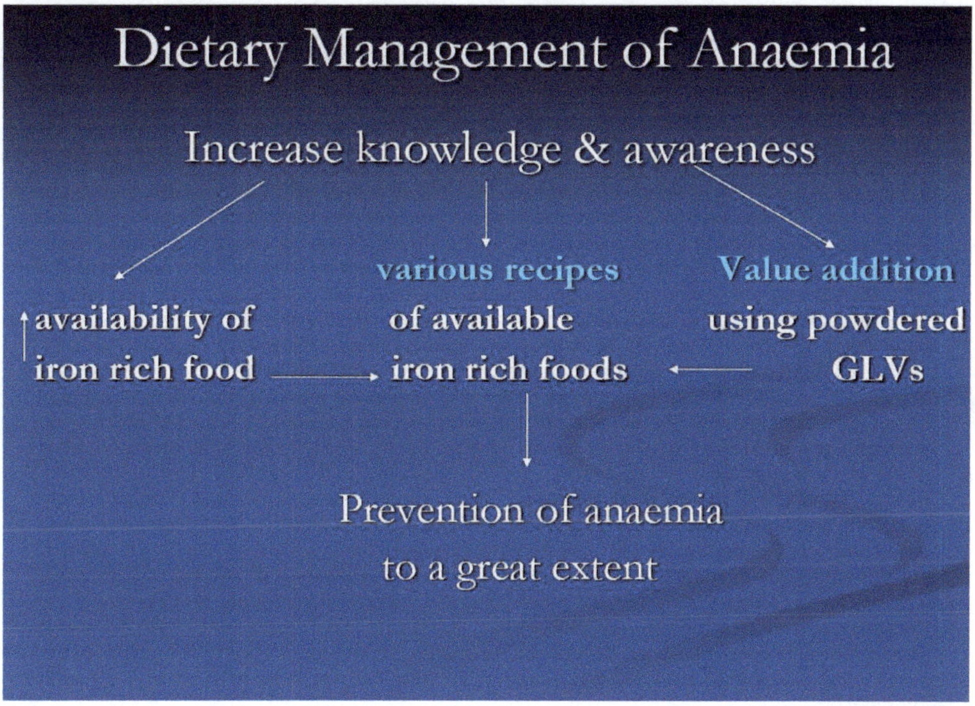

could be collected. Appropriate methods for drying GLVs were standardized for year long availability of GLV's. Standardization of variety of simple recipes from different GLVs resulted in a book of recipes using GLV's. Spin-off effects on iron status of other family members (young children, elderly adolescents etc.) were also measured. The programme reinforced the potential benefits of community based nutritional Anaemia control programme which was incorporated nutrition education in schools/colleges and community infrastructure to promote regular intake of iron/ folic acid-rich foods and promote intake of food which helps in absorption of iron and folic acid.

2.0 Empowerment and Sustainable Development

Empowerment of women demands innovative approaches, institutional development, awareness, education, socio-economic motivation and environmental soundness, which are complimentary to each other and lead to a sustainable package. Such development at village level can be sustainable only when social and economic processes are more equitable, efficient, and endogenous. Need is to weaken the culture of silence and dependency and promote the process of empowerment. Thus, the promotion of gender based equity in access to and control of productive resources; access to research and extension service benefits; enhancing women's participation in decision making process at all levels must be pursued in order to address the root causes of persistent poverty and food insecurity among rural women and the families they support. In this process, focus must be given for technological empowerment of women with skills and critical thinking that fosters a sense of self-reliance and ability to evaluate what is beneficial or detrimental to their interests as well as the factors

Box 13.2

The reproductive health status of women is greatly affected by the quality of sanitary protection and standard of hygiene maintained during menstruation. Initiatives towards improving the status of health of women in India have to be with a focus on rural/ semi-urban and urban women. The female population of rural India in the reproductive age-group of 15-45 years is 16 million, i.e. more than double of the number of urban women in the same age group. Even though there are various brands of sanitary napkins available at all urban markets, their costs are too high. On an average the cost of a branded sanitary pad varies from Rs.2.40 – 7/ pad. With such a cost profile these products suite the needs of the HIG and upper MIG strata women. For the LIG women this cost is too high to meet on a regular basis. High costs coupled with difficulty in accessibility, problems of disposability, along with traditional beliefs and taboos associated with the use of readymade napkins has led to most of the LIG women using cloth pads made at home. A survey to assess the level of need for sanitary napkins amongst women of low income category residing in urban slums in Delhi was undertaken. A total sample size of 500 working women both urban and rural in the age-group of 20-40 years was interviewed and based on the survey the task of developing a low cost product conforming to international standards of quality standards of absorbency, ability to withstand pressure after absorbency, water retention, flexibility, pH, disposability and microbiological parameters (bacterial and fungal counts) was developed. This technology has been found to be extremely promising and large scale field testing of the prototypes in rural areas of Maharastra show that the demand and acceptability of these napkins in rural areas is very high. Since the napkins are biodegradable disposability is not a problem. Keeping in view that farm women/construction workers have to use the napkin for much longer durations experiments on optimizing the thickness to suit their needs have also been taken up. The Department of Science and Technology is now taking up the task of setting up manufacturing units in different parts of the country both as a step towards entrepreneurship development of women and as a health intervention.

contributing to the degradation of environment and unsustainable development. Here, women's empowerment is operationally concerned with the genuine empowerment through education, ideas, through consciousness, through mobilization and participatory approach to improve their lives in a practical way, which would ensure a sustainable future for rural India. This is where sustainable development must have its roots. For this, priorities should be on activities which are easily manageable, require low external inputs, contribute to family nutrition, easily marketable, provide regular income and create assets of direct benefit to women. Technology must ensure sufficient production of food, fodder, fuel, energy, clean drinking water supply and fertilizers without spoiling and damaging the environment. Thus, drudgery should be reduced both at household related tasks and occupational tasks. This would energize rural women to move towards active and useful participation in other economic activities and in the overall technological transformation of the society.

More attention needs to be paid on gender sensitization of scientists and technologist with an aim to develop technology addressing women's needs. The need is to promote gender based equity in access to S&T based benefits as an integrated component for sustainable rural development. This will improve access to affordable and environmentally sound technologies and generate meaningful employment in the local economic structure.

2.1 Agents of Change-Voluntary Organizations

Women empowerment programmes in the country are run by the Central Government, the State Governments, the Panchayati Raj Institutions (PRIs), the Private Sector,Voluntary Organizations and Community Based Organizations. The tradition of voluntary action is deep-rooted in the country and has been a part of the national awakening and freedom struggle. Mahatma Gandhi and other great reformers of the country emphasized strong social and community action to remove social evils and catalyze development. The voluntary sector enjoys certain functional advantages, being community based, more accountable and capable of providing services at a lesser cost. More significantly, voluntary organizations have the flexibility to develop innovative projects based on local needs and resources in contrast to the standardization that normally characterizes governmental actions. The three most important contributions of VOs in national development, *viz.* innovation, empowerment and research and advocacy.

Since independence, this tradition of voluntary action has flourished and there has been a growth of voluntary organizations several of these are S&T oriented run by eminent scientists such as Dr M.S. Swaminathan, the father of green revolution in India. Young scientists have also taken the initiative to form voluntary organizations which are involved in demystification, demonstration and training in high end technologies such as biofertilizers, biopesticides, hybrid seed production, aquaculture. Voluntary organizations have done commendable work in different areas of nation building such as integrated rural development, watershed management and traditional rain harvesting, construction of community toilets, woman and child welfare even in remote tribal and hilly areas.

In areas of food and nutritional security, implementation of ICDS programme and midday meal programme of the government the voluntary organizations have an important role in creating awareness and dissemination of information. Over five hundred VO's are currently working with the Ministry of Science and Technology and are engaged in activities in sectors such as watershed development, social and water conservation, agriculture and cattle wealth, sanitation and drinking water, post harvest processing,health, promotion of Information Technology and Biotechnology.

2.2 Technology Choice

For environmentally sound sustainable development, technology choice can have a critical impact on many aspects of women's empowerment, especially the way we choose it, the way we innovate and design it and the way we deliver it. Experiences in the past showed that the introduction of technology and mechanization based on top-down perspective have favoured rich with capital assets and displaced millions of land-less producers, especially women from agricultural work. Therefore, technology and its scale must be carefully chosen to enable rural women to acquire and imbibe knowledge of technologies appropriate to their needs, priorities and rural development. An important point is that focus should be on diversification of labour use and not to the displacement of women labour. As in the case of agricultural implements, major factors to be considered are the characteristics of the tools such as handle, and size, based on the anthropometric data on women, the preferred body position, elimination of drudgery and economic enhancement.

A technology may be appropriate to a woman if it reduces their drudgery, improve their environment, improve their skills and enhance their comforts. Schumacher defines it as "the technology of production by the masses, making use of the best modern knowledge and experience that is conducive to decentralization, compatible with the laws of ecology, gentle in the use of scarce resources and designed to serve the human person instead of making him the servant of machines". According to another view, the technologies should be such as would upgrade traditional skills and capabilities; minimize fatigue and reduce drudgery; be innovative and capable of easy assimilation; generate significant and assured added value to existing methods of operation; generate employment and use local resources; need low capital investment and should result in low cost production of goods; be capable of replication and adoption and should blend harmoniously with existing environment for self-sustainable development of the rural women and community as a whole.

In rural scenario, choice and role of appropriate technology in relation to women and development became crucial in building up local capacity to devise solutions to tackle the identified problems to improve their quality of life. The emphasis should be to improve upon existing skills, provide managerial capabilities and to understand the science behind the processes/products. This will inculcate a scientific temper and make them more open to improved/emerging technologies for improving production efficiency and reducing drudgery in their day to day work. Women can participate in technology in three ways as consumers of technology, as producers of technology and change agents in dissemination of need based technology. Some of the emerging technologies which are directly beneficial are energy technology and small scale bio-technologies for sustainable agriculture, while, others like information and educational technologies offer lots of scope to overcome the barriers of illiteracy, lack of education and participation in development and adoption of such technologies. Livestock enterprise with appropriate production technologies also offers immense potential for social transformation of rural women's lives. This will help them to strengthen both nutrition and livelihood security. In terms of emerging technologies, biotechnology offers a powerful capacity to produce science based products like bio-fertilizers, biocide, microbial culture and vermi-culture for sustainable agriculture; tissue culture based production of fuel, fodder and commercial species of plants; fertility control vaccines for population control; improved production and productivity of cattle and buffaloes using embryo transfer techniques and conservation techniques to tackle narrowing genetic base to meet the needs of rural womenfolk in particular and community as a whole.

2.3 Technology Modulation and Transfer with Systems Approach

For effective transfer of technology, it is essential that various inter-related components of the technology, namely, "hardware" (materials such as crop variety), "software" (technical know-how, information), "humanware" (human ability)," organoware" (organizational and management aspects) and the final products (including marketing) are rendered accessible to rural women as consumer/end-users. In fact, a large number of technologies and institutions dedicating themselves for scientific and technological research work are available, but, there is a need for 'systems approach' to modulate and transfer technology to the field as user and environmental friendly and cost effective so as to have easy adaptability and replicability. Thus, focus of technology development and transfer would be to a problem solving approach rather than merely developing products or processes for the benefit of rural women and society.

An effective technology development and transfer process should, therefore, involve the following steps:

- Need assessment: need identification and technology gaps as perceived by women in the village;
- Identification of technology source (S&T institutions);
- Maximum utilization of local resources which can be sustained economically and ecologically; skills and markets;
- Participatory technology modulation/development which are environmentally compatible and socio-economically sustainable;
- Self-sustaining S&T base in rural areas (S&T field groups) to facilitate participatory planning, designing, execution and monitoring of chosen technologies: technology development, field testing, demonstration and training; and finally dissemination to meet the local needs;
- Setting up of viable S&T based micro-enterprise involving S&T field groups in rural areas and linking them with national S&T institutions/labs for adaptive R&D based on field generated specifications;
- Qualitative shift in production/distribution from household models to collective, networked "industrial" forms for higher productivity, incomes and reduced drudgery;
- Constant interaction between the innovator, the producer and deliverer or technology transfer group for technology assessment and further refinement; and finally
- Innovations should be in the spheres of both social behaviour and technology to bring about the desired ends of self-reliance, equity and sustainability.

To reach majority of those women who live in the vast areas of rural India, grass-root level voluntary organizations, with scientific and technological capabilities, provide the crucial link between emerging new developments in knowledge and technology, their modulation and dissemination to strengthen and diversify the rural economy using local resources and material as per women's needs. They can facilitate the articulation of felt needs on one hand, and also match those needs with potential source of technology. However, to make such intervention at village level, it is imperative to consider and emphasize contribution of target groups to the innovations and appropriateness of technology which are made based on their traditional knowledge system and perceptions. This will make them feel part of it as producer or consumer or end user of technology.

Awareness generation and technology capacity building through demonstration-cum-training among rural women about the importance of technology relating to their needs is also a necessary aspect for technology delivery system. The entire technology package will have to be developed based on women' needs, which, in turn must be evolved through a detailed scientific micro-planning exercise. Thus, appropriateness of technology delivery system through S&T base would revolve around effective education and learning process to facilitate the development of creative, innovative and autonomous capabilities in individual woman and women's groups through attitudinal/behavioural changes to promote evolution and transfer of environmentally sound technologies.

The voluntary organizations with a strong grass-roots base active in rural India represent the poor's interests. The development approach of several of these based on the above systems approach have resulted in sustainable models of activities, where application of science and technology have resulted in improved living conditions as well as new opportunities for gainful employment to rural poor, including women. For instance, in a collaborative programme between the Council for Advancement of People's Action and Rural Technology (CAPART), New Delhi, and the Department of Science and Technology (DST), New Delhi, CAPART has taken up large scale replication of some of the technology models as package of appropriate technologies evolved and proven by the latter through their partners *i.e.* S&T field groups along with collaborating S&T institutions through systems approach. In this endeavour, one of the most successful models recently developed is in food processing sector – involving women as 'primary stakeholders" at all levels as producer, as consumer and change agent from designing to adoption of cost effective technology package to suit local needs.

Case Study 2: Model: Post Harvest Processing of Horticultural Produce at Village level–A Success Story

In most of South Asia, the majority of poor women depend on agriculture for their livelihood. Likewise, data shows that in rural India, agriculture may continue to be the major economic activity in the near future. Therefore, need is being felt to give major emphasis on sustainable agriculture based employment generation. As agriculture is the backbone of our economy, so is the empowerment of women in view of their role in Indian agriculture. In this context, focus should be given to support women-managed rural production and marketing ventures in areas like horticulture, floriculture, and post-harvest processing in commodities.

Need Assessment and Conceptualization

Small farmers and other poor growers of fruits, vegetables and other horticultural produce suffer from the pressure to dispose off perishable produce, often at distress prices due to poverty and lack of access to storage or other preservation facilities. India is the second largest producer of fruits and vegetables in the world but processing is very low, approximately 3-5 per cent. It is also estimated that about 20-30 per cent of the total production of fruits and vegetables goes waste due to spoilage at various stages from collection to sale. Thus, not only valuable source of nutrient wasted, but its producers face impoverishment and consumers have to depend on high cost branded products from organized sector. As World Resource Institute points out, "When 20 per cent of harvest is lost, the actual crops loss is just part of the problem. Also wasted is 20 per cent of all the factors that contribute to producing the crop–20 per cent of the land used to grow the food and 20 per cent of the water used to irrigate it, along with the human labour, seeds, fertilizers etc. In other words, post-harvest food loss translates not just into human hunger and financial loss to poor farmers, but, into tremendous environmental waste as well". This is where, Science and Society Division under the Department of Science and Technology, launched a co-ordinated programme on Low Cost Processing and Preservation of Horticultural Produce during ninth plan period for four years duration. The programme was initiated with 'systems approach' to set up economically viable and self-sustainable production units having networking of poor/marginal growers and other weaker sections, especially women through appropriate technology package for value addition at village level involving S&T based developmental field groups and institutions.

These units established in the villages (based on field survey done to access the need and required S&T intervention), are close to the source of raw material and easily accessible to women growers. It was also ensured that local market potential should exist and range of raw produce should be available in order to run the units round the year.

Technology Package

The technology package includes some innovative techniques, suitably de-scaled and adopted equipments/machines to produce conventional products or intermediates like juice, pulp, puree, pickles etc. with a major emphasis on production of innovative dehydrated/osmo-dehydrated products such as onion and garlic powder, dried apricots/plums, apple chips, papaya candy and mango/jackfruit leathers etc. Given the focus on innovative work, and the perspective of adopting low cost energy saving technologies(such as agro-residue based flue-heated driers, zero energy cooling chamber for on farm fruits and vegetable storage based on the principle of evaporative cooling, lactic acid fermentation and other nutrient preservation techniques), major thrust under this programme was given on minimizing pre and post harvest losses; increased shelf life with quality control; controlled atmospheric and aseptic packaging, and emphasizing on drying and dried products to have all year round production and sustainable income ensuring maximum benefit to the women groups.

Systems Design

Each location comprises of mother unit at Tehsil level for drying, quality product production, packaging and marketing linked with several home based or cluster of village based sub-units for providing semi-processed produce to mother unit, together covering a cluster of, say, 10-15 villages. Adaptive R and D work, training in quality control and production process is carried out at mother unit, while, overall organization and supervision of the units is done by the S&T field group with a network of women. Training, an essential element of the programme was designed to be imparted at two levels: at institutional level by resource agency (Indian Agricultural Research Institute, Delhi) to core women group and at field level by trained women core group to network beneficiaries. It was ensured that at every level of operation, network of women is closely involved in production, processing, packaging and marketing promotion.

Impact Analysis

As a result of above co-ordinated efforts, seven full scale processing and production units have been established as common facility centres in different geographical locations in the country with technological back-up/training from the Indian Agricultural Research Institute, New Delhi. The programme was co-ordinated by the Centre for Technology and Development, S&T based field group based in Delhi to provide technical back-up as well as managerial support to participating partners in this programme. Each group has carefully developed low cost technology package, recipes and processing protocols through field trials based on traditional knowledge/techniques upgraded with modern S&T inputs, availability of raw material in the area and market demand of innovative products. Technologies of appropriate scale/type were introduced or wherever necessary adapted/standardized under field conditions with functional networking between different levels of production. Each Unit is having a core team of 4-5 trained women with on line quality control laboratory facility. These trained women as para-technologists are now involved in imparting training and technical know-how to other women of the area as on-going activity having multiplier effects. Each unit also has got requisite FPO license and is now self sustainable, being managed by women co-operative to process, preserve, pack and market the finished products. These innovative products are now being marketed successfully by them under the common brand name "Farmer's" ensuring maximum benefits to women growers/beneficiaries at the village level with an approach of value addition at source itself – thus reducing dependency on middle-men, reduction in post harvest loses, saving in transportation cost, and time. A new sense of accomplishment by way of new skills and earning independent incomes have given them the confidence to bring about other reforms in their lives. Many of them are now developing fruit belts around the units through better nursery plantations techniques for better production. It is estimated that about 70-80 families around each unit have benefited directly to have all year round production for self consumption and sustained income with minimum wastage. Many of innovative products: natural plum and apricot drinks, rhododendron squash etc. made from farm fresh produce by the caring and knowing hands of trained women are now available in the market. Each unit is estimated to have year-round production of 500 kg/day of finished products with monthly turnover of Rs.50,000 to Rs.1.00 lakh. Moreover, local processing has helped not only in extending the shelf-life of otherwise perishable produce, but also in generating local employment in rural production system.

Success and Spread Effect

Success of this programme lies in "bottom up approach" and linkages amongst Department of Science and Technology (DST), S&T based Field Groups, S&T Institutions and above all target group (Women) – from planning to implementation stage as partners, carriers and change agents in technology development and transfer in de-centralized way of functioning. This has resulted in long term sustainability by linking the women from small grower families in the area with the production and marketing system – involving their full participation at all levels, thus, making them self confident. Now, all units with trained women are serving as resource agencies to replicate this model in other areas in respective states.

As a village based S&T intervention, it is well suited to rural women as income generating activity based on sustainable use of local resources and material for large scale replication in rural areas.

Seeing the success potential and 'systems approach' in planning and implementation of the above programme, a UN agency–Asia Pacific Centre for Transfer of Technology, Delhi has taken the initiative in collaboration with DST, Delhi to replicate this technology model for large scale replication in the North Eastern States, and is therefore, a path-breaking effort to carry S&T for the socio-economic upliftment of poor rural women.

Box 13.3: Innovative Horticultural Products Developed By Various Field Groups

Sl.No.	*Fruit/Vegetable*	*Products*
1.	Apricot	Jam and Drink
2.	Plum	Drink
3.	Apple	Juice, Jam and Chips
4.	Bael	Squash
5.	Gooseberry	Preserve
6.	Papaya	Chutney and Candy
7.	Citrus Fruits	Lemon Cordial, Kenu Squash, Orange Squash and Lemon pickle
8.	Pineapple	Drink
9.	Tomato	Puree and Sauce Powder
10.	Rhododendron flower (Boransh)	Drink and Jelly
11.	Pickles	Yam pickle, Mango pickle, Lemon pickle, Jack fruit pickle, Mixed vegetable pickle
12.	Dried Powders	Turmeric powder, Dried Mango powder, Ginger powder

3.0 Conclusion: Gender Based Planning and Technology Literacy

Technology Models described above illustrates how scientific and technical interventions can improve the quality of life of women in rural areas. The experience has brought to light the fact that there is vast potential, intelligence, strength and ability in our women that can be tapped and used for sustainable development. This also shows that following factors/approaches are crucial for technological empowerment of women as well as to ensure sustainability of developmental programme at the grass roots level:

- Gender-sensitive planning at the local Panchayat level with gender/sex-segregated information to support S&T based developmental efforts in rural production system.

- Gender integrative participatory technology development/appropriation.
- Eco-restoration and employment generation through judicious use and transfer of appropriate technology package appropriated with involvement of women at all levels both as "change agent" as well as "consumer" of technologies.
- To deal with rural poverty, need is to deal with ecological poverty through participatory approach for ecological restoration.
- Proper recognition of the productive and domestic roles of women.
- To make such intervention successful, other areas of interest and support activities to provide basic facilities like drinking water, health and sanitation, nutrition, family planning, adult education and social security etc. should be taken care for overall sustainable community development to have a catalytic effect. With such support services, access to training and education or participation in economic and social life become a reality for rural women.
- Importance of S&T base in the actual area for effective technological intervention/ to develop location specific technology package taking into account the perceptions of women beneficiaries.
- Involvement of local women motivator as active "change agent" or "natural carrier" of technology through dove tailed approach of motivation through talks and audio-visuals, awareness build up through demonstration, and hands on the job training in relevant field.
- Strong institutional linkages between S&T field groups, R&D institutions with an extension machinery to disseminate the proven technological package at grassroots which empowers and enable rural women to seek local solutions and have face-to-face interaction between them and scientists for proper scientific know-how.
- Equal sharing of benefits amongst different stakeholders.

Participatory technological development and dissemination is a the mainstay for developing sustainable technology models for replication which will have an impact not only on the economic sphere, but also on the social, political, cultural, moral and environmental levels in rural production system. As in rural areas, the entire activities go around womenfolk and women's active participation should be ensured at every level from planning to implementation stage both as change agents as well as beneficiaries for the achievements of goals of equality and development and empowerment by changing social attitudes and elimination of all forms of gender-based discrimination–will be a guiding factors for any future development planning. The Rural Women's Technology Park and Golden Jubilee Biotechnology Park are models of S&T based empowerment which has demonstrated that careful planning with participatory approach and handholding are critical to success. These models also show that women know what they want from technology; the gap arises when technology developers do not know what is the requirement neither do they try to elicit information from actual users. The bottom up approach with active participation of actual users is the best approach to address innovative traits in women. However in new technology areas, the methodology can be a mixture of ' top down approach' to demonstrate the technology and then a ' bottom-up approach' for technology transfer. In a country like India having about 5,91,000 villages, each of them covered by one or other programmes of rural development, the need of the hour is to devise strong and well-knit technology delivery systems in identified areas and dovetail them with such programmes. This requires a systems approach regarding needs-identification, choice of technology, appropriate scaling, technology upgradation/ innovation, economical and ecological sustainability and optimum use of local resources with emphasis on technology literacy amongst women through mass communication for upgradation of skills and capabilities.

Women trained as para-technologist working in processing unit for value-addition at village level: Need based development and effective delivery of technologies as complete package is crucial for empowerment and sustainable livelihoods.

Women's Group in Uttarakhand working on Rhododendrons for value addition

Rural Women Technology Parks

A unique attempt to ensure women's empowerment through inputs of technology and social support. The objective of setting up these rural WTP's is to transfer proven technologies through a common platform, highlight issues related to women's technology needs, identify specific S&T interventions, and serve as a platform for technology information pool / documentation and to demonstrate live models in technology areas. The technology areas addressed in these Parks are:

- Land-based (agriculture, horticulture, plantation crop) appropriate agricultural practices, simple tools and implements to reduce drudgery, organic farming, biopesticides, vermicompost, mushroom cultivation, nursery techniques, energy plantation, hydroponics crop diversification, low-cost tissue culture and green house techniques, high intensity cultivation of cattle fodder, floriculture, chromaculture, cultivation of medicinal plants, nutrition gardens, seed bank, etc.

- Post-harvest technologies: Food processing, fruit and vegetables processing, management of household pests, protected storage structures.
- Natural resource conservation, augmentation, management and byproduct utilization: Identification, sustainable collection, use, genomic conservation and legal protection of natural resources, extraction of fibre from non-traditional plants.
- Forests based activities: Processing of non-timber forest produce for value addition.
- Water, Health and Sanitation: Basic health, hygiene and sanitation, treatment of animal diseases, water purification, water saving practices, harvesting and storage of water, nutrition, health packages for adolescents, women in the reproductive and post-menopausal age.
- Non-farm activities: Electrical, electronics, repair of household gadgets, computers, screen-printing, block printing, Handloom, wool weaving, fibre utilization.
- Energy Conservation: Energy efficient cooking means.
- Livestock Management: Fodder and feed management and improved animal sheds
- Marine and fresh water based activities.

The Department of Science and Technology has set up over a dozen rural women technology parks in different agro climatic zones of the country and is committed to setting up WTP's in all the districts of the country.

The Department of Biotechnology, Government of India has set up a Golden Jubilee Biotechnology Park for Women, the first of its kind in India, at Kelambakkam, 41 kms south of Chennai. This initiative seeks to use biotechnology for the uplift of rural women by providing opportunities for setting their own ventures. The scheme is a collaboration of the Department of Biotechnology, M.S. Swaminathan Research Foundation (MSSRF) and the Tamil Nadu Industrial Development Corporation (TIDCO). The project, was launched to commemorate the 50th anniversary of India's Independence, covers an area of 8 hectares and is managed by a society under the chairmanship of eminent agricultural scientist and architect of India's Green Revolution, Dr. M.S. Swaminathan. The Park aims at developing an integrated approach involving technology identification, incubation, dissemination, training and retraining, development of necessary techno-infrastructure through feasibility studies using the criteria of value addition and market demand. The park is managed by professionals with active stakeholder participation. The highest standards of environmental management will be adhered to in accordance with the United Nations Conference on Environment and Development (UNCED) dealing with environmentally-sound management of biotechnology. The Park acts as a platform for bringing together women entrepreneurs, scientists, financial institutions and industry. The first batch of women entrepreneurs have set up the industrial modules covering the areas of food, biotechnology, floriculture, medical diagnostics, pathological testing facilities, hatcheries for ornamental fish, herbal products, fortified food processing plants, Spirulina production and organic composting.

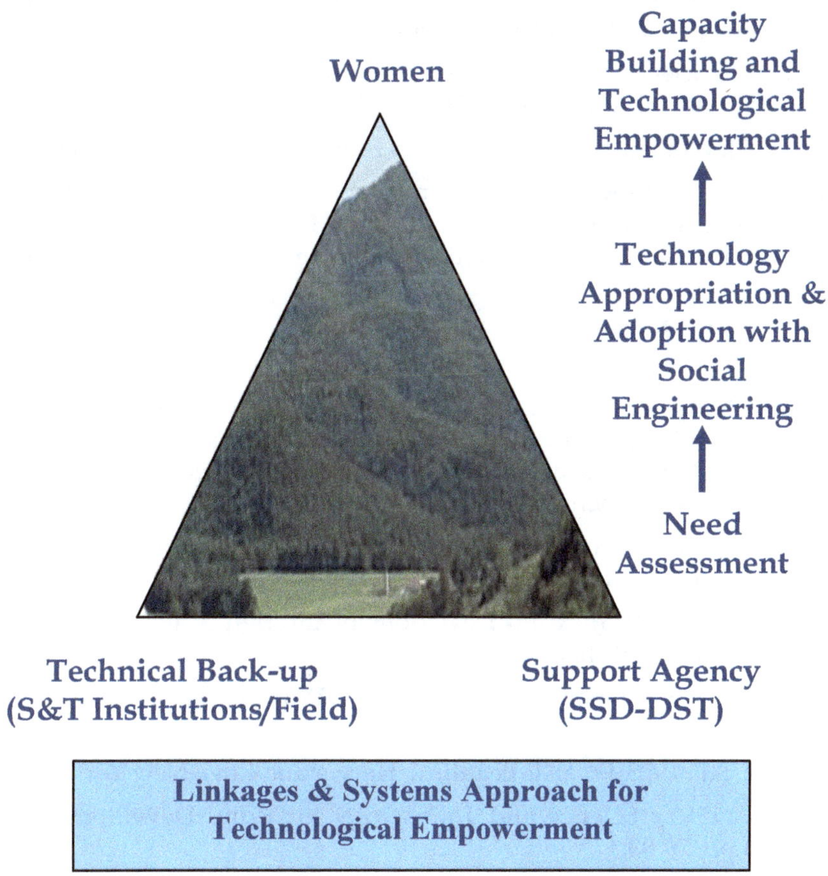

Systems Approach : Reaching the Un-reached

- Bottom Up approach with system design.
- Based on Traditional Knowledge System (TKS), local resources, material and skills: capacity building.
- Networking and linkages amongst people (Primary Stakeholder), S&T field groups, S&T Institutions.
- Need based Intervention: Technology Modulation, field testing and transfer to provide technological solutions to location specific problems.
- Participatory approach: ensuring backward and foreword linkages.

References

Agarwal Bina. (1995): Feminist Environmentalism or Eco-feminism. The Hindu, Survey of Environment, pp.1-14.

Agarwal, Bina. (1997) : Gender, Environment, and Poverty Inter-links: Regional Variations and Temporal Shifts in Rural India, 1971-91, J. World Development, 25(1), pp. 23-52.

Agarwal, S. and Kumar K. (1993): Science, Technology and Women for Sustainable Environment and Rural Development. Int. J. Sust. Dev. 1(3): 56-60.

Department of Science and Technology, Government of India (1995): Technology Models for Rural Applications, DST Publ., N. Delhi., p.115.

ESCAP (1997): Women in India. A Country Profile. United Nations, New York.

Khosla A. (1991): Role of Science and Technology in Sustainable National Development. In Proc. National Summit on Technologies for Environmentally Sound Sustainable Development, Development Alternatives, N. Delhi, India. pp.4-7.

Khosla A. (1999): Women and Environment, Development Alternatives Newsletter, 9(9), p.11-15.

Kwaschik, R., Singh R.B. and Paroda R.S. (1994): Technology Assessment and Transfer for Sustainable Agriculture and Rural Development in the Asia-Pacific region: A Research Management Perspective.(Research and Technology Development Service, SD Department, FAO).

Papa R. and Shanmugasundaram Y. (1996): Proc. Women and Emerging Technologies (Indo-British Seminar), British Council, Chennai.p.255.

Raghunandan, D. (1988): Rural Development through Science and Technology Application, Kurukshetra, xxxvii(1).pp 77-81.

Raj Rani (1997): Environmental Degradation and Women, Kurukshetra, XXLV(4&5), 57-59.

Schumacher, E.F. 1973: Small is Beautiful. Harver and Row, New York.

Sharma, M. (1991): Science and Technology for Rural Development. Yojana, 35(1&2),pp. 78-83.

Venkateswaran, S. (1992): Living on the Edge: Women, Environment and Development, Friedrich Ebert Stiftung, New Delhi

UNDP (1997): Human Development Report, Oxford University Press, New York.

World Bank (1991): A World Bank Country Report: Gender and Poverty in India, World Bank, Washington, D.C.

World Resource Institute, World Resources, 1998-99, Chapter 9.

Chapter 14

Women Empowerment to Women Capabilities: Bridging the Gap

Kavita Mehra
National Institute of Science Technology and Development Studies (CSIR), New Delhi– 110 012, India
E-mail: kavitamehra@yahoo.com

ABSTRACT

Though women constitute nearly half of the world population but even in the 21st Century, nowhere in the world they are placed at par with men either socially or economically. For this matter, 'women empowerment' issue has always remained very important for policy planners and social activist organisations. There are numerous discussions in the literature on definition of 'Women empowerment'? Paper elaborates the issue with reference to realization of women capabilities beyond poverty alleviation and welfare approach and tapping those capabilities with the use of S&T (Science and Technology) interventions. This fact has been illustrated through three cases across sectors and organisations. Since women play the major role in dairying and constitute more than 70 per cent component of labour force in this sector, National Dairy Development Board (NDDB) aimed at expanding the participation of women dairy farmers, and strengthening their capacity by making available the S&T information and practices in dairy. With this objective, NDDB initiated special programmes for women in the form of women cooperatives and giving them leadership training. There is another case of application of biotechnology by rural women in Himachal, a total new endeavour by women. This task of women employment (leading to their empowerment) has been achieved by Institute of Himalayan Bioresources Technology (IHBT), a Council of Scientific and Industrial Research (CSIR) laboratory in Himachal region for through Department of Biotechnology, Govt. of India funding. The third case depicts the role of voluntary organization, Self Employed Women Association (SEWA) demonstrated the capacity to deploy ICT and mainstream it effectively. SEWA's ICT efforts focus on the following aspects: capacity building, livelihood generation and security, and knowledge sharing. The above cases have been picked up from numerous Indian literatures on the ongoing work towards women empowerment, to show as examples of strengthening the existing capabilities and exploring the latent potential existing within women but lying unexplored. There is need to realize women capabilities in women empowerment programmes.

Keywords: *Women empowerment, Women capability, Science and Technology, Women cooperatives, Voluntary organizations, Plant Tissue Culture Technology, Information and Communication Technology (ICT).*

1.0 Introduction

In the current era of economy when human capital is the most important factor, participation of people irrespective of the gender is essential for economic and social development of the nation.

According to 2001 census, 48 per cent of total population consists of women in India. Participation of this component of population in full way is still a very big concern for various reasons. Though women constitute nearly half of the world population but even in the 21st Century, nowhere in the world they are placed at par with men either socially or economically. The nature and intensity of discrimination depends upon the socio-economic and cultural structure of the region to which the women belong.

A vast majority of Indian women work throughout their lives. Only a small number of women workers are officially recognized as workers and wage earners. Of the female labour force in India, more than 94 per cent are in the unorganized sector. Majority is concentrated in the primary sector in agriculture and animal husbandry, mainly as wage workers. Women mostly employed in the informal sector as unskilled, low-paid labour, unaccounted in national statistics. Women lack access to markets, and skills to meet changing market demands and financial management skills. This scenario is more true to rural women as the gender issue situation is graver in the rural areas. On the contrary, in the urban areas, women are working in all the sectors from traditional to modern and holding even top positions. The paper is dealing with the issue of women empowerment for rural women and the role of science and technology towards women capabilities building and strengthening.

First section of the paper briefly discusses the issue of women empowerment – the conceptual meaning and the process required. There is a brief mention of the role government and voluntary organizations etc. in the process of women empowerment. The following section is about the concept of the paper. Next section elaborates on the concept of paper through various cases. The concluding section discusses the impacts of the government programmes and initiatives having fruitful results as well as the need for change in mind set from women welfare programmes (in many cases but not all) to realize and unfold women capabilities and make avenues available to them to show their potential.

2.0 Women Empowerment

Empowerment in the real sense is the power that a woman is endowed with to decide for her and for others, to choose what to do and what not to do and finally to act and pursue upon her decisions, which leads toward organized action for social transformation.

Empowerment means changes in attitudes and skills; enhanced self-confidence supported by expanded knowledge. Reducing the constraints, which impede equal participation, can attain empowerment of women. *Empowerment* is a process that enhances the ability of disadvantaged individuals or groups to challenge and change existing power relationships that place them in subordinate economic, social and political positions (Agarwal, 1995). There is a difference between women's empowerment and women's welfare. Empowerment for women is fundamentally about changing social institutions and practices, about changing rules, norms and rights, and about changing the balance between women's obligations and responsibilities and their command over the resources needed to discharge these obligations and responsibilities. And in the context of development, empowerment cannot be given to anyone; it is a process that takes place wherein an inequality moves towards becoming equality. This is attained by reducing the constraints which impede equal participation and thus the inequality starts moving towards equality (Deshmukh-Ranadive, 2002).

Further in conceptualizing empowerment, she emphasizes that empowerment is a process and not a goal to be achieved. In this process the constraints that impede equal participation are reduced so that that the inequality starts moving towards becoming equality. These constraints are most often structural and connected to both the macro environment and the domestic environment that the women find her in. Empowerment has been identified as one of the key constituent elements of poverty reduction and as a primary development goal in recent literature.

Process of women's empowerment requires:

1. Taking into account of the existing endowment of women in terms of knowledge and capabilities.
2. Providing opportunities to women to acquire necessary information and training and to apply and utilize their upgraded capabilities.
3. Bringing the latent capabilities to the recognizable level through political and social forum.

In the literature, emphasis is there on economic empowerment of women through collective empowerment. To develop an effective framework for women empowerment is in the concept of 'Collective empowerment'. 'Women become empowered through collective reflection and decision-making'. What is meant is enabling women collectively to set their own agendas, to organize to help each other and make demands on the state for support and on society itself for changes. This understanding of women's empowerment is a good deal more radical than the more common approach *i.e.*, 'economic empower or getting women in the cash economy generally through self-employment or income generation' Young (1994). The empowerment of a single-woman or a number of women does not necessarily lead to change for other women in the same situation or community (Mathur, 2002). Collective empowerment, instead, refers to building strength through women working in a group that share common ideas and objectives.

Developing collectives or common goals using alternatives, non-hierarchical framework of organizing can set in place mechanisms for women empowerment that would have longer-term-social, cultural and political impact. This social mechanism has been applied in India, Thailand, Bangladesh and Philippines.

India has been described as a rich country with poor people. With abundance natural resources and varied and favourable climatic conditions, the country is a home of largest number of poor in the entire world. Almost half of India's one billion is deprived of adequate nutrition, safe drinking water, basic healthcare and education. The problem is compounded by the fact that more than 75 per cent of the population resides in rural areas in more than 700,000 villages (Chawla, 2002).

All over the world, women suffer from poverty disproportionately. According to foreword in Human Development report (1995), 'the most persistent of these disparities has been gender disparity'. Women still constitute 70 per cent of world's poor and two-third of the world's illiterates. They occupy only 14 per cent of managerial and administrative jobs, 10 per cent of parliament seats and 6 per cent cabinet positions. Much of their work remains unvalued, unrecognized and unappreciated.

Thus to begin with the process of development in India, one has to reach the rural women. The organization of production in rural household is neither discretely managed as production and consumption nor spatially separated as workplace and home. Women's work in the farm as extension of their domestic work around production seasons and their peak workload coincide with crop cycle.

2.1 The Role of Government of India in Women Empowerment

The first few plans followed a welfares approach and treated women as recipients of aid. It is only during the fifth plan a separate Bureau of Women's Welfare and Development (WWD) was set up in 1976 as part of the erstwhile Department of Social Welfare in order to intensity the country-wide efforts launched during the International Year of the Women (Pithadia and Sharma, 2005).

The sixth plan for the first time in India's planning history contained a separate chapter on Women and Development. To make the International Women's Decade a success it emphasized on three strategies via economic independence, educational advancement and access to health care and family planning. Thus with the beginning of International Women's Decade in 1975 a number of schemes were introduced and earnest efforts were made by the government to improve the status of women. Thus the Department of Women and Child Development being the national machinery for the development of women plays a vital role assisted by the Central Social Welfare Board and the National Institute of Public Co-operation and Child Development. Rashtriya Mahila Kosh under Department of Women and Child Development provides microfinance to poor women as a tool to promote self employment.

Considering it is not just a financing system, but a tool for social change, especially for women. Delivery is normally through Self Help Groups (SHGs).

2.2 Voluntary Efforts and Women Empowerment in India

There are number of voluntary organisations involved in women empowerment through initiating cooperative structures or taking up initiatives based on programmes. In most of the cases approaches have more towards poverty alleviation rather than using them as a competent human resource with latent capabilities. Self Employed Women Association (SEWA) is an organization worth mentioning in its role towards women self employment and empowerment.

SEWA is a trade union registered in 1972. SEWA, a member based organization of poor, self-employed women was one of the first organization to successfully rural women and women working in the informal sectors and now has 61 cooperatives, 400 self-help groups, 4000 saving groups over 530000 members in Gujarat – with its membership often encompassing all the women in the entire village. It is an organization of poor, self-employed women workers. They are the unprotected labour force of our country of the unorganized sector. SEWA's main goals are to organize women workers for full emploreliance. Full employment means employment whereby workers obtain work security, income security, food security and social security (at least health care, child care and shelter). SEWA organize women to ensure that every family obtains full employment. By self-reliance it means that women should be autonomous and self-reliant, individually and collectively, both economically and in terms of their decision-making ability (SEWA brochure, www.indianngos.com).

SEWA is both an organization and a movement. The SEWA movement is enhanced by being a sangam or confluence of three movements:

- The labour movement,
- The cooperative movement and
- The women's movement.

But it is also a movement of self-employed workers: their own, home-grown movement with women as the leaders. Through their own movement, women become strong and visible. Their tremendous economic and social contributions become organized.

3.0 Concept of the Paper

In this present paper woman empowerment strategies are dealt in terms of enabling women to realize their full potentials. The issue has been addressed based on *capability approach concept* (Nussbaum, 2000 and Garibaldo, 2002). The basic idea in this approach is not only to know about the person's satisfaction with what she does, but also about what she is in *position* to do (what or not she has the opportunities and freedom do use them). Women empowerment process needs to go beyond poverty alleviation and welfare programmes for rural women. Strategies should aim towards capabilities building to explore and utilise their full potential. They consist of greater access to knowledge and resources, greater autonomy in decision making, greater ability to plan their lives, greater control over the circumstances that influence their lives and finally factors which would free them from the shackles of custom beliefs and practices. One of the mechanisms discussed for empowerment of women based on exploring the capabilities and/or strengthening the existing ones is the role of Science and Technology in the form of information, training or the use of technology. The process leads to increased economic gains in the coded form and a self confidence as tacit contribution. It is believed that if women gain economic strength they gain both a visibility and a voice in the home, workplace and in the community which leads to community building.

4.0 Cases: Process of Empowerment

There are three different cases:

1. Strengthening the capabilities and organizing them in groups – Women Dairy Cooperatives
2. Role of government R&D laboratories in introducing women to new technology (Plant tissue Culture)-Exploring their capabilities with new resources in Floriculture Sector
3. Role of Voluntary Organization (SEWA) in introducing New Technology (Information and Communication Technology) to self employed women

Case 1: Women Dairy Cooperatives

In India, women play the major role in dairying. It is the women who usually look after the cattle. For expanding the participation of women dairy farmers, and strengthening their capacity, National dairy Development Board (NDDB) took a step by instituting 'Women dairy Cooperatives' (NDDB, 2001). This institution of cooperatives meant to serve as the basis of women empowerment through Dairying.

The case studies of Women Dairy Cooperative Societies from Ludhiana (Punjab) reveal out the role of dairy cooperatives towards women empowerment and community building (Mehra and Kalra, 2003). There are more than 800 milk cooperative societies whereas number for Women Dairy Cooperative societies exceeds 70. In Women Dairy Cooperatives women of the village manage the women cooperative society. The secretary and helper of the society may be a male or a female.

Members of the society pour their milk twice (morning and evening) in the society, the milk collection centre. The milk weighing and testing system is either manual or computerized in some societies. The fat content and the weight of the milk are done by the secretary at milk collection centre and the reading is entered into the pass books of the members. Collected milk is taken away by the vehicle of the Verka Dairy Plant (vehicle goes to the chilling centre which is around two kilometer away and from there the chilled milk goes to the Dairy Plant). At plant the SNF (solid not fat) content along with other observations is measured. After ten days, the payment is sent to the society for the milk on the basis of fat and SNF content and the amount of milk in litres, and is distributed to the members on the basis of amount contributed by each member. Profit of the Dairy as bonus and also annual price difference variation goes to society. The price difference remains in amount of milk collection in terms of kg (at society level) and litres (at plant level). Normally, 40kg.milk is equivalent to 41 litres. Thus, societies gain at the end of year. This part as well as the bonus is annually distributed amongst the members.

Role of Milk union at cooperative society level is to arrange training on their farms or to ake them to Anand in Gujarat. Training/information to dairy farmers is imparted on–Animal keeping and care, animal breeding-new techniques (Artificial Insemination for superior breed), selection of bull, delivery and post delivery care, factors affecting milk production, role of balanced diet for cattle, concept of quality versus quantity of milk etc., how to produce clean milk etc. Readymade feed formulations, better seed for fodder, medicines and mineral supplements for animals are also provided to farmers on cost basis. For solving problems of farmers about animals or fodder, consultancy is provided by the scientists. Thus the purpose of NDDB is in organizing the producers and milk marketing, inculcating awareness about best practices in dairying and availing sources amongst the milk producers for more quantity and better quality milk production in hygiene conditions. This aim has resulted in better income for women dairy workers. Economic empowerment leads to desired goal of women empowerment and community building. We can examine the impact of cooperative institution and scientific and technological knowledge imparting through some case studies.

Women Dairy Cooperative Society Ltd. (WDCS) at Jogi Majra, Ludhiana. The society started during 1989.Earlier the male members were running the above society in the 1980s. Due to some disputes among the members, the society was dissolved. After a period of time, in late 1980s, a group of women restarted the society and at present, it is one of the best performing women Dairy Cooperative Society.

Average Milk production per day over the years has increased from 220 litres to more than 1000 litres per day (by 2001) with the increase in membership as well. Society is able to sell substantial volume of ready made feed and mineral mixture obtained from milk union. It shows the gradual acceptability of recommended feed for their cattle. Members do get bonus also. Over the period society owns a building with a generator set, computerized milk testing/weighing facility and was planning to set up the chilling centre. In other words, women running that society very successfully which once became non functional by male members.

Bhumadi WDCS started by Harpal Kaur, a woman who wanted to avert the exploitation by milk vendors (middlemen). She managed to know the primary requirements for setting up a dairy cooperative. She got support from milk union in initiating the society. Women developed certain strategies within the society to invite more members and produce greater amount of milk. The initiative led to a great success and Harpal Kaur became a member of village panchayat. She is an excellent example of emerging leader.

In another WDCS at Jhandey, is the fruit of the seed sown by Mahila Mandal, an older widow who also wanted to overcome the exploitation by milk vendors. She donated her land to set up the society with the support of milk union. This WDCS teaches also the ways of pickling and papad making as well as established a thrift society.

NDDB's initiatives in terms of S&T based interventions have led to:

- More usage of recommended feed and other inputs for animal health care
- Artificial insemination technique is becoming more and more acceptable
- Using recommended allopathic medicines along with traditional formulae in animal health care
- Mineral supplements are given to cattle
- Use of milk testing gadgets
- Computerization of cooperative offices for record keeping
- Chilled storage facilities for increased shelf life of milk for longer transportation

Impact of Dairy Farmers on the Process of Empowerment of Women

The outcome of initiatives of women dairy farmers led to the economic empowerment of women due to:

1. Women in the cooperative themselves receive the payments and thus have achieved financial independence
2. Income has been raised and their standard of living has improved.
3. Leadership quality is emerging among them. On getting the opportunities, are able to take decisions and the latent talent is being revealed out. In some areas it has led to the setting up of Mahila Sangh (Women Centre) for imparting training on house hold requirements.
4. The women are transforming into entrepreneurs. The dairy as subsidiary occupation is changing into commercial activity. The financial gains have encouraged them to increase their farm sizes.
5. They have been found to be more alert to know the methods of farm improvement and being females, are able to give more warm care to their cattle
6. Their life is more organized. The regular visit to milk collection centre gives them an opportunity to organise their household work and to take out time to meet others. It helps in sharing their knowledge about the dairy farming too.
7. Animal health care knowledge through Union extension workers brings enlightenment in them to practice that knowledge, where ever applicable, in their daily life too such as hygiene, health care etc. So it is a source of education to them.
8. A matter of proud to the women. One could read the feeling of some achievement and contentment on their faces.
9. Impact on the Socio-Economic Status of the village children of the members of society are having higher education in varieties of fields, English M.A, PG in IT, MBA from USA. There are telephone connections with other modern gadgets.

Case 2: Plant Tissue Technology for Women for Floriculture Planting Material

Himachal Pradesh offers unique opportunities to develop cut flower trade because of abundant rain fall, bright sunshine and pollution free environment. In the past few years, quite a few growers have taken to commercial growing of flowers which include gladioli, chrysanthemum, carnation, tuberose, liliums, tulips, and orchids. Institute of Himalayan Bioresource Technology (IHBT) at Palampur, a constituent laboratory of Council of Scientific and Industrial Research (CSIR) has been engaged in developing the protocols based on plant tissue culture technology for the multiplication of elite planting material of flowers in demand (Sharma *et al.*, DBT website). Laboratory successfully demonstrates the beneficial use of adopting tissue culture raised plantlets (TCPs) in the field and has made the farmers to harden the TCPs themselves in their own fields under the supervision of IHBT scientists. This has resulted in greater excitement among people involved in multiplication of planting material through tissue culture as an enterprise. Thus, they are in a position to supply the culture vessels containing the plantlets of standardized as prescribed by IHBT and growers who are hardening these thereby are happy and confident this way completing the chain, from production to utilization of the produce based on technology adoption. Growers of the area got fully convinced about the superiority of the germplasm thus raised by tissue culture. This development led to awareness amongst rural women.

Keeping this in view, IHBT through Department of Biotechnology, Govt. of India sponsorship, women are being trained in tissue culture. In this unique approach rural women were enabled to use their time for establishing tissue culture units, without disturbing their daily routine. Under this project, about 40 rural women have been imparted training in micropropagating disease free liliums and orchids during three years period. In the beginning, women were invited to the institute to apprise them of the tissue culture technique in a simplified manner. A mobile laboratory with a custom designed laminar flow was used for training purpose. Later on difficulties encountered during incubation of cultures at their houses, two tissue culture units complete with the basic infrastructure were established one each at village Daloon under the aegis of Chinmaya Tapovan Trust and at Khundian under an NGO group Environment and Rural Awakening (ERA) by the scientists of IHBT. The women have become proficient enough to handle the plantlets and explants both of orchids and liliums independently. In order to ensure sale of their produce for financial returns, a group of small growers of region under the umbrella of The Kangra Floriculture and Horticulture Growers' Federation at Dhoron have been trained in hardening procedures only. The Institute has established its credibility by demonstrating the superiority of tissue culture raised plants in the fields to the growers of the area. This has led to a spurt in the demand for Tissue Culture raised plantlets from the growers. IHBT has competence in the areas of plant micropropagation and virus indexing. It is proposed to test the cultures periodically for viral infections so that they do not face difficulties during the sale of their produce.

Close working relationships exist between IHBT and the women associated with various NGOs operating at the village levels and the scientists and technicians visit their premises quite regularly.

Similarly, the farmers have also been regularly visiting the Institute for their problem solving and attending various training programmes organized from time to time. Efforts of IHBT are bringing a perceptible change in the floriculture trade in the area for ensuring production and availability of quality planting materials at affordable prices to the growers/farmers so that their dependency on imports is reduced and finally eliminated completely.

Case 3: Role of SEWA in Introducing Information and Communication Technology to Self Employed Women

SEWA has realised the potential of harnessing the power of information technologies in the context of the informal sector at much earlier stage. Its organisational structure and reach gives it the potential to harness such technologies effectively. In order to optimally utilise the power of ICT at the grassroots level, SEWA has launched various integrated development programmes and has demonstrated the capacity to deploy ICT and mainstream it effectively (Nanawati, 2006). SEWA has pilot tested various ICT tools for poverty alleviation, micro-enterprise development and even disaster mitigation, with remarkable success. SEWA's ICT efforts focus on capacity building, livelihood generation and security, and knowledge sharing.

Use of Mobile Phones

SEWA trains women to use the mobile phones, giving them loan for purchasing it. Women use their phone to find the prevailing prices in the market for trading their produce. This gives them the power to negotiate with the local trader and an escape from the middleman. They have formed a cooperative like seasame seed producers. The cooperative now sell directly to the main market without involving any middle men. Women like Jesuben, are now recognised as business women.

Video-Conferencing

It has played a very important role in supporting trade related activities of its members. When the price of procurement had to be decided, members used telecommunication facilities to directly communicate with the Sewa Gram Mahila Centre. Video-conferencing was a very effective medium of addressing the community issues and questions, is a much simpler and effective way of sharing information and communicating in such cases. SEWA members can not only freely talk to one another and discuss the project, the visual medium that video-conferencing provides helps in better communication.

Access to Electronic Medium

Satellite communication facilities are also used by cooperatives in all districts for sharing of ideas and as for discussions and issue resolution. Sat-com, as a medium of communication, has become very popular as it is available in all SEWA district centres. These facilities are operated by the members themselves, the programme being held fortnightly at the centres.

Women were taught how to operate an electronic weighing machine to weigh milk etc. They say "People trust electronic gadgets and hence felt that electronic machines would never lie". Laxmiben does the largest sale of milk in her village. Her income has increased many folds. Other simple tools such as calculators have also helped the members. Women operate their own accounts whether it is in the SEWA bank or through SEWA's ATM facilities. The use of these simple tools makes a huge impact in the small micro-enterprises of these women. These are very simple but very necessary for any person who is linked to the market. Exposure to these tools removes a lot of unfounded fears (related to ICT). The removal of these fears has helped in introduction of more complex technologies.

Advancing Steps in Technological and Management Inputs

The plight of the salt farmers is known to all. They work in the salt pans that are very far from their homes and when they need to send a message, buy a commodity or a replacement for a broken part of their machines they have to travel to the village. ICT interventions have been helpful to these farmers not only in improving their trades but also in improving their quality of life. Salt testing laboratories have been set up in the villages for these salt farmers. These labs have been equipped with the latest testing technologies. Training is imparted to the salt farmers to improve the quality of their salt. They receive technical inputs on how to make industrial salts which fetch a much higher price than the edible salts.

Computer Training

SEWA organises computer trainings for the women and their children in the villages of nine districts of Gujarat. The trainings are all about exposure and introduction to very simple computer applications that can help them in their trades help them communicate with one another and ease a lot of burden in their lives. Many women are trained in Excel worksheets. Some use Excel as a planning tool and others use Excel for maintaining their accounts.

These women do not use very high end technologies but their acceptance of technology broadens their outlook. In the districts of Kutch where girls are not encouraged to study, computer trainings for adolescent girls have grained popularity. The training gives them a better vision of the future. Many who exhibit greater talent undergo further training and then become teachers and run SEWA's Sanskar Kendras.

5.0 Discussion

Dairy case of women cooperatives is towards strengthening the capabilities of women dairy farmers by organizing them into groups – Women Dairy Cooperatives This process served dual purpose in development–organizing the dairy sector for its growth and the path for women empowerment. Strengthening of their capabilities of ongoing dairy practices along with providing assured market and other facilities has led to direct access of money in their hands. The economic empowerment has led to them towards women empowerment and its positive impact on the socio-economic status.

The application of biotechnology (plant tissue culture) case illustrates reaching the so far un-reached women capabilities. Efforts of IHBT are commendable in showing faith in 'capabilities' outside the context in which women were living and training them with the techniques of new biotechnological applications. This case is a true example of bridging the gap between women empowerment approach (a welfare kind) and realizing the unrevealed part of women capabilities, a path towards equality to men by giving confidence and economic empowerment to them. Simple rural women through S&T intervention have proven that there lies the potential but need to be explored to the optimum.

Role of NGO (SEWA) emphasizes that women are no different from men. Only need is to consider their potential, strengthen the capabilities and give them opportunities to express them. SEWA is totally dedicated towards and showing results in all spheres of entrepreneurship and use of modern tools and techniques. Efforts are bearing results in terms of employment, increases earnings, confidence, entrepreneurship, decision making and exposure to outwardly world.

6.0 Conclusions

Cases discussed are far away from 'women empowerment' based on 'women welfare' approach. It is about building capabilities of women in different spheres keeping pace with S&T based developments meaningful to women for better economic gains. Thus, there is need for more government programmes and initiatives of voluntary organizations to work 'with' women, which is a way to empowerment more and more S&T based applications need to be directed to women groups in rural areas. An utmost need is there to realize the latent capabilities and to explore those to the fullest. This process of women empowerment has far reaching positive impacts on the development of the society.

References

Agarwal, B. 1995. 'Gender, Property and Land rights: Bridging a critical gap in Economic analysis and Policy' in *Out of the Margin, Feminist Perspectives on Economics* (eds). Susan Feiner, Notburga Ott and Zafiris Tzannatos, Routledge, London and New York, pp.123-129.

Anonymous, 2001. *National Dairy Development Board*, http://www.nddb.org (2 February 2003)

Chawla, P. 2002. 'Sustainable development through empowerment of women. Paper presented in the *Economic Empowerment of Women* seminar, NISTADS (CSIR), New Delhi, 18-19 February, 2002.

Deshmukh, R J, 2002. *Negotiating Spaces for Women Empowerment*. Paper presented in the *Economic Empowerment of Women* seminar, NISTADS (CSIR), New Delhi, 18-19 February, 2002.

Garibaldo, F, 2002. *The Capability Approach and the Experience of Working*. Paper presented in the *Economic Empowerment of Women* seminar NISTADS (CSIR), New Delhi on 18-19 February, 2002.

Mathur, K., 2002. 'Women's empowerment and Development: Theoretical concepts and operational challenges. Paper presented in the *Economic Empowerment of Women* seminar, NISTADS (CSIR), New Delhi, 18-19 February, 2002.

Mehra, Kavita and Kalra, R.K., 'applying the Action Research Toolkits: Women empowerment through Dairy Entrepreneurship. In K.S. Gill and A. Jain (Series edts.), D. Brandt (EDT.) *Navigating Innovations, Enterprises and cooperative networks for Regional development* Volume I, Indian Research Press, 2003, pp.259-294.

Nanavaty, R., 2006. Self-Employed Women's Association (SEWA), India Improving trade, securing livelihoods, with ICTs. www.i4donline.net (16 November, 2008).

Nussbaum, M C, 2000. *Women and Human Development-The Capabilities Approach.* Cambridge University Press, USA. pp. 195-198.

Pithadia, V. and Sharma, A., 2005. Empowerment of women: Indian Context. www.indiatiamba.com/faculty_column (14 November, 2008).

Sharma, Madhu, Sood, A. and Ahuja, P.S.: Facilitating entrepreneurship amongst rural women of Himachal Pradesh through biotechnological application. dbtindia.nic.in/paper14.htm

UNDP, 1995. Human Development Report 1995, Oxford University Press, Oxford

Young, K. 1993. Planning Development with women: Making a World of difference, U.K., MacMillan Press, U.K. pp. 10-11.

Chapter 15

Science and Women Empowerment in Iran

Shokouh Navabinejad
Psychologist and President,
Women NGO's Network in Iran, Tehran, Iran
E-mail: navabinejad@yahoo.com

ABSTRACT

Women are key for a national ability to develop education is fundamental human rights, and one of the most important instrument in a achieving gender justice. Equal access to education for men and women in all fields of studies is main objective of millennium development goals too. During the last decades Iranian government has taken significant intervention measure to ensure education for all at all levels. Including importance of women access to science and technology, increase number of technical and vocational school for girls off over the country as well as the increase the percentage of girls entering the university in science and technology fields. In this article women empowerment through science and technology as well as strategic objectives and obstacles in Iran, will be discussed.

Keywords: *Women empowerment, Gender justice, Science and technical education, Strategic objectives, University enrollment.*

1.0 Introduction

The Islamic Republic of Iran has always taken education extremely seriously and has had great achievements in promoting education for all and making every effort to eradicate illiteracy.

Education is a fundamental human right and one of the most important instruments in achieving gender justice and safeguarding the rights of women. Equal access to education for women and girls is one of the most emphasized areas in the Beijing Platform for Action and one of the main objectives of the Millennium Development Goals. The Government of the Islamic Republic of Iran while understanding the importance of education as a tool for the development and sustainable growth of its society recognizes that provision of equal opportunities for women and men in society is one of the factors that promote achievement of sustainable development. The Government has taken significant measures towards realizing this goal. One the most important measures have been to incorporate gender perspective programmers and policies in the 4th National Five Year Economic, Social and Cultural Development Plan. Intervention measures to ensure education for all at all levels; pilot projects and reproductive health education have been effective and far reaching, in view of the widespread campaign undertaken by the Islamic Government of Iran.

In parallel with these national strategies, implementation measures and national mechanisms have been established to institutionalize Government plans to promote literacy, such as the establishment of the Literacy Movement Organization of Iran (LMO) in 1979). This organization consists of two levels, adults education and education for children. In primary and secondary education are areas, which have not been easily accessible by the Ministry of Education.

The main targeted groups in this plan consist of rural and vulnerable women. The Bureau of Women's Affairs in the Ministry of Education has also taken effective measures to eliminate illiteracy and discrimination among women. In the 3rd and 4th National Five Year Economic, Social and Cultural Development Plans official advancement of women's education and technical and vocational training has been reiterated, particularly in rural and disadvantaged regions. Such measures have lead to an increase in literacy rates from 64.7 per cent for women and 74.2 per cent for men in 1996 to 81.7 per cent for women and 89.2 per cent for men in 2001.

This report has been prepared based on the strategic objectives of Beijing Platform for Action in the area of education and training.

2.0 Strategic Objective 1: Ensure Equal Access to Education

Government Measures

1. Promotion of education for women particularly in deprived regions.
2. Increasing the number of tribal girl students at primary schools from 41,609 in 1999 to 63.365 in 2003 and at secondary level from 4,009 in 1999 to 10,312 in 2003.
3. Decreasing the gap in the literacy rate between men and women from 23.4 per cent in 1976 to 9.5 per cent in 1996.
4. Implementing a vocational training programme for 1,400 early girl dropouts by the CWP.
5. Encouraging 25,000 girls to return to school to complete their elementary education by the Ministry of Education and Training.

6. Organizing literacy courses for female inmates by the Organization of Prisons, in three major prisons in the country.
7. Increasing the number of girls attending boarding schools from 96,449 to 156,067, a measure which provides access to education for girls who live in inaccessible regions.

Obstacles

1. Geographically dispersed and inaccessible regions.
2. High costs of private education.
3. Employing ineffective structural adjustment policies.
4. Inappropriate traditional discrimination with regard to enrolling girls at primary and secondary schools.

Adult education has covered 3,177,699 persons during the years 1999-2002 over 82 per cent of whom were women. Other effective measures for bringing education to rural and distant regions is strengthening the central rural schools that provide access for several surrounding villages, provision distance, part time, correspondence, media and home-based education, establishing satellite schools and providing free meals emphasizing education for girls, provision of school costs and transportation expenses to secondary and high school students, and developing mobile technical and vocational training in rural regions.

3.0 Strategic Objective 2: Eradicate Illiteracy among Women

Government Measures

1. Developing research projects on the necessity of literacy for women and girls through numerous projects.
2. Increase in literacy rate of women, which has reached 74.66 per cent at the end of 2002.
3. Teaching practical use of education through publishing several books by the Ministry of Education.
4. Establishing the Literacy Movement Organization (LMO) in 1979, which has been active in regions that have been inaccessible to the Ministry of Education.
5. Establishing follow-up groups for continuing education of newly literate people in 1994 by the LMO with the main objectives of eliminating a return to illiteracy and promoting cultural reading and self–education.
6. Basic technical and vocational education by the LMO.
7. Special focus by the LMO on rural regions in particular educating rural women to the extent that 81 per cent of their activities target women.
8. Covering 92.42 per cent of women by literacy follow-up groups.
9. Increasing human resources in the literacy movement to 1,287 persons as trainers and administrative staff.

10. Training 30,000 rural women by the CWP in cooperation with the LMO towards increasing political participation of rural women.

Obstacles

1. Girls dropping out of school owing to economic and cultural reasons and early marriages at secondary and high school levels.
2. Poor cultural perspective of parents and their negative attitude towards girls' education in primary and secondary schools.
3. Lack of pre-school centers and inadequate legal follow-up for mandatory primary education for children.
4. Insufficient role models of educated girls in rural and urban regions.

The percentage of literate women over six years of age in 2002 was 78.2 per cent, with an increase of 5.3 per cent since 1996 and 120 per cent since 1976. There have been higher literacy rates among men, but the increasing literacy rates or women during these years indicate that the Government has achieved its goal of reducing the gap between men and women.

During the academic year of 2002-2003, 47.9 per cent of primary school students consisted of girls. Girls constituted 45.8 per cent of students at the secondary level and 50.7 per cent at senior high school.

4.0 Strategic Objective 3: Improve Women's Acccess to Vocational Training, Science and Technology and Continuing Education

Government Measures

1. Increasing the number of technical and vocational schools for girls to 8371 and art schools for girls from 357 in 2000 to 540 in 2003, and 1092 in 2006.
2. Implementing self-sufficiency programmes in rural boarding schools for girls in 214 boarding schools in 2002.
3. Implementing a plan for developing and diversifying girls' skill training and technical courses from 72 to 80 at high school level, such as accounting, information and technology and tourism.
4. Providing computer training to women employed in the educational sector.
5. Training 886 women trainers in 28 provinces to develop skill training of girls.
6. Implementing an evaluation plan on diversification of skill development and technical courses for girls in 8 provinces to improve quality of skill development and eliminate discrimination.
7. Reform and curricula programming in skilled, technical, and vocational training.
8. Providing computer courses for education and training of female staff of the Ministry of Education.

9. Organizing in service training for capacity building among head mistresses in tribal regions.
10. In crease the percentage of girls entering various fields of study in science and technology. (8.9 per cent in 1986 compare to 59.1 per cent in 2006)

The 3rd National Economic, Social and Cultural Development Plan states indicates that technical, skilled and in service training of government employees has integrated a gender approach with equal training opportunities. The 4th national Economic, Social and Cultural Development Plan commits the Government to provide equal access to education, particularly in less developed regions, emphasizing the promotion of knowledge, skills and increased productivity of human resources, especially for girls. The Government is obliged to implement education for all by providing suitable facilities for mandatory primary education. The Head of the CWP is member of staff of the Coordinating Caucus of technical and skill development training, established as part of the 4th National Economic, Social and Cultural Development Plan's Coordination Centre.

Table 15.1: Percentage of Girls Entering Different Fields of Study

Fields of Study	*1986*	*1991*	*2003*	*2006*
Experimental science	44.3 per cent	48.6 per cent	72.4 per cent	82.9 per cent
Human sciences	44 per cent	31 per cent	67.5 per cent	75.5 per cent
Mathematics and technology	8.9 per cent	15.4 per cent	47.1 per cent	59.1 per cent
Arts	61.2 per cent	52.3 per cent	86.9 per cent	58.3 per cent

Table 15.2: Total Number of Permanent Staff at Universities and Educational Institutions

Year	*Women*	*Men*	*Total*
1998	4099 (18.5 per cent)	18095(81.55 per cent)	22194(100 per cent)
2002	5232(19.13 per cent)	21936(80.7 per cent)	27168(100 per cent)

Future Actions

1. Ensuring equal access to public education.
2. Reforming cultural perspectives in the importance of girls' education.
3. Implementing strategies to ensure equal access to education and elimination of gender injustice.
4. Reforming public perspectives in enforcing mandatory education laws.
5. Paying attention to cultural reasons for girls' early dropouts.
6. Development and diversification in skilled, technical and vocational training for women.
7. Encouraging girls in disadvantaged regions in the direction of mathematics, sciences, and providing possibilities for girls to study electronics, chemical industry and agriculture.

8. Strengthening and updating skill and knowledge training of teachers in formal, long-term and short-term courses.
9. Precise planning for qualitative and just distribution of facilities and privileges in employment and welfare of the educational sector.
10. Ensuring complete access for girls in using training agricultural training centres.
11. Monitoring the enforcement of mandatory laws in education.
12. Compatibility of educational environment with emphasis on girls' psychological characteristics and developing life skills.
13. Training necessary skills in relationships for girls in establishing healthy, social relationships.
14. Developing projects to promote schoolgirls' health in the 4th National Economic, Social and Cultural Development Plan.

The Government of the Islamic Republic of Iran has taken serious measures to eradicate illiteracy, nevertheless there are still illiterate women, particularly in disadvantaged regions where literacy rates are so low that it provides suitable grounds to be discriminated against and remain silent against all violence or discrimination. Therefore, it is the Government's duty in public education to eliminate discriminatory conditions and cultural barriers for education of girls and women. Although the enrollment of women is high in human and social sciences at higher education levels, an equal presence of girls in subjects such as science and technology, technical and professional courses is still fairly inadequate. In the 3rd National Development Plan with the emphasis on policies in eradicating deprivation and increasing the number of girls' boarding schools and further emphasis in the 4th National Development Plan in future, it is hoped that these obstacles will be overcome.

Developing technical courses and the possibility of attendance by girls in courses such as electronics, chemical industry and agriculture and striving towards providing grounds for necessary and effective cooperation among organizations is among forward looking strategies of the Islamic Republic of Iran. Despite the Government's numerous successful efforts to bring education up to international standards, we can still aim for future advancement in this important area of concern.

References

National report on: Women Status in The Islamic Republic of Iran., Centre for Women Participation. 2006, pp.21-29.

National Report to the World Summit on Sustainable Development, Network of Iranian women NGOs, 2005.

National report by Ministry of Science, Technology and Research, 2007.

Chapter 16

Content Analysis of Web logs: Perspectives of Young Iranian Females

Zohreh Nosrat Kharazmi
Faculty of World Studies,
Tehran University, Tehran, Iran
E-mail: zkharazmi@ut.ac.ir

ABSTRACT

We live in the world where our geographic distances do not affect our interconnectedness and interdependence. Advanced communications and information technology link us all over the world regardless of our backgrounds or reluctance to involve in the process of communications. A significant aspect of today's world is globalization which as Robertson signifies is the mutual give and take of the global and the local (Barker and Willis, 2003). Simultaneously, the process of globalization is current phenomenon not only in the physical world, but also in the virtual world. (Ameli, 2002) So, to "grasp the flux", which means to perceive the world as it is now in the flow and flux (Meyer and Geschiere,1999), is the key element to empower the nations and countries to be successfully presented in the contemporary world. Women being a major part of the human capital of any society are important factors to take part in this grasping of the flux. This study explores the process of empowerment of women by presenting themselves in the cyberspace, thereby providing a proper potential for the nation to globalize their national values with their local identity. The paper analyses the features of Iranian national identity represented in weblogs written by Iranian young females, aged 18-25 years.The content analysis of these weblogs depicts the faces of Iranian identity which are more significant for young Iranian women and is presented through the virtual world as a process of making them globalized. The weblogs which are written in English, as the international language to support the process of transference of meaning internationally, are taken into consideration.

Keywords: *Iranian females, Globalization, Weblogs, Cyberspace, Cultural engineering, National identity.*

1.0 Globalization and the National Identity

Globalization is the phenomenon that we are experiencing consciously or unconsciously, and it shapes our lived experience in various aspects of life such as the political, the social and, much more significantly, the cultural.

What are the essence and elements of globalization is the question never answered completely and not always on total agreement. There are distinct approaches to globalization, some extremely far in two extremes of the spectrum.

Janet Morrison describes globalization as a number of processes through which the "products, people, companies, money and information move freely and quickly all around the world and the national boundaries are not very restrictive" (Morrison, 2002). There are approaches which take globalization as opportunities to improve the world of communications and a suitable area to get and share its benefits together.

Some socialists such as Herbert Schiller assume that globalization is the synonym of Americanization. He claims that the American transnational companies export American commodities and culture simultaneously. He takes the commodity which is exported from America to other countries, the same as the culture it exports (Storey, 2007). Globalization may be perceived as a force that reduces masses power, destroys cultures, weakens democracy and poses Americanization on others (Wolf, 2004).

Notably, the active part of the receivers of the cultural products and their potential resistance has not been considered in such an approach. So, taking globalization at par with Americanization or Westernization does not give importance to the other non-Westerns who affect the West and each other in the process of globalization. Giddens distinguishes that globalization is "one of fundamental consequences of modernity". So the essence of globalization is mostly Western, coming from Western cultures, but Giddens indicates that globalization "introduces new forms of world interconnectedness", something beyond a simple process of Westernization, Americanization or McDonaldization (Hopper, 2007).

This paper focuses on the third way balanced approach. The advantages and disadvantages of globalization are determined in different ways due to different factors. Here, multilateral connections and determinations are significant while forces, particularly the economic ones, are not deniable to determine who will win or lose more benefits in long term. Elliot and Charles argue that "The differences that make the world`s worlds are real" (Elliot and Lemert, 2006).

The sub concepts and various respects of globalization are actually very integrated and Collier and Ong depict that the global forms can assimilate to new environments to code contexts and objects in terms which can be controlled and valuated. But it's also notable that these global forms are limited and delimited by the technical infrastructure, governmental and value regimes (Ong and Collier, 2005).

Influencing factors and outcomes of globalization are traceable in various trends. But "culture is arguably the most direct way in which we experience globalization". Globalization processes affect the continuity or disruptions of cultures and "cultures are not immune to the processes of globalization" (Hopper, 2007).

Globalization does not necessarily mean that we are facing a dominant globalized culture. The cultures are globally in circulation and flow. Thus, Featherstone argues that the "globalization of culture" is the process under which "cultural integrations and disintegrations occur". We do not have a "global state" to own a global culture (Featherstone, 1990).

The globalized integrations and the blurring of national boundaries are parts of the globalization of culture. To be able to successfully manage the national cultural engineering we should know the globalization of culture quite well. And the great part of this understanding, say its corner stone, is to get the sources on which our distinct faces of national identity depend.

Historically, the national identity was constructed and defined by our attachments to the locality and place we lived in. It does not look as if identity formation, nowadays, takes place in the same way. Samuel S. Kim believes that the new phenomenon we encounter today is "Identity Surfing", that is, the experience of "multiple collective identities" which makes the individuals able to shift regularly from one group identification to another. If the national identity was once related to loyalty to land, tribe and family, the contemporary globalized world provides colorful loyalties, including "ethnic, gender, religious, cultural and linguistic, regional and even brand loyalty" to slide from one to the other specifically in different contexts (Kim, 2006). The unified, large-scaled collective social identity is broken today. It does not mean that the collective social identity faded out completely, but it lost the considerable homogeneity it had (King, 1991).

The second part of our understanding of the globalization of culture is to understand advanced communication and information technology which accelerates the processes of "Surfing" from one to the other identity. Communications and information technology specifically manoeuvering in cyberspace gave the chance to anybody, in any place, with any race and gender to present herself / himself under different masks that was totally impossible for her/him in the real world. Synchronization and Trans locality of cyberspace will be the concentration of the study which provides the opportunity to experience parallel times and places with a click in Internet, which could never be done in the physical world.

2.0 Globalization of Culture and Trans-National Identity

The cultural dynamics of globalization emerge from the interaction of numerous processes, flows, networks and interconnections, which are interpreted, experienced and contributed to by different actors and agencies in a range of cultural, political, and social environments and localities.

Such contexts may be regional, national, local, religious, institutional, historical and so forth, while the extent and nature of our engagement with these forms of globalization are similarly important. So, the heterogeneity and complexity of pervading tendencies of our globalization is not surprising. Globalization should not be viewed in the abstract; we experience its processes in tangible ways (Hopper, 2007). [Where does "begin? Clarify the starting sentence in quote.]

The globalization of culture may happen in various scales specifically "trans-national" and "trans-societal" ones through processes of exchange and flow of "goods, people, information, knowledge and images". This process of exchange is the fruit and simultaneously the amplifier of communication processes which on the global level has taken some autonomy (Featherstone, 1990).

As explained in the last section, the interconnectedness and interdependence, as an essence of globalization, are the features to be focused on. The features accelerated by new technologies, specifically in cyberspace, also cause new complexities of identity and the process of identification.

Therborne (2000) defines globalization as:

1. "A worldwide reach and connectedness of social phenomena," and
2. "A world-encompassing awareness of actors." In other words, a global spatialization of the social reality" (Ben-Rafael and Yitzhak, 2002).

In this definition, two points are attractive: first the "connectedness" and second "awareness of actors" about this mutual relationship. Then, representation of national identity in cyberspace as the core of the argument includes both cases. Females are connected to an extremely open-ended world in cyberspace and they alertly are participating in such a space to represent different aspects of their identity, including their national identity.

The constructing factors of globalization are beyond the scope of this paper. They are diverse and complicated as Anthony Giddens and James Rosenau depicts that the current political, economic and cultural flows and high levels of global interconnectedness root us in a combination of forces of "identity, capitalism, industrialization, the nation-state, technological and scientific developments, critical thinking and so on" (Hopper, 2007).But, the outcomes and results of globalized connections are eloquent about the presence of women of my country as they are representing themselves in cyberspace.

Bartelson (2000) believes in three consequences of the interconnected globalized world:

1. Transference; the interaction and exchange among social settings are intensified;
2. Transformation; the outcome of transference which is the phase of "emergence of changes at the level of the local"; and
3. Transcendence; the result of the previous phase, that is, "dissolution of the divide between inside and outside" (Ben-Rafael and Yitzhak, 2002).

Before the era of globalization, the geographic place and the cultural experience were so much intertwined that the "local, autonomous, distinct, well-defined and robust" cultural connections were absolutely meaningful. Then, identity was a "cultural belonging" which resourced from an "undisturbed existential possession", a kind of "continuity with the past". (Tomlinson, 2001). Globalization has changed different layers of identity. As a main component of our individual and collective identities, the national identity is also affected. "The previously unquestioned and direct links between identity, nation and state are now less clear" (Kim, 2006).

According to Pilkington, national identity is a social construction which has the potential to change and to develop. The best response to this identity in the globalized world is to accept "hybrid" or multiple identities (Giddens and Griffiths, 2006). Featherstone and Lash believe in the triumph of universal "disembedding", characteristic of "the informational order", the discourse that the global framework is much more decisive than the nation-states (Ben-Rafael and Yitzhak, 2002). Globalization has compromised the national identity's capacity "to maintain exclusivity of identity attachment" with its sharing in global market and the tensions and complexities introduced by multiethnic societies. The proliferation of new kinds of identity makes problems for the nation-state's hegemony over its people`s cultural attachment (Tomlinson, 2001).

Kelly argues that globalization has deconstructed any "naïve" understanding of locality and opened the horizons of investigation toward "creole culture, transnational movements and emerging diasporas" (Meyer and Geschiere, 1999).

John Kelly indicates that "to grasp the flux" requires us to correct the traditional views that "the world is a collection of nameable groups". And Appadurai argues that the social scientists have to struggle to fix "practices, spaces and countries into a map of static differences" (Meyer and Geschiere, 1999).

Thus, what we observe today is the experience of "a liquidation of the concept of culture", that is, culture is becoming more and more apart from "a play of pure forms and structured relations". For Fabian, culture is a process and is considered as the "dynamics of a 'developing code'...undoubtedly characterized by contradictions, oppositions and contrasts" (De Lame *et al.*, 2005). With the liquidation of culture Stuart Hall says the identities which "are about questions of using the resources of history, language and culture" are in "the process of becoming rather than being" (Storey, 2007).

The interaction of global forms makes "global assemblage" which essentially implies mobility, heterogenousity, instability and partiality in it (Ong and Collier, 2005).

Collier and Ong argue that these global phenomena have the capacity for "decontextualization and recontextualization, abstract ability and movement across diverse social and cultural situations and spheres of life" (Ong and Collier, 2005). Collier and Ong consider even ethnicity, as a significant cultural feature, whereas today it is something different from the traditional one. They believe that ethnicity today is not a localized model of a "place/people-bounded", and the "macro-historical and social systematic forces" are unreliable.

With globalization, the concept of "others" which was a main factor to define one's ethnicity is not effective any more and "others" are, rather, our counterparts. The "native point of view" also missed its meaning (Ong and Collier, 2005).

The paper analyses how national identity is affected by the globalized interconnectedness and "globalization of culture", while one of the most significant factors in the process is the communications and information technology and more significantly the cyberspace. Cyberspace has the potential to accelerate the process of "globalization of culture" particularly due to its three characteristics: translocation, synchronization and extremes of choices. So, the flux in cyberspace can be a vital cause of continuity and ability to globalize the identity of a nation.

3.0 Cyberspace and Iranian Cultural Engineering

The phenomenon we are facing today is not the process of a global state with emergences of a global culture. On the contrary, as Robertson distinguishes this relationship is bi-or even multilateral. We as a nation are influenced by the global features, which itself is hybridized, and simultaneously are able to affect the global phenomenon with our distinguished characteristics. Therefore, the management of this process is the art provides us the opportunity to have the creative point of globalizing our local values and practices.

It was in November 2001 when Supreme Leader of Iran forwarded, as part of national policy, the thesis of Iranian Cultural Engineering that comprised of the following ideas:

- Multilateral view of Iranian policy makers towards globalization and its different aspects
- Iranian national identity to be effective in the globalized world
- The capability of programming the culture in the long term. (http://www.iranculture.org/about/rahbar/mrahbar/b01-20.php, 2009)

Ayatollah Khamenei tried to attract attention to different levels of culture to be programmed: universities and educational system, youth culture, cyberspace, science and research and arts. (http://www.iranculture.org/about/rahbar/mrahbar/b01-20.php, 2009)

Many Iranian scholars have tried to interpret the Iranian Cultural Engineering and its potentialities to make the Iranian national identity stronger to resist the cultural imports of others and to be able to export its noble ideas to other nations and cultures.

Mohammad Nahavandyan, ex-president of National Centre for Globalization Studies, believes that the idea of Cultural Engineering had two layers: looking at culture as an organized objective system and an active approach to engineer the culture. He argues that in the globalized world the culture that is able to survive is the one that is capable of managing the rapid "dynamism" of its social, economic and cultural domains (Nahavandyan, 2007).

Nahavandyan also distinguishes three levels of cultural engineering as national, regional and global. He argues that the main components of cultural engineering are the areas which define our cultural identity and particularly plan the map of our national identity. He suggests seven areas to make an efficient cultural engineering as management of efficient researches.

- Ability to manage researches in internet
- Protecting and improving creative arts with regards to maintaining the intellectual property

- Developing immigration policies
- Taking Farsi language as the national language but prevent putting it as the opposite of others
- Improving different languages in the society that improves our whole global communications
- Improving human capital in the family institution
- Managing efficient institutions and infrastructure to bridge the intercultural gaps within the Iranian society (Nahavandyan, 2007).

Saeid Reza Ameli, the dean of Faculty of World Studies, University of Tehran and the theorist of "dual globalization" also believes that Cultural Engineering specifically becomes significant while considering globalization in cyberspace. He depicts that the society with a well-organized cultural engineering is able to win the intercultural gaps and conflicts. Ameli indicates that the "Love of the nation-state" is the key to make the country able to engineer the culture well. For him, immigration today does not need any physical travelling, because people immigrate virtually when a nation is not interested and absorbed by its system and identity. In the long run, this reduces the efficiency of the national culture and identity (Ameli, 2007).

Ameli is emphasizing that for long term and even with high expenses, the country that gives a narrative of itself based on honesty and multilateral angles of its advantages and disadvantages, avoids any fake representations like many contemporary countries and media, and will be the winner of the trust and loyalty of its nation to their national identity and maintain the people's pride in their national identity and humanitarian approaches (Ameli, 2007).

Cyberspace has had various services to process of globalization. The features of cyberspace are fantastically expanded, as well. But, from many characteristics, there are translocalization, synchronization and widespread choice. Particularly in the analysis of web logs written by Iranian young females these are highlighted when Iranian young females are representing different aspects of their national identity in a new space of the cyber.

Translocalization is defined as "the processes by means of which migrants in their host societies build up new forms of locality and communality based partly on their own cultural backgrounds and partly on their opportunities and sociocultural contexts of the host society in terms that interconnect the different localities" (Kupiainen *et al.*, 2004). The same meaning today is applied in cyberspace. The potentiality to make new communities and to give new meanings to the concept of the local is increasing in the fresh space of cyber.

Cyberspace provided various opportunities for women to do business improve their political participation and fight for more equality (Armitage and Roberts, 2002). Many women who are deprived of their individualistic freedoms support the new contextualization in cyberspace in which they are accessing such freedoms.

In cyberspace, as Bernard Sharratt reminds us, there are also essentially ambivalent images which take shape in the "dual sense of both representation and misrepresentation" and both "reflection and fiction" (Robins,1996). Thus, women are able to compensate part of their real world misrepresentations, isolation and limitations in cyberspace to make new images and communities. Network communications is much easier to be managed in cyberspace specifically internet which is nowadays accessible to more and more women. Translocality helps women to experience common experiences of different places and localities through cyberspace. This is while the time is compressed in cyberspace and synchronization lets women just click and share and exchange their ideas and ideals with others from different time zones and even ages. The translocality and the synchronization are themselves the sources of choices, a kind of infinite choices, to select the components of their identity, find the similar minds in Cyberspace and cooperatively work to strengthen that identity.

Part of the identity is the national one which can be narrated and distinguished from the "other" in cyberspace. A great deal of cultural engineering, as the national policy of Iran, is applicable in Cyberspace. Women would be empowered as they have the capacity to present themselves and their national identity through cyberspace that will greatly influence the process of globalization of Iranian local values and national identity.

4.0 National Identity Contents of Young Iranian Females on Weblogs

The author went through the list of Iranian weblogs, with locations inside and outside Iran, as depicted in the website: http://blogsbyiranians.com/, with a total of 355 ones in the list. Of these, only 10 writers identified as female 18-25 years old. In these 10 weblogs, all posts sent from 2004-2009 were thoroughly analyzed, which reached 1780 posts.

These posts were analyzed by content and of two levels:

- The national identity and
- Iranian women identity as reflected in the Weblogs.

In the first level, the significance of Iran land and nature, national identity and culture, potentiality of Iranian culture to have efficient international and inter-religious dialogue and interactions, Iranian customs and style of life, Iran history and historical architecture and Farsi language were taken into consideration. Some sub-categories such as Iranian food or family life which were appreciated by writers were set into a category like Iranian customs and style of life.

Iranian young females are representing their national identity components in various posts. For instance, Narges, 25 years old, writes "I was pretty much grateful that I had been born in Iran and have grown up there", or Iranian Girl believes "Iran is moving toward the goal of equality, justice, freedom, welfare, education, strong economy, etc". There are posts which express the ability of Iranian women specifically; due to language learning to have efficient communications like Elham Binai writes "Basically, we are all the human beings with the same potential to be a good or bad human being. The important thing is to realize the positive and negative aspects.

Then try to increase the positive aspects and reduce the negative aspects". Similar phenomenon is depicted in literature, music, cinema, arts, and the historical background and architecture. Elham Binai depicts that "Iranians are proud of their poetry heritage...if you are non-Iranian, please do research (about) Saady..." or about its historic cities Iranian daughter says "Whenever I walk there (in Shiraz) I feel glorious and proud..."or "I'd like to write about Cyprus the Great ....Cyprus chart of human rights...".Most webloggers are appreciating Iranian customs such as Norouz, Yalda, etc. Antidrown asserts that "Norouz has been celebrated for at least 3000 years and is deeply rooted in the rituals and traditions of the Zoroastrian religion" or Iranian daughter believes "Iranian along with the nature celebrate birthday of spring". Farsi language is appreciated by many webloggers specifically those who live abroad. Unitari writes in her weblog "I've noticed that Farsi is much less biased than English...you don't differentiate between a third person (she) and a third person (he)" or Spreading Love in Ripples, a 20-year-old resident of Canada, thinks "Speaking of which (Farsi), I'd like to write a story 'The Glossary of useful Persian Terms and Phrases'..."

Out of 1780 identified posts the number of posts with references to Iranian national identity is 215. The graph below is presenting the percentage of posts relating to the first level we discussed.

The second level of analysis relates to the Iranian women identity reflected in weblogs. At this level, the significance of activities and movements to improve the conditions of women, especially women's education and ultimately the professional situation and emphasis on the women's employment are extracted from the identified posts and analyzed by content.

Antidrown, a 20-year-old resident of US, narrates the activities of women in Iran such as "There are many NGOs all over the country that help women and push for women's rights". She further narrates "When some American friends see an Iranian woman in a scarf, they see a weak woman...just look into the eyes of an Iranian woman ... That's audacity...audacity is righteous and self-confidence".

In addition to all these impressions in weblogs, what is very much significant is the feedback on women education and improvement of skills which lead to self-confidence and development of women's rights. An Iranian girl asserts that "I've been always interested in knowledge and science". Improving different skills, such as driving, tour management, calligraphy and specifically learning languages are all named as skills to pave the way of women progress. Iranian girl supposes the language skill as the accelerator of "touch with different ideas and thoughts". The percentage of posts with the sense of appreciation of education is significantly high while it does not look to be just with the purpose of getting a good job or income, but the essence of education.

About jobs, the independency and the opportunity to apply the skills and the ultimate fortunes they make, gives much significance to them. Antidrown eloquently says "This is what a feminist look like: I'm open-minded, I'm strong, I like Iran, I'm artist, I'm engineer, I like languages, I like to be perfect, I like to know everything, I like to be a PhD student, I'm getting married this summer".

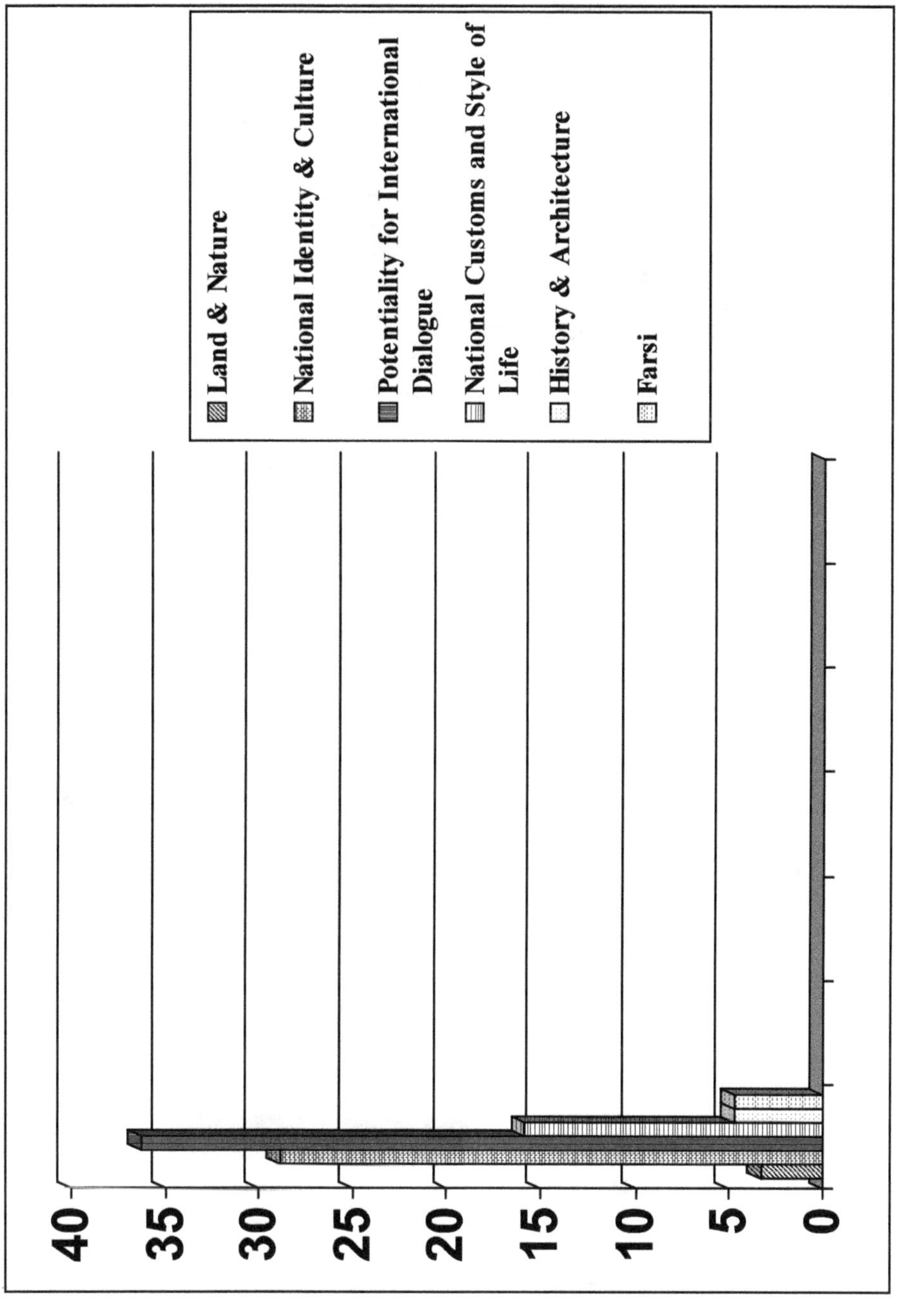

Chart 16.1: Iranian National Identity in Weblogs

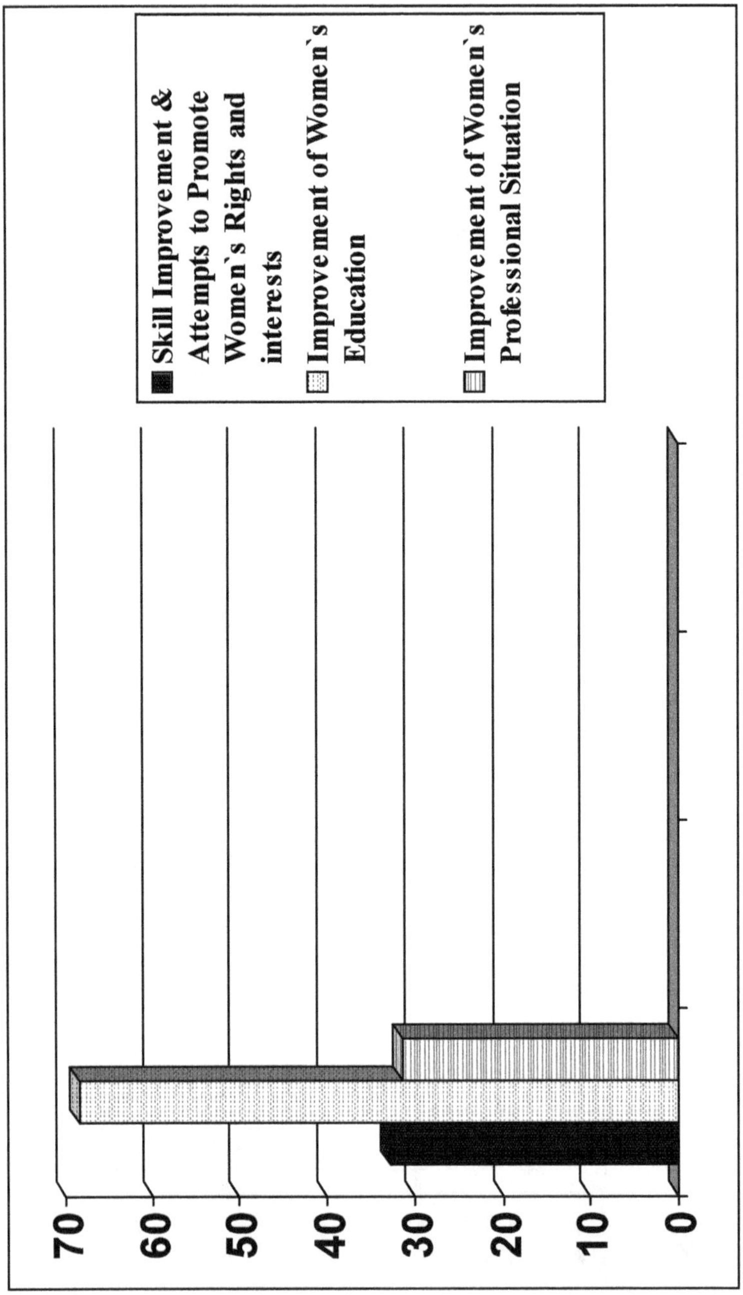

Chart 16.2: Significance of Women Affairs in Weblogs

As mentioned the total number of identified posts is 1780. The number of posts with references to Iranian national identity is 73. The graph below is presenting the percentage of posts relating to the second level we discussed.

5.0 Conclusions

The dominant narratives of the globalization in Iran have usually had a left socialistic and Marxist entity and globalization has been taken as the neocapitalism which is very much pro Americanization rather than a multilateral opportunity to globalize our localities (Ameli, 2007). So, it is very controversial inside Iran and similarly outside it, to investigate whether any country is losing its localities to Americanization or it has the chance to maintain it cultural sovereignty inside its borders and to globalize its local values and thoughts in the globalized world.

In the present study, I was to communicate that empowering the Iranian woman to present herself in the new worlds in the age of globalization, via the modern technologies of communication, is a great space to introduce and to globalize her Iranian values.

Iran that is passing from a traditional to a modern state can take the advantage of its woman's ability to answer for her needs and wants in new spaces such as the cyberspace to globalize her identity and values. Iranian woman, simultaneously, is offered the opportunity to have more choices to narrate her new forms of identity in a new bricolage of her tradition and modernity.

In analyzed weblogs, I came to the conclusion that Iranian young females, beyond their localities, utilize translocalization, synchronization and widespread choice of the cyberspace and represent their sense of agreement or their criticisms on various features of being an Iranian female. The emphasis they put on women education rather than just having jobs and gaining money and the belief of the potentiality of Iranians for international dialogues were the most significant characteristics of their posts. Thus, women while active in social spheres are efficiently influencing the future cultural engineering of Iran both inside and outside borders. It seems from the weblogs of the young Iranian females that a kind of new world based on Iranian traditional culture, including arts, land, language, architecture, customs and life style, with an articulation of more modern criteria such as the promotion of the women's education, rights and jobs looks the new ideal map of identity for the future Iranian female life.

References

Ameli, S., 2002. Globalization, Americanization and British Muslim Identity, London, ICAS Press, 138.

Ameli, S., 2007. The Very Challenge of the Cultural Engineering Is Taking the Only Narrow Pathological View, Cultural Engineering and Globalization, Vol. 10 and 11, 21 (1), 25(2).

Armitage, J., Roberts, J., 2002. Living with Cyberspace: Technology and Society in the 21st Century, London, Continuum International Publishing Group, 111.

Barker, C., Willis, P., 2003. Cultural Studies: Theory and Practice, SAGE, London, 174.

Ben-Rafael, E., and, Yitzhak, S., 2002. Identity, Culture and Globalization, Leiden, BRILL, 7(1), 8(2), 8(3).

De Lame, D., Arnold, H., 2005. A Hill among a Thousand: Transformations and Ruptures in Rural Rwanda, Wisconsin, Univ of Wisconsin Press, 21.

Elliot, A., Lemert, C., 2006. The New Individualism, the Emotional Costs of Globalization, London, Routledge, 154.

Featherstone, M., 1990. Global Culture, SAGE, London, 1(1), 1(2).

Giddens, A., Griffiths, S., 2006. Sociology, Oxford, Polity, 868.

Hopper, P., 2007. Understanding Cultural Globalization, Oxford, Polity, 9(1), 2(2), 3(3), 9(4).

http://www.iranculture.org/about/rahbar/mrahbar/b01-20.php (accessed 15 Dec, 2008)

Kupiainen, J., Sevänen, E., Stotesbury, J., Knuuttila, S., 2004. Cultural Identity in Transition, New Delhi, Atlantic Publishers and Distributers, 352.

Kim, S. S., 2006, the Two Koreas and the Great Powers, Cambridge, Cambridge University Press, 36(1), 35(2).

King, A D., 1991. Culture, Globalization and the World System, Minnesota, University of Minnesota Press, 45.

Meyer, B., Geschiere, P., 1999. Globalization and Identity, London, Blackwell Publishing, 3(1), 3(2), 3(3).

Morrison, J., 2002. The International Business Environment, New York Pal grave, 31.

Nahavandyan, M., 2007. Cultural Indicators for Global Competition. Cultural Engineering and Globalization, Vol. 10 and 11, 13(1), pp.14-15.

Ong, A., Collier, J. S., 2005, Global Assemblages, Blackwell Publishing, London, 11(1), 12(2), 11(3), 248(4).

Robins, K., 1996. Into the Image, London, Routledge, 7.

Storey, J., 2007, Inventing Popular Culture, London, Blackwell Publishing, 112(1), 80(2).

Tomlinson, J., 2001. American Council of Learned Societies, Cultural Imperialism, London, Continuum International Publishing Group, pp. 88-91(1), 271(2).

Wolf, M., 2004, why Globalization Works, London, Yale University Press, 13.

Chapter 17

Economic Empowerment of Malaysian Women through Science and Technology: Identifying Issues, Providing Solutions

Poh Sze Choo
147 Cangkat Delima Satu, Island Glades,
11700 Penang, Malaysia
E-mail: p.choo@cgiar.org, pohsze@gmail.com

ABSTRACT

In Malaysia, girls no longer lag behind boys in access to education right up to the tertiary level although women's enrollment in technical and vocational schools is still low. Women's entry as employees in the professional and technical category has increased from 28.2 per cent in 1957 to 44.6 per cent in 2000. The rate of increase in women employees in this category was even faster than the increase in male employees especially from the 1980s onwards. However, very few women are in top managerial and decision-making positions; in the year 2000, only 10.4 per cent of the 3.23 million working women in Malaysia were in this category. Although Malaysian women in science have progressed tremendously in economic advancement since the 1950s, women still encounter challenges which restrain them from greater economic contribution. Among these challenges are: sex stereotyping; attitude towards women and work; male domination and gender discrimination at the workplace. These challenges can be overcome by: creating awareness on the importance of gender equity; avoiding stereotyping; making the workplace more female friendly; providing role models; promoting image of successful women in science; recruiting male advocates; providing mentoring and fellowship facilities and making the teaching of science more female-friendly.

Keywords: *Women in science, Empowerment, Challenges, Sex stereotyping, Gender discrimination, Cultural norms, Solutions.*

1.0 Introduction

In Malaysia today, girls have equal opportunities and choices in education as compared to boys.

This is due largely to the Malaysian government's commitment to education and their investment in an "Education for All" programme since the 1950s. By 2005, Malaysia had already achieved gender parity fulfilling the target set in the Millennium Development Goals of the United Nations (UNICEF, 2005).

Malaysia has continuously been investing in education in her 5-year development plans. The proportion of development expenditure on education more than doubled between 1980 (7.5 per cent) and 1990 (15.3 per cent) and then nearly doubled again by 2003 when the amount increased to 25.9 per cent (UNICEF, 2005). The programme to provide universal education has been so successful that enrollment of girls in primary, secondary and tertiary education has now caught up with boys and in some instances even better than the boy's. Despite the improved educational attainment of Malaysian women, gender differences still exist in enrollment in science and technical subjects, and the education they receive has gender stereotypes that perpetuate gender inequality (Ministry of Women and Family Development, Malaysia, 2003 abbreviated to MWFD in subsequent references).

Although girls and women do not lag behind boys and men anymore in education, there are still some gaps when it comes to careers, promotions and retaining women in some science related careers. Women in science have made considerable progress in the past two decades, but women holding top positions in science are too few to count. This lag is due no longer to education but to some myths and glass ceilings that women need to shatter.

With an economy increasingly based on technology and science, we must maximize the talents of both women and men. Women bring unique perspectives and ideas different from men's. With the combined skills of both women and men, we can better address societies' challenges with greater creativity and innovation.

This paper examines some of the Malaysian statistics on the educational status of girls and women as well as the economic contributions of professional women especially those in scientific endeavours. Myths regarding women's ability in science and the glass ceiling that blocks women's upward mobility in the workplace are reviewed. The paper attempts also to provide solutions to enhance women's contribution to science, and to retain women in science in the workplace.

2.0 Enrollment of Students in Malaysian Schools and Universities

The enrollment of female students at the primary school level had increased by more than 3½ times during the period 1957-2000 (MWFD, 2003). In 2005, the primary net enrollment/attendance ratio for girls was 95.3 per cent and 95.1 per cent for boys, indicating more girls than boys were completing their primary education (UNICEF, 2005). In secondary schools, the enrollment of women increased by more than 36 times to reach a total enrollment of 985 692 students in 2000 (MWFD, 2003). However, women's enrollment in technical and vocational schools was still low, comprising 30 per cent of the total enrollment (MWFD, 2003).

In 1959, female undergraduates comprised 10.7 per cent of the total student enrollment in the University of Malaya which was then the only local university in Malaysia, but increased to 51.3 per cent of the total enrollment in local universities by 2000 (MWFD, 2003). In 2001, women constituted 52 per cent of the bachelor's students in computer science and 65 per cent in information technology in the University of Malaya (Lagesen, 2008). For the master's course, women comprised 43 per cent of the enrollment and 39 per cent in the PhD course. Lagesen noted that although the enrollments appear to look like a leaking pipeline, it may not be since many of the male PhD students were non-Malaysians.

3.0 Malaysian Women in Science Careers

Women's entry as employees in the professional, technical and related workers category (which includes engineers, scientists, doctors, lawyers, accountants, technicians, teachers, nurses, social workers, performing artists and religious workers) increased from 28.2 per cent in 1957 to 44.6 per cent in 2000 (MWFD, 2003). The rate of increase in women employees in this category was even faster than the increase in male employees especially from the 1980s onwards (MWFD, 2003). The MWFD report shows that the gender gap seemed to have narrowed tremendously among male and female professionals in selected services. In 1996, about 54 per cent of the professional staff in legal firms were females while for accountants the male-female ratio for professionals was 5.1: 4.9, and for dentists 5.6: 4.3. Females comprised 30 per cent of the doctor population. There were very few female engineers, surveyors and architects in the 1996 survey. In the year 2000 only 10.4 per cent of the 3.23 million working women occupied top managerial and decision-making positions (Mohd. Ariff, 2006).

Data collected in 2003 from Universiti Teknologi Malaysia, Kuala Lumpur campus showed that the ratio of women to men engineering educators was 1: 3 (Abdul Majid *et al.*, 2004). In their report, Abdul Majid *et al.* (2004) documented that the core engineering departments were dominated by men and in some departments (Marine Engineering Department; Remote Sensing Department; Microelectronics and Computer Engineering Department; Geomatic Engineering Department) there were no female engineering educators. Women educators were mainly employed in supporting departments such as the Administration and Services Department, Science Department and the Preparation Centre for Japanese Technical Course):

- Civil Engineering Department: 8 women; 9 men
- Electrical Engineering Department: 28 women; 25 men
- Mechanical Engineering Department: 7 women; 16 men
- Administration and Services Department: 40 women; 17 men
- Science Department: 37 women; 13 men
- Preparation Centre for Japanese Technical Course: 6 women; 2 men.

Women comprised a great majority of the key administrators and computer science faculties in two major universities in Malaysia (Othman and Latih, 2006). The staff profile in 2006 in the Science Faculty, University of Malaya is shown below (Dr. Sasekumar, pers.com.):

University Malaya

- Vice Chancellor (VC): Male
- 4 Deputy VCs: 3 Males; 1 Female

Faculty of Science

- Dean: Male
- 4 Deputy Deans: 2 Males; 2 Females
- 5 Heads of Departments: 2 Males; 3 Females
- Professors: 10 Males; 12 Females
- Assoc Professors/Lecturers: 19 Males; 21 Females.

The staff profile indicates that at the highest level of management (VC and Deputy VC), the ratio of male to female staff was 4: 1. There were slightly more women among the teaching staff (Professors, Associate Professors and Lecturers) with a ratio of 36 females to 31 males. The strong representation of women as Deputy Deans and Heads of Departments places University of Malaya ahead of many other tertiary institutions in the Islamic world. In the Middle Eastern countries, a study conducted by Hassan (2000) showed that "few women in universities and research institutes are presidents, deans, department chairs, directors of institutes, or heads of divisions or laboratories".

4.0 Challenges to Advancement of Women in Science

Although Malaysian women have progressed tremendously in economic advancement since the 1950s, they still encounter challenges which restrain them from greater economic contribution. Among the bottlenecks are the following:

4.1 Sex Stereotyping

Fewer Women than Men Succeed in Science and Math Careers Because of Innate Differences

The stereotyping of women and their innate inability to do science and math has adverse implications, and may doom women to fit the mold cast out for them. Lawrence Summers, the immediate past president of Harvard University created an uproar and intense debate over the above stereotyping.

It is generally accepted that men and women show no differences in general intelligence. The perceived difference in the innate ability of men and women in science can be due to many reasons.

According to Bob Triver, differential parental investment in their children may create highly specific predictions about when you should expect sex differences and what they should look like (Pinker, 2005). Nancy Hopkins, a biologist at Massachusetts Institute of Technology did not argue that there are no differences between the abilities of men and women but said that social factors do affect women's performance (Bombardieri, 2005). A report in The Economist (2006) reported that males and females of any age are equally good at computation and understanding of mathematical concepts and that men and women are equally good at navigating but using different styles.

Let us take the case of computer science/information technology education and career as a case in point and see how people fit into the stereotype mold created for them. In many western countries, the low enrollment of women in computer science courses is a well-known problem (Lagesen, 2008). Lagesen noted that these countries coded computer science as a "masculine" endeavor, and this seemed to affect the number of women taking up a "masculine" subject. In Malaysia, however, computer science is not tagged as "masculine", but is deemed as a suitable course and career for women because it is associated with office work which is perceived to be "safer" then work located outdoors. It can be seen from here that it is not the lack of innate ability of women to handle the complexities of computer science, but rather it is the gendering of computer science as "masculine" which makes it so formidable for women in the west.

Women computer science students/professionals interviewed in Malaysia perceived engineering (especially civil engineering) as "masculine" because engineers need to work a lot more outdoors which women perceived as dangerous, and women prefer not to work under the hot sun and with workers who are men (Lagesen, 2008). Women interviewed by Lagesen did not doubt their ability in handling engineering courses but instead did not choose the course because they did not find the working life of engineers to their liking.

4.2 Attitude Towards Women and Work

There are still many conservative Malaysians who believe that women's place is in the home and their primary responsibilities are domestic and child-bearing duties. Women receive little help from their spouses in terms of these duties and working women are often referred to as "superwomen" who have to juggle between work in the office and work at home. The dual workload of women often produces stress which may undermine women's health, their work performance and family harmony (MWFD, 2003). Very few women in Malaysia pursue a career path to become successful professionals, and many opt for early retirement once their husbands are financially sound and can afford a comfortable family life (MWFD, 2003). As aptly put by Professor Elizabeth Bartholet, "women who have made it to the top have overwhelmingly surrendered family and having children while men who have made it to the top have not, because they have spouses who are taking care of the family (Levine, 2005).

4.3 Male Domination and Discrimination at the Workplace

Jobs that are traditionally dominated by men like engineering, surveying and architecture may be deliberately avoided by women. In most fields of science, work is almost always done in a group.

When the workplace is dominated by men, the subtle gender differences in working styles, the subtle cues, and unspoken standards of performance may create tensions for the minority women in the group (Scientific American, 1998). As for computer science as a career, the larger number of women in computer science in Malaysia seemed to make the culture at the computer workplace more welcoming to women (Wajcman, 2004).

The glass ceiling blocks women from career advancement, especially when it comes to promotion to higher level posts (MWFD, 2003). Glass ceilings are a result of negative gender stereotyping and prejudices against women which may give rise to gender discrimination. The few women who have broken through the glass ceiling are exceptional examples; there had been cases where frustrated senior public service women employees had taken optional retirement when unfairly bypassed for promotion (MWFD, 2003).

5.0 Recommendations to Empower Women in Science

5.1 Creating Awareness on Importance of Gender Equity

Malaysians in general are still not very aware of the concept of gender equity and why it is important to work towards one. Many women, especially those from the older generation have low self esteem and have a self-sacrificing attitude. The cultural norm views men as head of households and women hold subservient positions. Men and women must recognize that women have rights to many of the things that are currently denied to some of them, for example equal pay for equal work, education and training, and the right to asset ownership.

From an early age, boys must be given training on domestic chores so that in manhood they are able to share the housework in the family. Even men too must be sensitized to help their spouses and lessen the latter's workload. Programmes that encourage the participation of both men and women can help to increase communication between the sexes and encourage a more even division of childcare responsibilities (UNICEF, undated).

5.2 Avoiding Implicit Gender Biases

Employers must ensure that the workplace adopts a gender-sensitive culture. As employers or employees, we must try also to get rid of the implicit gender biases which most of us harbor. On average, managers are less likely to hire a woman than a man with identical qualifications, are less likely to ascribe credit to a woman than to a man with identical accomplishments and, when information is scarce, will more often give the benefit of the doubt to a man than a woman (Anon., 2006).

5.3 Making the Workplace more Gender Friendly

In fields dominated by men, care must be taken to ensure that the workplace is woman friendly: cracking man-only jokes, expecting women to do all the informal duties in the office and harboring unspoken women stereotypes may all create a feeling of discomfort and discontent among women.

Women are usually less combative, less aggressive and less self promoting and these characteristics should be understood and not used against them (Scientific American, 1998a).

Characteristics, namely assertiveness and single-mindedness, have often been singled out (without scientific basis) to relate to scientific creativity yet these traits are stereotyped to be unacceptable for women (Anon., 2007).

5.4 Providing Role Models

One way of enhancing women's participation in science is to provide role models for women to identify with and to build up their self-confidence. The frequent stereotyping that women lack the ability to do science erodes their confidence and women often doubt their own ability to handle science subjects. School career councilors play crucial roles here to encourage and to ensure these young women on the verge of choosing their career that they have the capability.

In Malaysia the high number of female computer science professionals often act as role models to their younger female students. Lagesen (2008) interviewed many female students who said that they were inspired by their female teachers and the female head of Networking Department to take up computer science. The various winners in the "L'Oréal for Women in Science" awards in the international and national categories have undoubtedly served as role models for thousands of young women scientists pursuing Life Sciences in Malaysia.

The L'Oréal Group, which is the sponsor of the "L'Oréal for Women in Science Program", has clearly set a good example for other companies to follow. The L'Oréal Group believes in the value of gender diversity and 65 per cent of its researchers in the life sciences department are women, and women comprise 55 per cent of its researchers overall (Goh *et al.*, 2008).

5.5 Promoting Image of Successful Women in Science

Ms. Jennifer Campbell from the L'Oréal Group believes that communication is key, and to enhance women's participation in science, "we need to promote the image of successful women scientists who are being acknowledged, rewarded and valued for their contributions" (Goh *et al.*, 2008).

Malaysia has done its share to encourage and promote women scientists through the "L'Oréal-Unesco for Women in Science National Fellowships" which is a collaborative effort funded by the L'Oréal Group and supported by the Malaysian National Commission for UNESCO in partnership with the Academy of Sciences Malaysia, the Higher Education Ministry and the Science, Technology and Innovation Ministry. The yearly event and winners of the fellowship awards are given wide coverage in the media to acknowledge their contributions and to inspire other women scientists.

5.6 Recruiting Male Advocates

There will be greater success in the programmes to empower women in science if there are more male advocates (Goh *et al.*, 2008). Without male colleagues' support, influence and lobbying, the proposition becomes much more daunting. Malaysia's past immediate Prime Minister, Dato' Seri Abdullah Ahmad Badawi, is a good example of a male advocate who is very supportive of the working women's plight and have implemented several women-friendly regulations such as flexi working hours and no-pay maternity leave up to 5 years in the Malaysian civil service.

5.7 Providing Mentoring and Fellowship Programmes

Providing young women in science with the right skills set is very important, and this can be done through mentoring and fellowship programmes. The Gender and Diversity (G&D) Pilot Fellowship Programme (which has now developed into the African Women in Agricultural Research and Development (AWARD) Programme) of the Consultative Group on International Agricultural Research (CGIAR) tapped into existing world-class resources for the career development of women in science, agriculture and development. It is a good programme to emulate by the Centre for Science and Technology, NAM to empower women in science in the non-aligned and other developing countries. The G&D Pilot Fellowship Programme helped junior women scientists by providing a two-year mentoring relationship with a senior scientist in her field. This programme supported professional growth in both scientific expertise and people management, thus facilitating the development of female science leaders and strengthening their respective institutions (Goh *et al.,* 2008).

The G&D Programme under the CGIAR also provides training courses (Women's Leadership Course and Women's Negotiation Skills Course) to build up the leadership capacity and managerial effectiveness of women leaders, scientists and professionals who work in the CGIAR Centres and other international and national organizations.

To benefit women in science in the NAM countries, The Centre for Science and Technology, NAM can perhaps kick start some of the projects undertaken by the G&D Programme. A special unit for women in science can perhaps be set up to facilitate and co-ordinate work pertaining to the empowerment of women in science.

5.8 Making the Teaching of Science more Female Friendly

Science professors were found to be the most conservative in terms of sticking to traditional teaching techniques which have traditionally worked best for those who have become scientists (Scientific American, 1998b). Teaching of science must therefore be made more female friendly.

Furthermore, the pursuit in the improvement of science education is much less controversial than projects that might be construed as providing an unjustified advantage to a special interest group (Scientific American, 1998b).

6.0 Conclusions

According to the Human Development Report 2007/2008 of the United Nations Development Programme, Malaysia ranked 65 out of 93 countries in the Gender Empowerment Measure (GEM). The Report shows that females comprised 40 per cent of the professional and technical workers. Malaysia has attracted many women professionals in some scientific fields such as computer science, information technology and dentistry. However, women still lag far behind in fields like engineering, surveying and architecture. Providing more encouragement, mentoring and fellowship programmes, female-friendly teaching techniques and role models from senior female engineers and scientists may help to alter the situation and further empower women in science.

Acknowledgements

I am very grateful to Mr. Ahmad Wafi Harussani for assisting me to get my application endorsed as Malaysia's representative to the Workshop. Thanks are due to Mr. M. Bandyopadhyay for advising me to apply for sponsorship and to NAM S&T Centre for providing financial support. I wish also to thank Dr. Sasekumar who has always given me help so willingly whenever requested.

References

Abdul Majid, F. A., Md Salleh, L., Jusoh, M. and Borham, A. J. 2004. Educating engineers: Contribution of Muslim women in Universiti Teknologi Malaysia. In: Contribution of Muslim Women Enginnering Educators of UTM to the Society, Universiti Teknologi Malaysia. http://eprints.utm.my/3125 (accessed 12 November 2008).

Anon. 2006. Beyond bias and barriers: fulfilling the potential of women in academic science and engineering. Committee on Maximizing the Potential of Women in Academic Science and engineering, National Academy of Sciences, National Academy of Engineering and Institute of Medicine. 346pp.

Bombardieri, M. 2005. Summer's remarks on women draw fire. The Boston Globe. January 17, 2005.

Goh, A., Recke, H., Hann-Rollins, D. and Guyer-Miller, L. 2008. Successful women, successful science. Gender and Diversity CGIAR Working paper 48, pp.120.

Hassan, F. 2000. Essays on Science and Society: Islamic Women in Science. Science Vol. 290, no. 5489, pp.55-56.

Lagesen, V. A. 2008. A Cyberfeminist Utopia? Perception of gender and computer science among Malaysian women computer science students and faculty. Science, Technology and Human Values, 33(1), pp.5-27.

Levine, A. 2005. Bartholet and Sommers debate innate gender differences. http://www.hlrecord.org/home/index.cfm?event=displayArticlePrinter Friendly&uStor... (accessed 30 October 2008).

Ministry of Women and Family Development (MWFD). 2003. The progress of Malaysian women since independence 1957-2000. Ministry of Women and Family Development, Kuala Lumpur.

Mohd Ariff, M. 2006. More needs to be done in empowering women with ICT. Bernama.com. http://www.bernama.com.my/bernama/v3/news.php?id=226875 (accessed 20 November, 2008)

Othman, M. and Latih, R. 2006. Women in computer science: No shortage here. Communications of the ACM, 49(3), pp.111-114.

Pinker, S. 2005. The science of gender and science: Pinker vs Spelke. Edge the Third Culture: An Edge Special Event. http://www.edge.org/3rd_culture/debate05_index.html (accessed 15 November 2008).

Sasekumar, A. (pers. com.). 2008. Institute of Biological Sciences, University of Malaya, 50603, Kuala Lumpur.

Scientific American. 1998a. Digging deeper: the present status of women scientists suggests avenues of further research. Excerpted from a paper prepared by Gerhard Sonnert of Harvard University for the New York Academy of Sciences Conference "Choices and Successes: Women in Science and Engineering. April 27, 1998.
http://www.sciam.com/article.cfm?id=diggingdeeper&print=true (accessed 29 October 2008).

Scientific American. 1998b. Selling women on science: market forces alone won't eliminate inequities. Further intervention will be needed. Excerpted from a paper prepared by Gerhard Sonnert of Harvard University for the New York Academy of Sciences conference "Choices and Successes: Women in Science and Engineering. April 27, 1998.
http://www.sciam.com/article.cfm?id=digging-deeper&print=true (accessed 29 October 2008).

The Economist. 2006. The conundrum of the glass ceiling. The Economist-Print Edition. 3 August 2006.

UNICEF. Undated. Involving men to empower women.
http://www.unicef.org/malaysia/support_6065.html

UNICEF. 2005. Early investments in education pay off for Malaysian girls today. UNICEF Malaysia Media Centre.

Wajcman, J. 2004. Technofeminism. Cambridge, UK. Polity, pp.160.

Chapter 18

Empowerment of Women through Science and Technology Interventions in Myanmar

Khin Htike Htike Lwin
Technological University (Kyauk Se),
Department of Technical and Vocational Education,
Ministry of Science and Technology, Myanmar
E-mail: htikehtike88@gmail.com, most22@myanmar.com.mm

ABSTRACT

The performance of Myanmar women in the field of Science and Technology will be discussed. The Union of Myanmar is the second largest country in Southeast Asia, with the total land area of 261,228 square mile. Myanmar enjoys favourable climatic conditions and a rich diversity of topography. The population of the country is estimated at 54.3 millions with the population growth rate of 2.02 per cent. In Myanmar, there is no gender discrimination especially in the education sector. Moreover, the legal system provides equal rights to women and men in all areas. As for the area of education, the literacy rate of women is 73 per cent. With regard to health care profession, 50 per cent of women are doctors. Under the Ministry of Science and Technology, there are two implementation programmes, such as; Human Resources and Research and Development in Ministry of Science and Technology. In order to develop Human Resources Sector, Ministry is

- Extending and Upgrading Technological Universities and Colleges, Computer Universities and Colleges and Government Technical Institutes.
- Offering special courses and new specializations
- Providing advanced teaching aids and laboratory equipment
- Providing up-to-date reference materials
- Providing up-to-date machines and other facilities to be used in practical workshops.

Hence, MOST has already established 30 Technological Universities, 25 Computer Universities, 4 Government Technological Colleges, 7 Government Technical Institutes and 13 Technical Training Schools all over the country. Most of the heads of above Universities and Colleges are women as I am the Principal of Technological University (Kyaukse). They also take part in research work. There are more female teachers than male teachers in every University. Furthermore, the number of girl students is more than boy students. This means that women can work hardly like men in MOST. In addition, there are more female doctors than male doctors in health card sector in Myanmar. Therefore, women are working in leading roles to carry out research and development of Myanmar.

Keywords: *Empowerment, Science and Technology, Research and Development.*

1.0 Background

Union of Myanmar is a country with population about 54.3 millions where various national races are living in unity. The population growth rate is 2.02 per cent.In Myanmar 50.3 per cent of the total population are women. Thus the women outnumber the men.

The status of Myanmar women has always been high since the days of the Myanmar Kings. They enjoy equal rights as men. They do not take the name of the father or husband after marriage. Her name is her own and the name is known from birth to death unless she changes it of her own accord. With regard to education, the literacy rate of women is 73 per cent.

There is no gender discrimination in the education system. The government is making concerted efforts to promote the education status of women and children especially in the rural and remote border areas. With regard to health, the government places special emphasis on family health, the most basic and integral unit of society. The health coverage is extended to the rural and far-reaching border areas.

Myanmar is changing to the market economy and as such, not only the government sector, but also the private sectors are playing an important role. As such, more women are entering the private sectors, contributing a significant labour force. The legal system of Myanmar provides equal rights to women and men in the area of business and commerce. With regard to women in the profession, women outnumber men in the field of education and nursing, while 50 per cent of women are doctors.

In Myanmar family, there is no preference for boys. Both boys and girls are equally loved. Some may attach more values for girls. There is a Myanmar saying: When a daughter is born parents' usually say, "Assurance of an additional dish on the dining table". In Myanmar family, the husband and wife share equal household responsibilities. The husband provides the financial needs and it is the woman who manages the family decision making in providing food, clothing and schooling etc.

The women may go out to work for the social development; they still have the major responsibility to look after the family. The two responsibilities must be balanced for a woman who wishes to lead a harmonious and happy life both at home and in the society. One must be careful not to go against the cultural norms and values attached to our families.

2.0 National Machinery for the Advancement of Women in Myanmar

The government is striving towards the emergence of a peaceful, modern and developed nation and it attaches a high priority for the child's development and the advancement of women.

A Myanmar delegation attended the Fourth World Conference on Women, held in Beijing in 1995. The outcome of the Beijing Conference was the Beijing Declaration, highlighting the Beijing Platform for Action basing on the twelve critical areas of concern for women's advancement.

The government is committed to the Beijing Platform for Action and to implement it in accord with the national political, economic and social objectives. The government also recognizes the important role of women force in the national development.

As a fulfilment of the commitment made in Beijing, the government formed a National Machinery for the Advancement of Women, namely: The Myanmar National Committee for Women's Affairs on 3rd July, 1996. The patron of the committee is the Secretary-1 of the State Peace and Development Council and it is chaired by the Minister for Social Welfare, Relief and Resettlement, because, this Ministry is designated as the focal point for Women's and Child's affairs. The Committee members are the deputy ministers of the ministries related to the women and child welfares and representatives from the notable Non-governmental organizations. It is a high-level inter-ministerial policy making committee on women's affairs including the girl child and instrumental in providing the mechanism for intersect oral collaboration and coordination.

The working committee was formed on the 7th October, 1996 and was followed by the formation of State/Division, District and Township level (grass-roots) Working Committee for women's Affairs.

This committee are the operational committee. Their main objective is to systematically implement the programmes for the advancement of women. The working committee at the central level identifies the nine "critical areas of concern for the advancement of women in Myanmar.

The operational committees work alongside with the government departments, the NGOs and the communities.

The First critical area of concern was the need to form the National Machinery for Advancement of Women. It was formed on the 3rd July, 1996, namely, The Myanmar National Committee for Women's Affairs (MNCWA). The 3rd of July, the founding day of the MNCWA has been designated "The Myanmar Women's Day" with honour, in 1998, for the committees' formation is a historic undertaking.

The second area is to promote the educational status of women and children through formal education (at least to complete primary education), non-formal skills-based education and vocational training programmes especially for those who cannot afford to pursue higher education, to gain easy access to employment and to enhance income-generation.

The third area is uplifting of fitness and family health, focusing on the reduction of the infant and maternal mortality.

The fourth area is to protect women and children from violence, trafficking and abuses.

Violence is present in Myanmar but it is not a major issue. The research findings on the marital component of Domestic violence revealed that the magnitude is comparatively low, more mental violence than physical, the commonest causes are, inadequate financial provision for the family, excessive and habitual drinking, disharmony with in-laws in the extended families and adultery. Interviewers also found out that women are not aware of the existing laws that protect them.

The fifth area is women and economy. Myanmar women have been active in both fields of agricultural and commerce since ancient days. But now as Myanmar is integrating more into the world's market economy, more Myanmar women have become involved in the business sector, where they enjoy equal status as men. The economic sub-committee enhances rural women's income generation potentials by providing such courses as sewing, knitting, weaving and cooking. Non-traditional income-generation activities like auditing, computer science, driving, English language speaking, fruit carving, costume jewellery-making and secretarial courses are given to gain easy access to employment. Credit and loan scheme.

The sixth area is women and culture. As Myanmar opens its doors to the world, foreign influences have entered and to a certain extent, Myanmar cultural norms have come under pressure and this is now a concern to society. Therefore the cultural sub-committee, through video and radio plays, has made concerted efforts to preserve and safeguard the cultural heritage and national characters of Myanmar society. The sub-committee encourages young Myanmar women to uphold Myanmar cultural norms, to love and respect the country and its people, to honour and value parents, teachers and elders, to cherish family and society and safeguard their honour and dignity.

The sub-committee encourages young women to participate in the Traditional Performing Arts Competition which is held annually, and sponsors fashion shows on National Attires. Talks by prominent literary figures, summer courses on religion, culture and tradition are held at schools and in communities.

The seventh area is women and environment. Myanmar women are very much aware of the existence of the National Commission for Environmental Affairs, its policies and issues. Every year, to welcome the national labour month, tree planting ceremonies are organized through out the country by the Myanmar National Committee for Women's Affairs. Saplings are planted to honour the Myanmar Women's Day which falls on the 3rd of July. The auspicious day is celebrated by the entire mass of Myanmar women down to the grass roots levels. Myanmar women are also engaged in environmental sanitation activities, prevention of home air pollution, water pollution and in the increasing the availability and accessibility of safe water supply, under the supervision of the State and Divisional Working Committees for Women's Affairs.

The eight area media in Myanmar gives a positive image of women in her varied roles as mother, friend and life-partner. Thus the media gives women an equal status as men. Gender sensitive issues are being made aware to the media personnel. The activities of Myanmar National Committee for Women's Affairs are regularly featured in periodicals, television and radio for public awareness of their aims, strategies and efforts and what they have accomplished. Some of them are produced and directed by women, themselves. More women are encouraged to take-up journalism and some have reached decision-making levels in the media. Women are also trained to enhance their skills, knowledge and access to information technology. National media are used to support education and training programmes for women. The ninth area is the girl-child. There is no discrimination against the girl-child in Myanmar.The government attaches a high priority to the development of children because they are the potential leaders of tomorrow. The Child's Law was promulgated in 1993 to implement the U.N. Convention on the Rights of the Child. Myanmar ratified the Convention and it became a member state.

A national mechanism for implementing the rights of the child was established, namely a National Committee, State and District Committee on the Rights of the Child have been formed to protect and safeguard the child's right, including the girl-child. All the critical areas stated above are also concerned with the development and protection of the girl-child.

The department of social welfare, in collaboration with the NGOs and INGOs is responsible for the welfare of disadvantaged children and children in especially difficult circumstances which also includes girl-child. Special training programmes tailored to their needs are provided at the respective institutions.

The Myanmar National Committee for Women's Affairs (MNCWA) held the First and Second National Women's Conference in November, 1988 and June 2001 respectively. At the First Conference, the National Plan of Action for Advancement of Women was discussed and adopted. At the Second Conference, the works implemented by the State/Division Working Committees for Women's Affairs, at the grass-roots levels were reviewed, the constraints encountered were discussed and the future Plan of Actions recommended.

The MNCWA also observe the International Women's Day and celebrated "The 3rd July Myanmar Women's Day", every year. Myanmar hosted the Regional Consultation on Women's Health and Development Initiative: Country Profiles, organized by the Ministry of Health, the Myanmar National Committee for Women's Affairs and the South East Asia Regional Office of World Health Organization in 1988. Regional Workshop on Violence Against Women and the Role of the Health Sector was held in Yangon, Myanmar in 1999, sponsored by the Myanmar National Committee for Women's Affairs, Ministry of Health and the World Health Organization.

3.0 Ministry of Science and Technology

Ministry of Science and Technology carry out the development of Science and Technology for promotion of industrial production contributory towards the National Economic Development plans.

With the aim of enhancing successful implementation of development programmes in Science and Technology, the Ministry of Science and Technology was established on 2nd October 1996.

The Ministry of Science and Technology is organized with the following departments:

1. Myanmar Scientific and Technological Research Department (MSTRD)
2. Department of Atomic Energy (DAE)
3. Materials Science and Materials Engineering Research Department (newly established)
4. Department of Technology Promotion and Coordination (DTPC)
5. Department of Advanced Science and Technology (DAST)
6. Department of Technical and Vocational Education (DTVE)

3.1 Myanmar Scientific and Technological Research Department (MSTRD)

The "Myanmar Scientific and Technological Research Department" was First established in 1947 and was reorganized as the "Union of Burma Applied Research Institute"(UBARI) in 1954.It was later also known as the "Central Research Organization "(CRO). The present name" Myanmar Scientific and Technological Research Department"(MSTRD), was adopted in May1994,under the Ministry of Science and Technology was Formed, MSTRD stood as the sole department conducting multi-disciplinary research activities in major scientific and technological disciplines for the industrial and economic development of the nation. The head-office is located at No.6,Kaba Aye Pagoda Road, Yankin Township on a (33) acre site. The "Pilot Plant Complex"(Hmawbi) is also one of the sub-departments under the MSTRD.

Objectives

- Carries out research works in order to utilize the national resources more abundantly for the development of the national economy.
- Carries out research works in order to establish more industries based on advanced scientific and engineering techniques.
- Conducts applied research for the domestic production after acquiring the modern scientific and technological know-how from foreign countries.
- Conducts scientific and technological research activities for the benefit of the nation.

Activities of MSTRD

The MSTRD has (9) research and development departments and (6) service departments. The service departments not only assist the technical departments of MSTRD but also offer technical services to other government departments, according to their requests. Moreover these departments are participating in HRD programmes of MOST, by supporting and supervising the thesis of post-graduates and Ph.D candidates.

The Technical Research and Development Departments

Applied Chemistry Research Department

The department performs basic and applied research works in Chemistry. The main technological development projects deal with the production of finished and semi-finished products from indigenous agriculture, forests, marine and mineral resources of the country.

Ceramics Research Department

The research and development activities include the test-production of finished commodities of ceramic and clay materials from the natural and mineral deposits of the country.

Food Technology Research Department

The department conducts research not only on the chemical and biological methods of food preservation but also on various food-processing technologies.

Metallurgy Research Department

This department is concerned with the processing of indigenous mineral ores, from laboratory scale experiments through pilot plant test runs to industrial productions.

Pharmaceutical Research Department

The department is involved in the development of drugs and pharmaceutical with a view to commercial exploitation from indigenous as well as foreign medicinal plants. It is currently carrying out research on Black Creeper, Quinine Succiruba, Ginseng, Indian' telegraph plant, Indian Gentian, Me-Ky-Ru, Red Reish, Winter Cherry, and Indian Valerian. The Present research works involve in the production of Anti-malarial drugs and food-supplement tonics out of indigenous pharmaceutical resources.

Physics and Engineering Research Department

This department is concerned with testing and determining the physical and engineering properties of materials. It is carrying out Research and Development works on solar and wind energy.

The department has a Non-Destructive Testing (NDT) laboratory and Provides Technical Services for the non-destructive testing of iron, steel, cement and concrete materials and electrical cable wires.

Pilot Plant Research Department

The research activities of the department cover the upgrading laboratory scale experiments to pilot scale test runs. Now research project, "Purification for industrial waste water by anaerobic digestion method for producing the methane gas and using the residual water as fertilizer" has been being carried out..

Polymer Research Department

The Department performs research in the field of polymer synthesis, polymer physics and processing of composite materials. Now research projects, "Preparation of fibre optics" and "are being carried out.

Pulp and Paper Research Department

Pilot Plant Complex (Hmawbi)

It was established in 1976 on the basic of successful research projects by the Indian Government' sand UNIDO Aid. In 1980 it commenced to operate in pilot scale production. Now, the Pilot Plant is involved in the production of toilet paper, straw pulp and activated carbon are produced in pilot scale too.

The Technical Service Departments

Analysis Department

The department has the responsibility to analyze samples from and issue the analyze reports to government organization. The samples analyzed are drinking water, industrial and agricultural water as well as waste waters. Agriculture and farm produce, food, feed, adulterants, additives, contaminants, coke, coal, cement, cosmetics, drugs and pharmaceuticals, materials, ores, alloys, industrial raw materials, finished and semi-finished products of industries are also among the samples analyzed.

Fine Mechanics Department

This department is an aggregate of fine mechanics, electronics, electrical, optics and blowing sub-department. The R&D works under-taken by various research institutions, government departments, cooperatives and private enterprises are assisted by this department in the maintenance, repair and fabrication of scientific instruments and laboratory glass equipments.

Standards Department

The standards library of this department is the centre for providing information regarding standards and specifications and maintain a large collection of international and national standards and related publications. Attached to the library is a metrology laboratory where standards of national physical units as well as international units are maintained, providing references for calibration and testing purposes of various weights and capability of department by using the measuring instruments of Metrology Laboratory donated by Mitutoyo Corporation(Japan).

Technical Information Department

The department is an established source of scientific and technical information. The library has textbooks, pamphlets, reports, periodicals and reprints, etc.

Mechanical Workshop

The workshop provides services for the fabrication installation and repair of scientific and industrial processing equipments.

Administrative Department

This department is concerned with the administration and welfare of personal, technical planning, material, budgetary and financial matters of the organization.

Department of Atomic Energy

Myanmar's interest in nuclear energy for peaceful purpose is longstanding. As early as 1956, An Atomic Energy Division was established in the Union of Burma Applied Research Institute (UBARI).

The UBARI was recognized as the Central Research Organization (CRO) which again got a chance to the new name of Myanmar Scientific and Technological Research Department (MSTRD). Until 1997, the AEC existed all along as a division under the UBARI,CRO and MSTRD. In 1997, DAE was established as a new separate Government Department by combining the Atomic Energy Research Department and the Research Policy Direction Board.

Myanmar joined the International Atomic Energy Agency (IAEA) in 1957. It has been an active member of that organization and has benefited from scholarships and training programme of the IAEA. Myanmar nationals have been trained by the IAEA in the application of nuclear technology for peaceful purposes. Myanmar acceded to the Non-Proliferation Treaty(NPT) in December 1992.

In April 1995, Myanmar signed the Safeguards Agreement with the IAEA. In this age of globalization it is imperative that developing counties such as Myanmar must explore all possible avenues to promote development for our peoples.

The Myanmar Government is striving to acquire modern technology in all fields, including maritime, aerospace, medical and nuclear. In Myanmar the application of nuclear technology remains limited. The constraint is due mainly to the fact that isotopes have to be imported.

In the circumstance, when the new ministry of Science and Technology was set up in October 1996, ways and means to overcome the constraints to the peaceful use of nuclear energy were explored.

Department of Atomic Energy (DAE) was established as Department under the Ministry of Science and Technology in 1997 to carry out research, development and training in the field of atomic energy and to ensure the safety of the radiation sources and the protection from nuclear radiation hazards. With these new establishments Myanmar efforts for development and applications on nuclear science and technology have started DAE.

The DAE has its roots in the Union of Myanmar Atomic Energy Centre (UBAEC) established in 1955 as a division under the Union of Burma Applied Research Institute (UBARI). The DAE have been trying to develop and promote peaceful uses of nuclear technology with the help of IAEA. The DAE is composed with four departments:

- Radiation Protection Department
- Radiation Application Department
- Reactor and Isotopes Department
- Administration and Finance Department

The current strength of those departments is about 200 employees of which 25 per cent are trainees.

Objectives

The main objectives of DAE are as follows:

- To carry out research works for the development of nuclear technology in the country
- To carry out research, development and training in the field of atomic energy
- To protect radiation hazards or to implement nuclear radiation protection
- To coordinate with Government and Private Sectors for their nuclear technology applications and promotion

Facilities

Radiation Measurement Laboratory

This facility is equipped with radiation detectors, spectrometer for high resolution spectrometry, X-ray fluorescence analysis (using both isotope sources and tube excitation), air samples for environmental monitoring.

Instrumentation Laboratory

This laboratory was established as part of the Agency assisted TC Project for Nuclear Instruments Repairs and Maintenance. It is provided with accessories equipments and tools required for a repair work. It is also capable of fabricating some nuclear related electronic modules.

Radiation Dosimetry Laboratory

The equipment used are TLD card readers and monitoring devices also provided by the Agency. It is providing dosimetry services for more than 440 radiation workers from 35 different government departments, but only occasionally for the public sector. The DAE is extending personnel dosimetry services to the radiation workers on nation level. The number of radiation has increased during the last three years. The TLD badges are distributed to the workers to wear during their duty hours and checked the exposure dosage for every two months. Dosages of workers are recorded.

Gamma Irradiation Facility

A 12,000 curies Cobalt gamma chamber (Gamma 5000 from BRIT, India) has been installed in 2000. It is providing irradiation for sterilization of tissues from Tissues Bank (also an IAEA undertaking). Irradiation for research including agriculture is just beginning.

Non Destructive Testing Laboratory

An agency assisted project to establish an NDT laboratory started in 1995-96 in MSTRD. The facilities include Cs-137 Gamma radiography, X-ray radiography, ultrasonic, magnetic and dye-penetrate testing.

Current Activities

The current activities of DAE are as follow:

- Services for monitoring of radioactivity;
- Preparing Atomic Energy Regulation in order to accomplish the provisions of Atomic
- Energy Laws;
- Insurance of licenses for importing and re-exporting of radiation sources;
- Radiation Dosimetry Services for radiation workers in the country
- Inspection of used radioactive sources and enforcement of regulatory compliance
- Determination of radioactivity in minerals and elementary analysis of ores by XRF
- Maintenance of nuclear instruments, calibration, and testing of nuclear and testing of instruments from various governments
- Sterilization of tissue grafts
- Determination of types and nature of minerals and compounds by XRD
- Study of radiation effects on various materials using Co-60 Gamma sources
- Technical cooperation with IAEA in case of expert visits, fellowships and training, scientific visits and participation in internal conferences

Conducting national seminars and training courses for promoting new nuclear techniques

- For training and education, senior and junior scientists' conducts teaching and guidance for graduate and post-graduate students related with nuclear technology.

Material Science and Materials Engineering Research Department

(Newly Opened)

Department of Technology Promotion and Coordination

In order to become an industrially develops nation, Union of Myanmar has to build many Industries. This process of industrialization requires importing machineries from abroad or construction of machineries locally. Establishment of the Department of Technology Promotion and Coordination was approved by the Government Cabinet Meeting (26/98) potentials in the construction of machineries which are requires for the Industrial Process of Union of Myanmar.

Department of Technology Promotion and Coordination is organized with the following sub-departments:

1. Mechanical Technology Promotion and Inspection Department
   Mechanical Power Division
   Mechanical Production Division
   Mechanical Measurement Division
   Mechanical Inspection Division
2. Chemical, Metallurgical and Textile Technology Promotion Department
   Chemical Technology Division
   Metallurgical Technology Division
   Textile Technology Division
3. Electrical Technology Promotion Department
   Electrical Technology Division
   Electronics and Automation Division
   Electrical Inspection Division
4. Design and Construction Department
   Design Division
   Construction
   Division
5. Administration and Finance Department

Mineral Research and Development Centre (ELa) which was previously under Myanmar Scientific and Technological Research Department was transferred to Department of Technology Promotion and Coordination on March 1, 2000. This Mineral Research and Development Centre is located about half mile form Ela village, and is 2 miles away form Yangon-Mandalay Motor Road. The centre is located on 346.69 acres of land and centre itself 1155 feet long and 689 feet wide. The Mineral Research & Development Centre is organized as follows:

- Administration Department
- Basic Research Department
- Applied Research Department
- Technical Service and Consultation Department

Biotechnology Department was transferred to Department of Technology Promotion and Coordination on August 16, 2001. The Biotechnology Department is organized as follows:

- Environmental Biotechnology
- Biochemistry
- Microbiology
- Molecular Genetics

- Plant Biotechnology
- Plant Tissue Culture
- Industrial Biotechnology
- Aqua Technology
- Animal Biotechnology

Department of Advanced Science and Technology

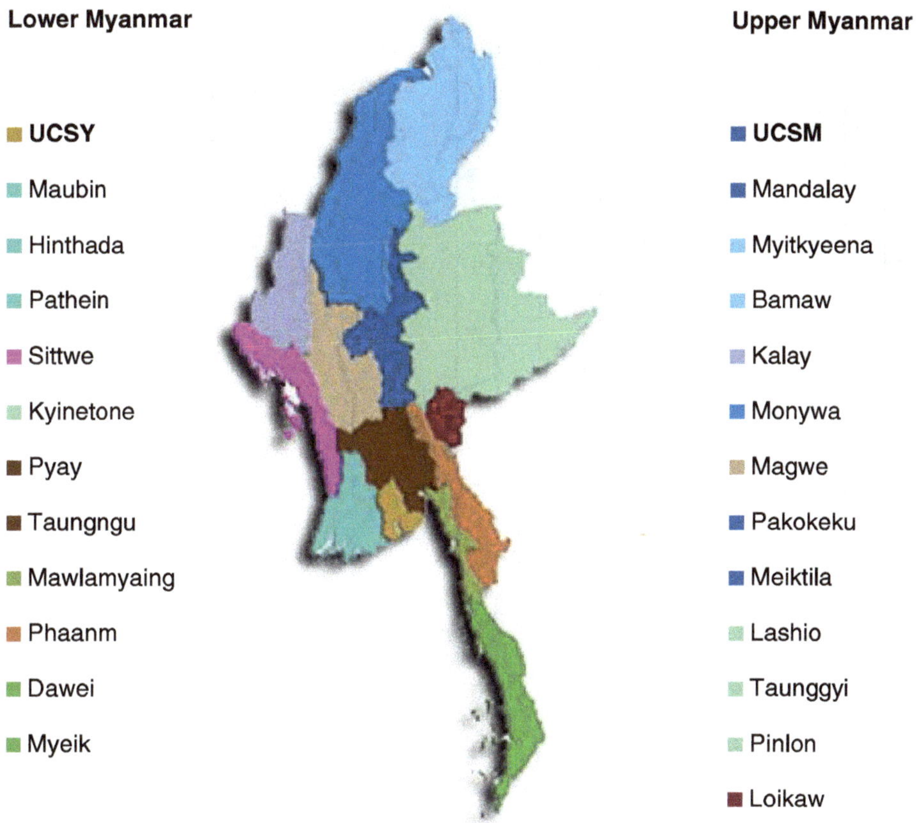

Department of Technical and Vocational Education

The DTVE has 26 Technological Universities, 3 Government Technological Colleges and 6 Government Technical Institutes. Technological Universities:

1. Pyay Technological University (PTU)
2. Technological University (Myitkyina)
3. Technological University (Banmaw)
4. Technological University (Mandalay)
5. Technological University (Kyaukse)
6. Technological University (Meikhtila)

Sl.No.	Level	Engineeering			B.C.Sc			B.Sc			B.A			ECO			Acadamic		
		M	F	Total	M	F	Total	M	F	Total	M	F	Total	M	F	Total	M	F	Total
1.	Bachleor	611	877	1488	25	272	297	240	494	734	117	783	900	13	272	285	8	44	52
2.	Master	542	428	970	65	557	622	121	696	817	17	226	243	–	6	6	–	–	0
3.	Ph.D	129	246	375	5	178	183	13	215	228	–	1	1	–	–	0	–	–	0
4.	Diploma	92	378	470	22	140	162	–	2	2	–	2	2	–	–	0	–	–	0
5.	Pre–Diploma	414	602	1016	23	146	169	–	–	0	–	1	1	–	–	0	–	–	0
	Total	1788	2531	4319	140	1293	1433	374	1407	1781	134	1013	1147	13	278	291	8	44	52

Sl.No.	Level	Law			Aero			wd&pmäef			Agri:			Others			Total		
		M	F	Total	M	F	Total	M	F	Total	M	F	Total	M	F	Total	M	F	Total
1	Bachleor	2	9	11	6	–	6	2	3	5	4	6	10	74	76	150	1102	2836	3938
2	Master	–	1	1	38	25	63	–	–	0	–	–	0	68	191	259	851	2130	2981
3	Ph.D	–	–	0	4	2	6	–	–	0	–	–	0	3	7	10	154	649	803
4	Diploma	–	–	0	1	–	1	–	–	0	–	–	0	–	3	3	115	525	640
5	Pre–Diploma	–	–	0	–	–	0	–	–	0	1	2	3	–	–	0	438	571	1189
	Total	2	10	12	49	27	76	2	3	5	5	8	13	145	277	422	2660	6891	9551

7. Technological University (Lashio)
8. Technological University (Taunggyi)
9. Technological University (Kyaington)
10. Technological University (Pinlon)
11. Technological University (Loikaw)
12. Technological University (Taunggu)
13. Technological University (Monywa)
14. Technological University (Pakokku)
15. Technological University (Kalay)
16. Technological University (Magwe)
17. Technological University (Sittwe)
18. Technological University (Hinthada)
19. Technological University (Pathein)
20. Technological University (Maubin)
21. Technological University (Hmabi)
22. Technological University (Thanhlyin)
23. Technological University (Pha-an)
24. Technological University (Mawlamyaing)
25. Technological University (Dawei)
26. Technological University (Myeik)

Government Technological Colleges

1. Government Technological College (Mohnyin)
2. Government Technological Colleges (Sagaing)
3. Government Technological Colleges (Myinchan)

Government Technical Institutes

1. Kyaukpataung
2. Yenangyaung
3. Yamethin
4. Thandwe
5. Wakema
6. Chauk

Chapter 19

Empowerment of Nepalese Women through Science and Technology Intervention in Nepal

Kayo Devi Yami

Chief Science Faculty, Nepal Academy of Science and Technology,
P. O. Box 3323, Kathmandu, Nepal
E-mail:yamikd@yahoo.com

ABSTRACT

Nepalese Government has been giving priority to all those programmes which can alleviate poverty since 7th five year plan. More than 30.8 per cent of the total population is below poverty line. Poverty persists in condition where human resources are undervalued where as land and material resources are over valued. People are poor because they have no assets – no land, no livestock, no houses and often no access to education. Their only assets are time and labour. The challenge is therefore to enhance the economic value of these assets, a goal that could be achieved through transition from unskilled to skilled work and making them aware of the commercial value of their skill and their local resources. Ministry of Environment, Science and Technology is the policy making Government organization whereas Nepal Academy of Science and Technology is the apex body of Science and Technology and Women in Science and Technology is an organization whose members are all well qualified and skilled women power ready to deliver their scientific knowledge to needy ones. Of the total scientific manpower of the country (28,134) 15 per cent is the women scientist and technologists. A pilot project was launched with the tripartite collaboration of these three organizations. The main objective of the project was to improve social economic conditions of marginal and disadvantaged group of rural population by generating entrepreneurship and employment. In 2008 Bode village of Bhaktapur Municipality was selected as a bio village where four training and demonstration programmes were launched *viz.* mushroom cultivation, biomass briquetting, use of biopesticides and apiculture, and environment awareness and solid waste management programme. The participants were all farmers with marginal lands. More than 80 per cent of the participants were women farmers. All of them participated with great enthusiasm. Since the project attracted much enthusiasm such biovillage establishment has been proposed to replicate in rest of the village development centers of the country. The initiation of such project not only empowered women villagers but also gave opportunity to qualified women Scientists and Technologists of Nepal who otherwise did not have access to deliver their research capability from their lab to land.

Keywords: *Poverty, Empowerment, Biovillage, Mushroom, Biobriquetting, Biopesticide, Training.*

1.0 Introduction

Nepal is a land locked agricultural country situated between two Asian giants, China in the north and India in the south occupying an area of 141,181 sq km. Only 21 per cent of the total area is used for crop cultivation and 40 per cent is covered with forest. Paddy, wheat, maize and pulses are the major crops. The total population of the country is 26.4 million, male population accounting 50.1 per cent and female population being 49.9 per cent. Rural population is 85.2 per cent. The annual growth rate of the population is 2.25 per cent (2001 census).The per capita GDP at present is only US$ 383. Human Development Index rank 142. While Nepal consists of unlimited natural resources such as water, flora, timber, hydropower, and scenic beauty, it has an extremely fragile environment. It is often affected by severe flooding, landslides, famine and faces environmental problems such as deforestation because of an overuse of wood for fuel. These often have negative impacts on people in rural areas who depend highly on nature for their survival. It must be noted that the majority of people, especially those in the rural areas, depend on land and agriculture as an economic activity.

The status of women of this country shows 86 per cent women live in rural settings, 63 per cent never attended school, and the economically active female population is 40 per cent. The proportion of females in professional/technical occupation was only 15 per cent (population monograph, 1995). Characterized by rural farm based economy, 91 per cent of the economically active female population is engaged in agriculture. The last decade had observed a marginal shift of female economic activities from agriculture to non agricultural sectors like tailoring, weaving, knitting, fashion designing, beauty parlor, travel agency, handicraft production, food processing, advertising and secretarial services.

Among those involved in gainful economic activities, two third of women are in manufacturing business and remaining in service sector (Shrestha et. al 1999). With more than 80 per cent people living in rural areas the target of science and technology for empowering women needs to bc rural based.

Nepal is rich in biodiversity and natural resources because of the varied topography ranging from tropical to alpine climate. Inaccessibility of 75 per cent Of its land covered with high hills and mountains has benefited in the production of organic fruits, tea, coffee, herbal plants etc.

Technology intervention is required in order to exploit the commercial value of these available local resources.

There is a serious requirement of capacity development of the local people and integrating with scientists of urban and international agencies so that they are technically upgraded to make them capable in exploiting technology, commercializing their skill and local products. A well established knowledge based society needs to be established to empower these people specially women folk through Science and Technology.

2.0 Weaknesses, Obstacles and Challenges to Economic Empowerment of Women in the Nepal

Poverty Status

Nearly one third (30.8 per cent) of the total population live under the poverty line as per the Nepal Living Standard Survey 2003/2004.

Social Structure

Nepal was a Hindu country till last year and more than 75 per cent of the total population is Hindu hence the influence of Hindu philosophy dominates the mentality of the society. Although the status of women in Hindu philosophy is conflicting the dominant view regards them as procreators of an heir for the family lineage. They are generally placed at a lower status than their male counterparts. This is vividly depicted in the literacy scenario (Yami and Yami 1991). Of the total literacy rate of 65.5 educated female make up only 42.8 per cent. Under the Hindu culture strong division of labour exists between males and females. Males, considered stronger and superior are expected to handle enterprising jobs which place them in outdoor activities, while females are expected to look after the household, thus internalizing their work. This attitude is well reflected in the education pattern of women: the majority takes art subjects particularly home science in academic sector. The legal structure follows male in hiring ancestral property which automatically makes less economic empowerment of women in order to mobilize their ancestral property in vital decisions like mobilizing the property for loan from bank to start business entrepreneurship etc.

Despite the overall involvement of women in various fields, the benefit of their work does not reach them. Women work day and night to produce a product only to lose it to mates when it comes to dealing with cash. Although women have access to land, forest, animals etc they rarely have formal ownership of these resources which access to other resources and information. They can't use land as security to obtain loans, they need to invest in their farms or develop small, innovative business.

Various Science and Technology tools to support such businessmen are also lacking. This phenomenon not only robs the dignity of work of women but also alienates them from their own work. Hence the urge to improve their produce never arises unless they are made part and parcel of the S&T programme.

Education

Literacy rate of the country according to the 2001 population census is 65.5 per cent male, 42.8 per cent female with the total per cent of literacy as 54.1 per cent. As education in Science and Technology subjects require time, dedication and money. Majority of women from rural villages do not get access to scientific and technical education that can help modernize their professions and increase their productive output. Hence women qualifying master degrees in Science and Technology fields are mostly from urban cities. Despite these constraints, women with basic scientific and technical education are increasing over the years. These groups have entered technical carriers such as teachers, engineers, doctors, nurses, pilots, agriculturists, entrepreneurs etc.

3.0 Science and Technology in Nepal

With the political change of 1951 greater number of S&T institutes have become accessible to the public. The 7th plan (1985 – 1990) gave S&T a sectoral status and recognized as an essential vehicle for the development of the nation. The 9th plan (1997-2002) aimed at integrating S&T with daily lives of people. Hence the development of Science and Technology in Nepal is still in infancy stage as compared to that of our neighbouring countries. Some of the major institutions that deal with S&T and women are as follows:

- Nepal Academy of Science and Technology (NAST) established in 1982
- Women in Science and Technology in Nepal (WIST) established in 1992
- Federation for Business and Professional Women in Nepal (FBPWN)
- Ministry of Women, Children and Social Welfare established in (1995)
- Ministry of Science and Technology established in (1996)
- Department of Women Development
- Gender Sensitizing Units in National Planning Commission and other line ministries established in (2005).

4.0 The Participation of Women in S&T Related Activities

Science and Technology which demands laboratory works, more concentration and field visits is often considered incompatible with the welfare of the family. Women's participation in the field of technology is further hindered as this sector requires students to go abroad and this is discouraged in our society for female students. However, with the introduction of technical education in few subjects like Medicine, Engineering degrees within the country women are now getting more opportunities to enter these fields. It is encouraging to find that despite all the social prejudices against women, the number of female students opting for S&T is increasing.

In 1996 there were only 8,236 scientists and technologists in the country among which only 992 were women. By the year 2006 the total number of S&T manpower had gone upto 28,134. Among them only 15 per cent were women scientist and technologists. It has been recorded that maximum women were from medical and biology areas. There are 60 PhD women S&T women power in the country till now among them 20 are from chemistry followed by Botany and Microbiology.

By subject as shown in the following table, engineering showed the highest manpower (39.97) followed by Natural Science (22.30) and Medical Science (19.56 per cent) (Bajracharya *et al.*, 2006).

However participation of women in engineering field is only 6.6 per cent. Maximum number of women participation is in natural science and medical science.

Table 19.1

Sl.No.	*Subject*	*Women S&T*	*Men S&T*	*Total*
1.	Agriculture	138(4.6)	2862	3000
2.	Botany	534 (47.1)	599	1133
3.	Chemistry	232 (27.07)	625	857
4.	Computer and IT	242 (30.90)	541	783
5.	Engineering	742 (6.59)	10,514	11,256
6.	Env. Management	18 (21.69)	65	83
7.	Food technology	23 (11.61)	175	198
8.	Forestry	34 (4.26)	764	798
9.	Hydrology	20 (7.66)	241	261
10.	Mathematics	55 (4.29)	1,225	1,280
11.	Medical Science	1,742 (32.56)	3,607	5,349
12.	Meteorology	5 (6.49)	72	77
13.	Microbiology	84 (43.07)	111	195
14.	Natural Resources	4 (19.04)	17	21
15.	Pharmacy	48 (31.99)	102	150
16.	Physics	17 (1.45)	1,151	1168
17.	Statistics	14 (1.57)	775	889
18.	Zoology	167 (26.25)	469	636
	Grand Total	4,219 (14.99)	23,915	28,134

Source. Nepal Academy of Science and Technology (2006).

Traditionally agriculture and animal husbandry has been the main domain of women activities. It is disheartening to find that an almost negligible number of women have entered this field. Because of such less involvement of women the traditional methods of cultivation still linger on. Also we find that only 4.26 per cent women are qualified in forestry. Forestry and animal husbandry being interrelated, depletion of the forests means starvation of cattle's which leads to further deforestation thus leading to an unending vicious circle of destruction. Nepal being an agricultural country, there is plenty of scope for the role of women in the development of S&T in agriculture, animal husbandry, forestry etc.

Women in S&T Academic Institutions

Most of the educated female manpower opts for going to teaching line because of its comparative advantage in time saving which can be of immense value for looking after children and household affairs. There are four universities, the oldest public university being Tribhuvan University. In 1999 there were 105 professors, 294 readers, 950 lecturers in Institute of Medicine, Institute of Engineering, Institute of S and T, Institute of Forestry and Institute of Agriculture and Animal Science (TU Today, 1999). Among them there were only 15 women Professors, 47 women Readers and 284 women Lecturers (Shrestha *et al.*, 1999). There are altogether only 60 women PhD holders in S &T till now. Under Tribhuvan University there is only one public Institute of Agriculture and Animal Science to support development in agro industry. Due to the lack of opening for job opportunities there is much fresh female manpower passing master degree in S&T areas that are doing part time teaching job in public and private S&T institutes. Unemployment is the major problem.

About 40 per cent of the literate manpower is unemployed. Therefore majority of them try to go abroad for pursuing further study and for getting jobs.

Women in S&T Research

Majority of S&T institutions of Nepal are located in the capital itself. In most of these institutes women as well as men are least involved in R and D activities. The reason could be due to the negligible amount (<1 per cent of GNP) of investment on S&T research and development. The infrastructures are not adequate for doing modern research work. Hence most of the scientific manpower is concentrated in conventional research works only. A survey in 2006 showed that the number of women scientists in different research institutions were 42 in Department of Plant Resources, 33 in Nepal Agricultural Research Council and Department of Agriculture, 29 in National Food Research Laboratory, 15 in NAST, 13 in Nepal Drugs Limited, 5 at Department of Mines, 2 at Department of Quality Control and two each in Department of Water Supply, Department of Archeology, Ministry of Forestry, Forensic Science Department and Horticulture Department. Majority of them are engaged in Biotechnological research works including tissue culture, bioprospecting, and molecular characterization of medicinal plants.

Women in S&T Entrepreneurship

Nepalese women trained in science and technology has demonstrated extraordinary entrepreneurship in starting businesses that utilize their training as well as their business acumen.

This generates important technology related services that increase the economic productivity of the country and also increases employment. The survey showed 20 entrepreneurs have started over 10 businesses ranging from tissue culture, floriculture, mushroom culture, biofertilizer production, alternate energy, biodiversity inventory preparation, medicinal herb cultivation and production and computer and IT services. Some women without proper knowledge are involved in S&T related business ventures like nurseries, apiculture, mushroom cultivation, seed production, vermicomposting etc.

Professional societies like Women in Science and Technology (WIST) and Federation for Business and Professional Women in Nepal are actively engaged in training women manpower in job oriented trainings like repair and maintenance of electrical appliances generally used by women, repairing of watches, mushroom cultivation, biofertilizer production, preservation of fruits and cultivation, conservation and use of medicinal plants, water purification and sanitation etc.

Many of these trainings have led to household entrepreneurship development. Business incubation centre has been established under Ministry of Industry however, it is not focussed for women particularly.

5.0 Government's Effort to Empower Women

The present Government of Nepal has made it mandatory to allocate 33 per cent seats for women from ministerial level to constitutional assembly members down to decision levels of various institutions.

In this fiscal year alone the Government has allocated 13.94 per cent of its total budget for the overall development of female population of Nepal including health, education, training, entrepreneurship development etc. In the social empowerment sector a new law has been introduced because of which now children can get nationality certificate by their mother's name also.

6.0 Establishment of Information Communication Technology Centres

Nepalese rural areas have resources with commercial values such as medicinal plants, organic fruits like apple, orange, grapes, tea, coffee, honey which are in great demand in foreign countries.

Inaccessibility is one of the basic characteristics especially in mountain areas. It implies lack of physical access to markets, higher transport costs, and limited information about markets.

They go waste because women lack knowledge of preserving, packaging and juice extraction methods.

A serious programme and strategy with sufficient government financial, technical and expert support programme needs to encourage through high policy from Nepal government.

Nepalese women also have wealth of knowledge regarding the use of medicinal plants in health most of which have scientific base that are passed orally from one generation to another.

Most of this knowledge have never been researched or documented. ICT tool could be used to preserve this knowledge and use for the benefit of human beings.

Nepalese government is aggressively launching establishing Tele Centres (approximately 400 numbers) and take care of such problem including road and transport network. Although these tele-centres are not specifically to empower women these centres should also be used for:

Bringing together women farmers, women agriculturists, scholars, development workers, service providers, networkers and advocates, and of course mothers, daughters, sisters.

Drawing nation and world's attention to women's immense contribution in sustainability life in harsh and fragile environments, their roles, responsibilities and rights as nurtures of their families and communities.

Integrating women in Science and Technology so that women's voice are heard not only inside house and also outside, at a much larger scale through the use of ICT.

Training them in cultivation, conservation and use of medicinal plants, preservation of fruits and food products they produce and get them into market connectivity.

Since poverty, illiteracy, lack of computer knowledge and language barriers are among the factors impeding access to ICT infrastructure particularly for women special trainings are required to address these problems.

7.0 Schemes and Programmes including Training Courses on S&T Suitable for Economic Empowerment of Women

Bio-village Development Pilot Project

The inspiration of launching Bio-villages in Nepal came from the Bio-village development programme introduced by the Father of Green Revolution of India, Dr. M. S. Swaminathan in South India.

Bio-village model for rural development addresses key issues like natural resource conservation, productivity improvement and poverty eradication. What makes the Bio-village Project different from contemporary development pathways is its embrace of job-led economic growth rooted in the principles of ecology, equity, energy efficiency and employment generation. A simple example like the practice of using the straw that has served as the growth medium for mushrooms can be used as a natural fertilizer for the cultivation of organic vegetables or for growing cut flowers that could be taken as one such example of the ecologically efficient techniques that can be passed down through the Bio-village Project.

Eco-farming could be another programme, in which chemicals and capital—the building blocks of modern farming—are replaced with knowledge and biological inputs like vermiculture, bio-fertilizers and bio-pesticides; this in turn creates new eco-jobs in villages. Thus more and more avenues will be opened for rural non-farm employment based on marketing opportunities. New opportunities for earning a living are devised through analyzing a family's resources. As a result, landless laboring families take to household mushroom cultivation, ornamental fish-rearing, coir rope-making, rearing small ruminant animals under stall-fed conditions and other enterprises which are within their means.

The tripartite collaboration between Ministry of Environment, Science and Technology, the policy making Government organization, NAST, the apex body of Science and Technology and Women in S and T, an organization whose members are all well qualified and skilled had launched Bio-village programme in Bode village of Bhaktapur Municipality last year. The main objective of this pilot project was to improve social economic conditions of marginal and disadvantaged group of rural population by generating entrepreneurship and employment. The reason for involving women scientists was to transfer their scientific knowledge from their LAB TO LAND and let them directly interact with farmers to help them in solving their problems.

This pilot project comprised of the trainings, demonstrations and implementation of following programmes.

Biomass Briquetting Program

Briquetting technology has been gaining popularity among the rural population and many NGOs and INGOs are also involved in the promotion of Beehive briquetting technology. Beehive briquettes have some advantages such as smokeless and efficient combustion, clean kitchen environment, comparative low cost, etc. The main aim of this programme is to substitute the traditional firewood and expensive imported kerosene for cooking purpose, consequently, to mitigate the indoor air pollution problems due to the burning of traditional firewood and at the same time making the use of this important industrial/domestic energy resource.

Beehive briquetts are made from a mixture of charcoal of agriculture waste (Banmara) and clay. briquette. Such briquettes are smokeless, efficient, and comparatively cheap.

Mushroom Cultivation

It is a training programme for making them capable to produce Pleurotus mushrooms on their own.

This programme is aimed to provide a proteinous food as poor farmers' nutrient supplement and to increase the income of poor farmers through mushroom cultivation. The poor farmers can adopt the low investment technology to achieve the high income.

The resources of seeds are available within the country. It is very appropriate technology for Women and landless farmers.

Biological Control Program

Although Nepal do not have industries for producing pesticides a large quantity of these chemicals are imported and used indiscriminately by the farmers mostly in accessible agricultural lands.

Realizing the health hazards of the use of these chemicals on human health biopesticide training programme was envisaged. These biopesticides are pesticides that come from natural sources as animals, plants, bacteria, and certain minerals.There are many advantages in using biopesticides, from both an environmental and business aspect as they are biodegradable, environmental friendly and are safe to bees, ladybeetles and other beneficial insect. There is also a shorter re-entry period for fields sprayed with biopesticides while chemical biopesticides have a much longer reentry period-one to three days-during which nothing can be done in the field.

Biological control is an awareness generation and training programme to promote the use of eco-friendly plant extract as biopesticides.

The local farmers were made aware on the use of extract from locally available plants like Lantana, Neem, Eupatorium, Acorus, Melia etc.as biopesticide.

Apiculture Training Program

Honey provides energy as much as 3.5 kilo-cal/kg and requires no digestion, but only a slight digestive action. Ayurvedic system of medicine uses honey as a major ingredient for the preparation of tonics, vitamins and herbal medicines. In more recent years it has been scientifically proven that honey is very effective in rapidly cleaning up infection and promoting healing. It increases appetite; helps to control gastritis and can also give relief to allergies, sinus, arthritis and asthma. In many areas of the Himalayan region where sugar is not easily available honey is used as a substitute of sugar. Honey produced in high mountains are fetching a lot of money in international market especially in Japan and European countries.

Environmental Awareness and Solid Waste

This programme focused on the need for environmental protection, awareness and trainings to rural villagers for preserving and restoring the environment.

It is an awareness programme to make rural people aware of Environmental pollution and its control, health and sanitation issues and training on management of solid waste and composting/vermicomposting.

There were 65 participants all of them farmers from \bode village. It is interesting to note that more than 90 per cent of them were women. After getting training in these four fields all the participants expressed in developing small scale entrepreneurship and have asked the Government to provide some sort of support for it. Therefore we have proposed to the Government to replicate such programme in 3915 village development centres.

8.0 Conclusions

Greater involvement of women in S&T is easier said than done, as it involves the restructuring of the whole socio-cultural framework. Women scientists hold triple responsibilities of reproduction, household and employment. Nepalese cultural division of labour defines their primary role as to look after home and bring up children.

Despite all the above mentioned problems, the situation is not frustrating for women; rather a positive trend is emerging for girls and women scientists and technologists. Gender disparity is slowly diminishing and most of the parents have a strong commitment to educate their daughters and wish them to stand on their own feet. The past attitude that technical fields such as engineering, medicine are for male only is slowly disappearing. Politically also concept of male supremacy is fading away slowly with the mandatory introduction of 33 per cent seats for women in every Government nomination from ministerial level, constitutional assembly down to institutes.

Recommendations

First of all, the cultural hurdle needs to be overcome, *i.e.*, the culture of discrimination against women which inhibits women from gaining education.

The misconception of the inferiority of women which is deep rooted in our society and not just the institutions offering S&T must be dispensed with. There is ample evidence, particularly in relation to women in sports that could be used to challenge this misconception.

Special incentive package should be created to encourage more girls to enroll in Science and Technology fields.

In a developing country like Nepal where the fulfillment of the basic necessities of the people is the most urgent task, the role of S&T in general to implement the national programme is obvious.

Since women remain at the root of fulfilling the basic needs of the family, they should be the first target group for any S&T programme.

Increase investment in S&T, increase budget in R&D so that more jobs could be created reversing the BRAIN DRAIN into BRAIN GAIN.

Motivate women scientists and technologists by awarding them for their achievements in research works.

Integrate more women in national development by giving them more opportunities to participate in decision making processes at local, regional and international levels.

Acknowledgements

I would like to acknowledge NAM S&T centre for kindly sponsoring me to participate in the workshop. I also would like to thank IROST for the local hospitality they provided. Finally I would like to thank my parents who are no more there, for educating me and my five sisters in five different fields of science and technology.

References

Bajracharya, D., Bhuju, D.R. and Pokhrel, J. R., 2006. Science Research and Technology in Nepal. UNESCO Kathmandu Series of Monographs and Working Papers No. 10, pp. 39-43.

Central Bureau of Statistics, Nepal. 2001. Population Census of Nepal.

NPC. 1986. The seventh plan (1985-1990). National Planning Commission, His Majesty's Government of Nepal, Kathmandu.

NCST. 1996. Women scientists and technologists of Nepal. National Council for Science and Technology, Kathmandu.

NPC. 1998. The Ninth Plan (1997-2002). National Planning Commission, His Majesty's Government of Nepal, Kathmandu.

Nepal Living Standard Survey 2003/04, Statistical Report Vol I. Central Bureau of Statistics, Nepal.

Shrestha, G., Shrestha, J., Joshi, S.D., Sharma T., and Yami, K.D., 2001. A Status Paper on Women in Science and Technology, RONAST Status Paper No. 2.

TU, 1999. Tribhuvan University Today, Tribhuvan University, Kirtipur, Kathmandu.

Yami, K. D. and Yami, H. 1991. Role of women in the development of Science and Technology in Nepal. Proceeding of the Conference on the Role or Women in the development of Science and Technology in Third World, (3-7 October, 1988, ICTP, Trieste, Italy), pp. 496-498.

Chapter 20

Various Mechanisms Used by "Vidatha" Technology Transfer Programme for Economic Empowerment of Rural Women through S&T Interventions

Jayasamara Gunarathne,
Technology Transfer Unit, Ministry of Science and Technology, Colombo – 3, Sri Lanka
E-mail: jayasamarag@gmail.com, secretary@most.gov.lk

ABSTRACT

Sri Lanka is an Island located at the bottom of the Indian sub-continent, has a women population of 54.4 per cent, given high priority for women in many development initiatives. "Vidatha", a programme under Ministry of Science and Technology which is the main S&T Transfer Programme in the country, has no exemption from having women empowerment as one of their main Objectives. Vidatha works as a network that provides facilities for rural communities through its 250 Vidatha resource centres around the country. Each centre is equipped with internet facilities, machinery for particular area, packaging equipment and S&T Officer to assist rural communities in selecting technologies, operate machinery and help in any other request made by them. Various technologies developed by multiple sectors are transferred to rural communities through this vast government programme where women are also economically empowered through promoting entrepreneurship. The programme does not stop by merely giving the technology but also assists in post production stage by guiding on Quality control helping them to get quality assurance certificates.

Empowerment of women through S&T interventions is not an easy task within rural communities of strong cultural bonds. We have specifically emphasized small and medium technologies that can be easily handled and machinery that can be operated and maintained by women without many complications. Our transfer programmes are designed to utilize existing materials and surplus localities for sustainability of the projects and maintain the bio-diversity and a special emphasis is on understanding the social conditions of different communities. We believe that this programme has helped rural women at least in understanding their strength, that they can contribute towards national development let alone projecting themselves as strong entrepreneurs.

***Keywords**: Technology transfer, Mechanism, Improvement, Empowerment, Entrepreneurs, Quality.*

1.0 Introduction

Sri Lanka has a women population of 54.4 per cent (Central Bank of Sri Lanka, 2006), which obviously makes women empowerment a shaping factor for any development programme. The population of Sri Lanka is estimated to be approximately 20 million, the outcome of a five fold increase that has been recorded during the past century (Report, Department of Census and Statistics, Sri Lanka, 2007). The women population being higher than male population may be due to the fact that the life expectancy and living standards of women are comparatively higher within the region. (Country Gender Assessment, Asian Development Bank report, Manila, Philippines, 2004) The average life expectancy for women is 76.4 yrs. and 71.7 yrs. for men for the period (2001-2006), which depicts a sex differential of about 5yrs., in favour of women (Report from Department of Census and Statistics, Sri Lanka, 2007).

However only 36 per cent of labour force is women in the year 2006 which still shows less employment among women even though it is believed that economic independence will empower women both socially and economically. Women unemployment being twice as much as male unemployment, many Institutions have focused their programmes specifically to strengthen women participation in income generation activities.

"Vidatha" means giving Science and Technology (S&T) for free. This programme falls under Ministry of Science and Technology which is the main S&T Transfer Programme in the country and it has no exemption from having women empowerment as one of their main objectives for the programme. Since its inception in 2005 out of the total 3270 entrepreneurs, 1709 happen to be women making this programme successful in terms of women economic empowerment through S&T Interventions.

Vidatha works as a network that provides facilities for rural communities through its 250 Vidatha resource centres all around the country. Each centre is having computer and internet facilities plus machinery needed for particular area, packaging equipment and S&T Officer to assist the rural communities in selecting technologies, operate machinery and help in any other request made by them.

Various technologies developed by Government institutions, universities, research organizations, non-government organizations and private sector are mainly transferred to the rural communities through this vast government where women are also economically empowered through promoting entrepreneurship.

The programme does not stop by merely giving the technology but it also assists in post production stage by guiding on Quality control. All Vidatha entrepreneurs products are tested free of charge for Quality Assurance Report and then the Ministry assists in acquiring Good Manufacturing Practices (GMP) certificate as the first stage of standards and then assists them to acquire Sri Lankan Standards (SLS) in the second stage. Altogether more than 50 entrepreneurs have been successful in getting GMP for their products out of which around 30 are women entrepreneurs.

The mechanism used by Vidatha Technology Transfer programme for Technology Transfer and empowerment can be broadly divided into four mechanisms.

- Technology transfer for localities.
- Technology Transfer for Improvement.
- Development of new Technologies.
- Technology Transfer of Innovations.

In every mechanism the social context will play a major role in technology transfer. Even if we have developed the ideal technology, without understanding the social impact and their different perceptions, the technology transfer becomes useless. Therefore a deep consideration of social component is vital for the success of the technology transfer mechanism.

Technology Transfer for Localities

Introduction and Transfer of technologies to use existing raw materials, harvest surplus, unused materials of a particular area. This is important for sustainability of many projects since the materials used are from localities and readily available and women can participate in their production as home based economic activities since the material costs are not very high.

Technology Transfer for Improvement

Transfer technologies to improve existing traditional technologies for a particular production process. The importance of this is to improve the productivity and also many women who are reluctant in changing their traditional livelihood can benefit from this.

Development of New Technologies

Develop and Transfer technologies for needs and requisitions from community. Some rural women have innovative thoughts on various products but lack technological knowledge for production, and society has also restricted them of getting knowledge on inventions. Therefore, developing of technologies for them will make them encouraged for further involvement and economic empowerment.

Technology Transfer of Innovations

Transfer of Technologies of Scientific Innovation and research. Many Innovations lies in research centres and this will help communities reach the technologies.

Having different mechanisms for technology transfer has given many advantages for the programme. We were able to adapt to many situations since we were equipped with different mechanisms. An overview of each of the mechanisms and trying to take examples for each mechanism will help us improve our method of operation. It may also be noted that the case study selected may not be involved in just one mechanism but sometimes a combination of two methods.

2.0 Sri Lanka: An Overview

Being centered at the edge of the Indian sub-continent, Sri Lanka's location has given the island a unique character both physically and socially. Myths and legends have evolved around her name from the earliest of times. Her temperate climate, lush forests, abundant animal life and cheerful people have exerted a fascination on voyagers and wayfarers from ancient times.

Sri Lanka's recorded history begins with the sixth century B.C. but there is evidence to show human settlements well before that time within the island. The "Mahavamsa" contains the country's recorded history, which is a unique record spanning 22 centuries. It is a long tale of an agri-based civilization which even today remains as Sri Lanka's most enduring hallmark. The early settlers were Indo-Aryans and with the introduction of Buddhism by emperor Asoka of India, Sri Lanka became the seat of Theravada Buddhism, a position which it retains even today. The form of rule was a benevolent monarchy and under its canopy grew a great hydraulic civilization based on the cultivation of paddy and supported by a network of tanks which made Sri Lanka self sufficient in her staple diet, rice. On this economic basis flourished a magnificent superstructure of palaces, temples and places of learning whose ruins still testify to a great civilization which was the perfect blend of the material and the spiritual.

Sri Lanka which was known variously as Taprobane, Serendib, Zielan and finally Ceylon has been referred by such early travellers as Marco Polo (1293) and Ibn Batuta (1334). Their accounts of the country are of a paradise isle of unequalled splendour. Later chronicles such as the Englishman Robert Knox who belonged to the 17th century have written of a courtly people and he has described the Sinhala farmer as being fit to be a king if washed of his mud.

Foreign invasions, most notably Sri Lanka's subjection to colonialism by the Portuguese, and Dutch and the British, were to drastically transform the country's face. This experience gave rise to parliamentary democracy, the rule of law and modern social cultural structures. Sri Lanka is a land of hospitable, literate people which has periodically changed government through ballot, an ancient country that is nevertheless perfectly attuned to the newest technology of our age. With its pivotal geographical position and the tropical climate, Sri Lanka has been known as a favourite tourist destination for a long time.

Today, having a population of around 20 million, Sri Lanka is ruled by a democratic government elected through ballot. Even though agriculture is the main economical sector, many service sectors and other exports such as gem and jewellery hold significant portion within its economy. High literacy rate of 92.5 in 2005 and the life expectancy of 74 years have been the key-factors of its social status.

In spite of the small ethnic conflict in the (a smaller part of) omit Northern part of the country the bulk of country's population consisting of Singhalese, Tamils, Muslims and Burghers continue to live in harmony in the rest of the country.

The present government has launched a successful humanitarian assistance programme in northern part of the country in bringing a peaceful solution and people today have placed hope in this venture and await an end to the conflict which will also help for a rapid development within the Northern Sri Lanka.

3.0 Sri Lankan Women: An Overview

Within the traditional social system of the country the women were given an opportunity in decision making specifically as a mother within a household. Her main duties comprised of raising children, and looking after the household matters. She occasionally helped with the agricultural work. The change of the ruling system and the education system has changed some of her duties but yet within the household the mother holds a special place in making decisions.

Sri Lankan women have a relatively better status than women in many other developing countries[1]. The Constitutional provision of equal rights without discrimination on the grounds of sex, the ratification of the Convention on the Elimination of All forms of Discrimination Against women (CEDAW) 1981, universal franchise and equal rights to contest elections in 1931, and equal access to free education and health services in the 1940s have all contributed to promoting gender equity. Thus the free and equal access to health services has increased the high life expectancy of 74 against the 72 for men depicting the better status of the women in the country.

Economic independence is a crucial determinant of social and economic empowerment of women. For this very reason, more and more women enter the labour force. Women depend largely on the informal sector for economic survival. It is noted that occupational segregation and wage disparities have narrowed down substantially and women have entered many fields once dominated by men (Report from Department of Census and Statistics, Sri Lanka, 2007).

The free education system in the country has given many women the opportunity to have a better education paving the way for more employment opportunities. But the unemployment among women which is 14.2 per cent in 2004 was relatively higher than male which is 6.3 per cent. The cultural role of women in a society was the main reason for unemployment among women even though female percentage is high at secondary education level.

Table 20.1: Unemployment Rates by Education Level and Gender

Level	*No Schooling*	*1-5 years*	*6-8 years*	*9-10 years*	*GCE or Degree*	*All Levels*
Men	3.7	8.3	9.6	12.2	14.7	9.9
Women	1.6	10.2	21.2	29.6	34.1	20.8

Source: Country Briefing Paper, Asian Development Bank, 1999.

Women economic empowerment is a very sensitive aspect that has to be dealt with considering the social context of the country. The promotion of entrepreneurship among women can make women economically independent. Practice of women entrepreneurship is essential both from the point of view of equity and full utilisation. Though women constitute roughly half of the world population, till recent times their contribution in official terms has been negligible. 'It is reported that 2/3 of the world's work is done by its women, 50 per cent of the food is produced by them, yet the world pays them 1/10 of its income and allows them to own mere 1/100th of its property'. This inequity has to be reduced and one of the ways to do it is by encouraging women to become entrepreneurs (Banerji and Sen, 2000).

The recent trends in entrepreneurship development have helped in improving living standards of women in the country. The alarming figure of unemployment rate compared with male has also focused attention and many government-funded projects are implemented identifying the importance of this issue in National Development.

4.0 "Vidatha" Technology Transfer Programme

The Ministry of Science and Technology was established as a separate cabinet Ministry in 1994. In general, the word "technology" means, "a scientific method of achieving a practical purpose". Technological change means the technical knowledge used in the production of capital and machinery. Technological progress comprises of creation of skills, new means and methods of production, which can produce goods and services at less cost of production. It is the level of technology which acts as an index of economic development in the country. Technology includes machinery for production, housing, health, education, communication, and many other elements (Banerji and Sen, 2000).

The Ministry has attempted to use S&T Interventions in promoting entrepreneurship. "Vidatha" meaning giving away science and technology for free is the Ministry's Technology Transfer Programme that was introduced to

- Retain and promote the work of scientists and technologists in this country itself,
- Utilize their research findings and technology for development activities of the country,
- Up-lift the economic level of the rural people,
- Make arrangements to exhibit to the country the creativeness and skills of rural people and utilize their resourcefulness for the country's benefit, and
- Ensure the quality of the products from the village reaching the market and thereby to ensure a stable income for rural folk.

Vision

To utilize the full potential of S&T to improve the quality of life of all Sri Lankans, with emphasis on hitherto neglected villages.

Mission

To harness the natural and human resources in the village for development through the transfer of science and technology to the rural community, thereby reducing unemployment and improving the quality of life.

Vidatha Resource Centres

The Vidatha Programme is executed within the rural areas through Vidatha resource centres that were established to identify the resources and needs of the community for economic development, to draw the attention of Science and Technology Research Institutions to provide the technology needs of the community, and to demonstrate and transfer the appropriate technology for village economic development.

Every resource centre has machinery that can be used by the community members free of charge and the people will have the opportunity to utilize the resources during the initial stage of business. The resource centres also act as sales promotion centres where the products are displayed and also assist in increasing the demand

An important role of the resource centre is the coordination between entrepreneurs and quality assurance certifying Institutions in assisting for entrepreneurs to get Quality Assurance certificates in different stages. The first stage is the Good Manufacturing Practices (GMP) Certificate which is issued by the Sri Lanka Standards Institution (SLSI). The next stage is the Sri Lanka Standards (SLS), the main standard certificate issued by SLSI.

Vidatha Societies

These will be established at village or Wasam level by the divisional coordinator. Even though anyone can join these societies it will be targeted at those who wish to develop or improve a micro, small or medium enterprise. Every centre will have a coordinator and a computer operator to assist the rural communities. They also gain the opportunity for training free of charge to start an enterprise, establish interrelations with other entrepreneurs and share knowledge and experience.

5.0 Technology Transfer Mechanisms

Within Vidatha technology transfer programmes the mechanisms used can be broadly divided into four mechanisms,

- Technology transfer for localities.
- Technology Transfer for Improvement.
- Development of new Technologies.
- Technology Transfer of Innovations.

A special emphasis is stressed on the social aspect in every technology transfer mechanism

During the community awareness and transfer period the monitoring and evaluation component is strengthened. The post production Management component comprises the quality control measures as well as marketing of the products.

Technology Transfer for Localities

In many parts of the country the harvest surplus is wasted due to lack of proper post harvesting technologies. There are also many raw materials that are not properly utilised. In some cases some fruits are not used at all for any production and therefore merely get wasted. Therefore introduction of technologies is important to utilise these raw materials in industries and this mechanism deals with transferring such technologies for rural communities (Figure 20.1).

This is important for sustainability of many projects since the materials used are from localities and are readily available. Especially women can participate in these productions as home based economic activities, since the material costs are not very high and they can also utilise the machinery at resource centres during the initial stages of the production.

In this mechanism the need assessment plays an important role since the appropriate technology is selected based on the assessment. Community awareness is essential for sustainability of the project since the participation for production may largely depend on the attitude change through the awareness campaign.

Case Study–1

Damayanthi Rathnadivakara has started manufacturing coconut toffees, jam and cordial at a very small scale in her house as a self-employment. Most of her products were sold to neighbouring shops and houses. One day she visited the Vidatha resource centre within her area and her potential was identified to improve the production if she were exposed to improved technologies.

Most of her productions utilised the unused fruits within her area. She started with the dehydration of fruits such as pineapple, papaya, biling, nelli and mango that was getting wasted within the area. Thus her earlier production of jam and cordial was improved using the latest technologies given by Industrial Technological Institute (ITI). Now she was able give a variety of products to suit different needs and also to utilise the wasted fruits within the area.

In increasing her demand for production the GMP certificate played a major role. Ministry of Science and Technology along with Sri Lanka Standards Institute was able to provide her with GMP Certificate once her production process was tested free of charge by ITI. Her productions were displayed at various trade exhibitions organised by the Ministry of Science and Technology. She was in need to improve and increase the production to meet the increasing demand. Hence she also took Rs. 10,000.00 from Vidatha loan scheme to expand her business (Figures 20.2–22.4).

She started the business with the help of her family members but today she also has 3 paid helpers within her business. The significance of her business is utilization of wasted fruits within the area. Vidatha programme continues to support her in business and her strong determination has made her into a successful entrepreneur (Figures 20.5 and 20.6).

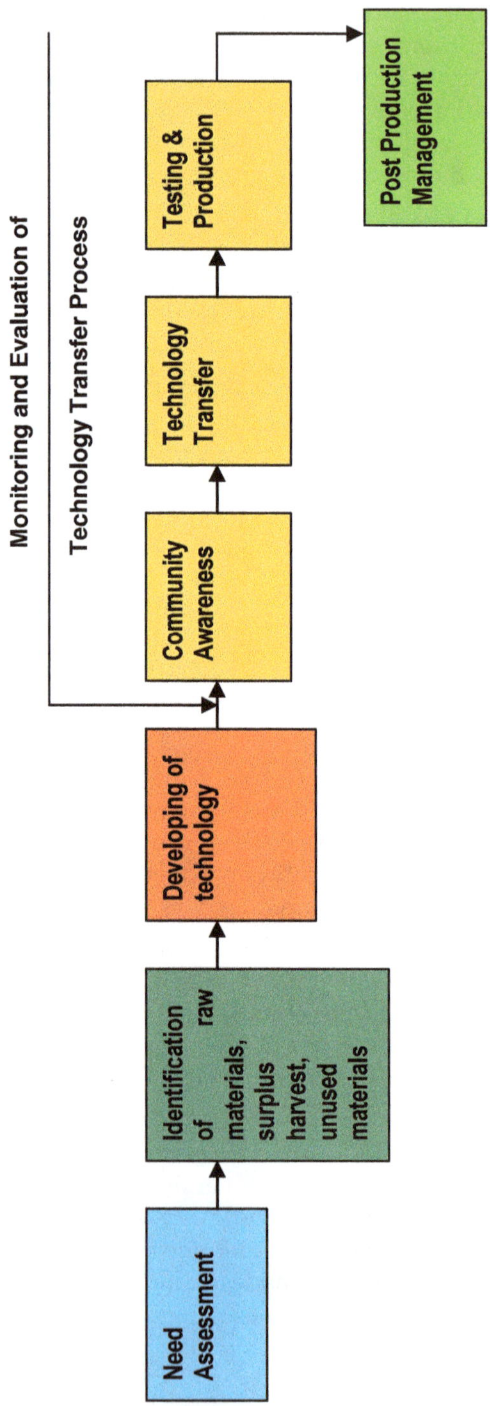

Figure 20.1: Steps in Technology Transfer for Localities

Figure 20.2: Production being Tested

Figure 20.3: Some of her Productions

Figure 20.4: During the Production

Figure 20.5: Damayanthi getting her GMP Certificate

Figure 20.6: Displaying Production in an Exhibition

Following is the improvement in her income,

Table 20.2: Increase in Damayanthi's Income

Year	*Monthly Income*	*Annual Income*
2006	6000	72,000
2007	10,000	120,000
2008	14,000	168,000

Case Study–2

A women's group of 5 at Bandaragama was selected by the Vidatha Resource Centre to introduce the technology of the dehydration of Jack and other vegetables that are wasted within the area. All these women come from low income rural families and a home based economical involvement will help them to improve their economic status.

The main production is dehydration of jack and breadfruit that are being wasted in the area. The programme will help to increase the household income, and also to utilise the wasted material in the area that bring down the material cost for production providing for a better profit.

The group is utilising machinery at the resource centre and dehydrated jack is produced directly aiming the export market. Since their production quality was improved they were able to get a long standing order from an export company to supply their products. The GMP certificate has helped them in getting many orders for their items within local market as well (Figures 20.7 and 20.8).

The average Income Generation of the group is as follows,

Table 20.3: Income Generation of Vidatha Women's Group

Description		*Amount Rs.*
Monthly cost	550.00 X 30 days	16,500.00
Daily income	50.00 X 30	1,500.00
Monthly income	1,500.00 X 30	45,000.00
Monthly profit		28500.00
Profit for one member	28,500.00/5	5,700.00

2. Technology Transfer for Improvement.

The transfer of technologies has improved existing traditional technologies for a particular production process. There are many local industries using traditional technologies which make the production process slow and energy consuming. This does not help in having good manufacturing practices. Since some families are engaged in their traditional technologies, developing improved technologies will benefit them to develop their industry without facing the difficulty of shifting to a different livelihood altogether.

The importance in this is to improve the productivity and also many women who are reluctant in changing their traditional livelihood can benefit from this.

The important factor for this mechanism is the study of the existing traditional technology. Some final products largely depend on their method of production and new technologies tend to change these. Therefore, careful consideration should be given to keeping the traditional quality of the product in developing technologies and try to improve the standards as much as possible. The community awareness is also critical in this since some may not be willing to change the traditional technology and to build the trust towards the new technology.

Figure 20.7: During the Production

Figure 20.8: Some of her Productions

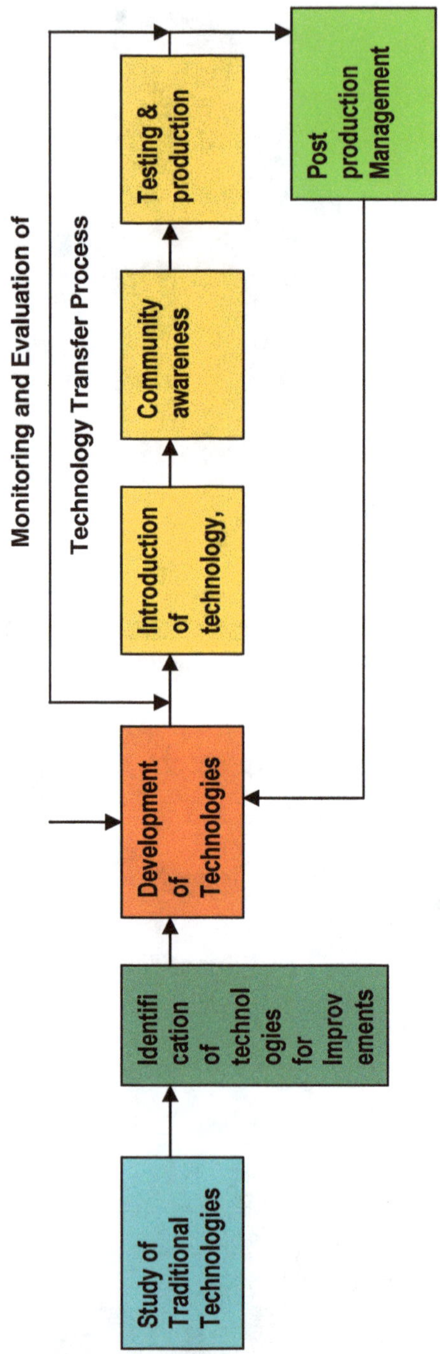

Figure 20.9: Steps in Technology Transfer for Improvement

Case Study

Thamara Dhamayanthi has been manufacturing coconut oil at home using the traditional technologies, which takes a lot of time for production thus bringing down the efficiency. In 2001 the white coconut oil production technology was introduced to her and she has purchased a small machine for Rs. 650 and started manufacturing white coconut oil. She started with manufacturing hair oil which she introduced to market, a 60ml bottle for Rs.30.00 and a 100ml bottle for Rs. 55.00. She was able to sell around 100 bottles per month.

In 2003 she was able to purchase a machine for Rs.6500.00. With this improvement she could increase her production to produce 500ml bottles priced at Rs.90.00 and was able to sell around 50 bottles apart form her previous monthly sales. By 2005 she has improved her enterprise to buy two more coconut producing machines, another machine for Rs.10, 500.00 and a large scale machine for Rs. 65,000.00.

The new machines have allowed her to start the manufacturing within separate premises instead of within her house where she started. Subsequently she was able to increase her production to 20L per month and also the oil prices has been increased to 100ml for Rs.65.00, 500ml for Rs120.00 and 750ml for Rs.350.00. Her products were taken by many supermarkets since the quality is of a very high standard (Figures 20.10 and 20.11).

Figure 20.10: During Thamara Dhamayanthi's Production

Figure 20.11: Large Scale Machine

She was able to increase the standards of the manufacturing process along with her enhancement in finances, which is an important criterion for starting a home-based industry. Vidatha resource centre has assisted her throughout her improvement from printing the label up to marketing the product. She was able expand her market through many trade exhibitions organised by Vidatha programme. Apart from the many customers she has already established she was able to start her own sales centre recently (Figures 20.12–20.15).

Development of New Technologies

This involves the Development and Transfer technologies for needs and requisitions from the community. Some rural women have innovative thoughts on various products but lack of technological knowledge for production and cultural conditions restrict them from getting knowledge on innovation. This is the reason why they have not started any production. Therefore developing of technologies for them will encourage them for further involvement and economic empowerment.

The community discussion and identification of needs is vital for the success of this mechanism. The developed technology should be simple and easy to handle by the community as well (Figures 20.16).

Figure 20.12: Her Products

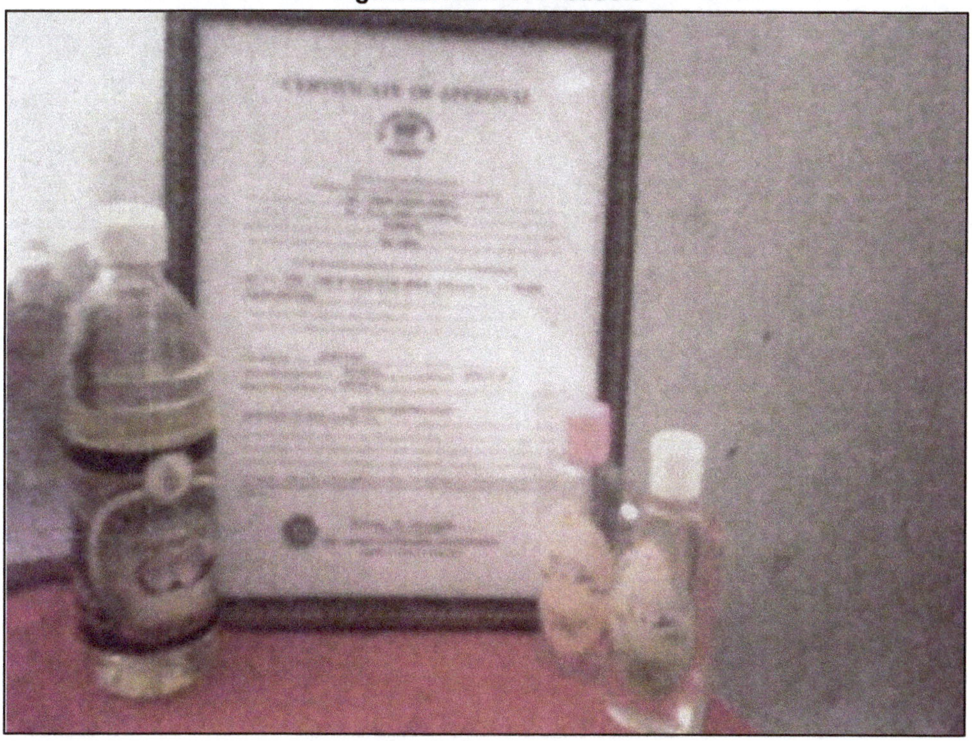

Figure 20.13: Her GMP Certificate

Figure 20.14: Tamara Damayanthi's Products and Achievement

Figure 20.15: Tamara's New Sales Outlet

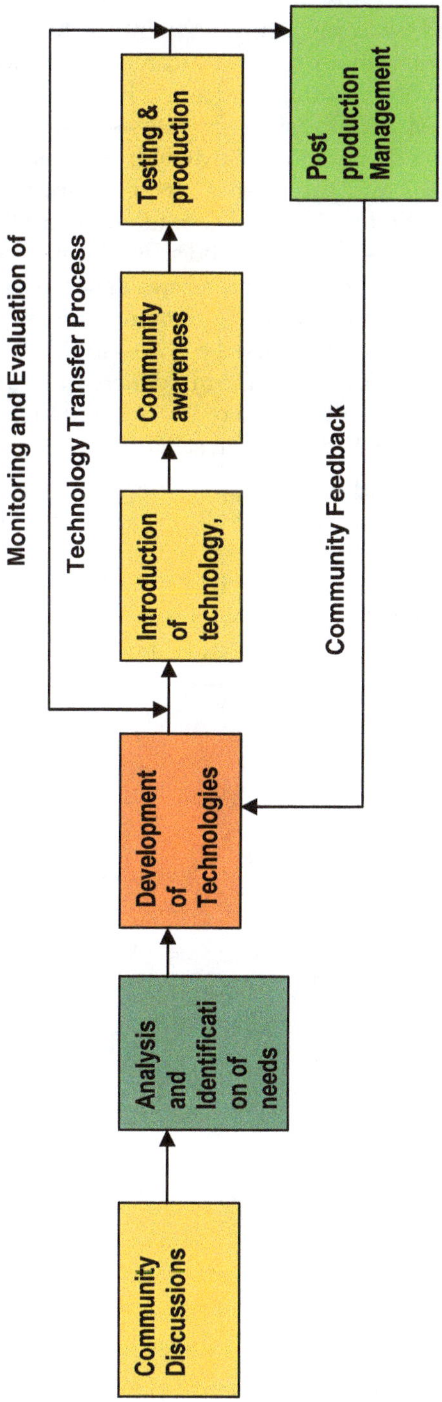

Figure 20.16: Steps for Development of New Technologies

Case Study

Malani Premalatha used to supply string-hoppers for the neighbouring houses and shops as a self-employment that provided income for her household. One day she came to Warakapola Vidatha resource centre and her major problem in her self-employment was identified. Most of the string hoppers she supplied to shops remain in the shops and they get wasted very soon. The dehydrating machine at the resource centre was introduced to her and the unsold string hoppers were dehydrated. Understanding the success of this technology, she wanted to start this as a commercial production and at the beginning she dehydrated string hoppers using the machine at the centre and then supplied all the Vidatha centres as samples. This proved to be a very successful product and she was able to get many orders. Her products were also introduced at many trade fairs organised by the Vidatha programme and the demand for the project thus increased. With the rapid increase of the demand she had the need to have her own machine and a loan of Rs. 50,000.00 at an interest of 13 per cent was given to her through the Vidatha loan scheme.

The Ministry of Science and Technology along with Sri Lanka Standards Institution has been able to award her with the GMP certificate for her manufacturing process. ITI has done all the quality testing free of charge for her production in cleaning and sanitation, personal Hygiene, product Information, pest Control, Waste Management, Packaging and labelling and storage. Once she had the GMP certificate, she was able to get many orders from super markets (Figures 20.17 and 20.18).

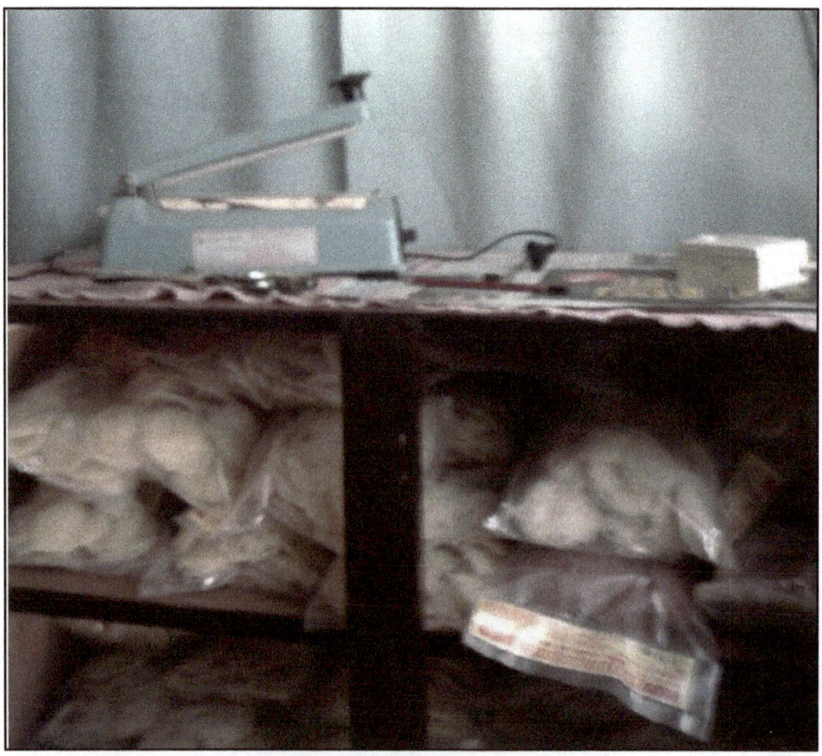

Figures 20.17: Her Production

She being a mother of 4 with a very low income level today has developed a business with the help of her family named "Laksiri Production Centre". Dehydrated vegetable biriyani is one of her products with a very high demand. Most important factor in all her products is that they are 100 per cent natural productions (Figures 20.19 and 20.20).

Figures 20.18: During the Production

Following is a summary of increase in her income.

Table 20.4: Malani's Income Increase

Year	*Monthly Income*	*Annual Income*
2006	5000	60,000
2007	9,000	108,000
2008	12,000	144,000

Figures 20.19: Production being Monitored

Technology Transfer of Innovations

Many research organizations have developed innovative technologies for productive and efficient productions; but a proper mechanism that identifies the social needs and understands the cultural barriers needs transfer of technologies of

Figures 20.20: During the Production Process

scientific innovation and research. Therefore this programme has mechanism for technology transfer of innovations for the use of communities (Figures 20.21).

In this mechanism a vital component is the selection of communities for technology transfer. The success of the production depends on the adoption of it by the community and proper implementation.

Case Study

The technology to manufacture instant herbal drinks such as belli, ranawara and Iramusu was developed by ITI. This innovation was important since making of these herbal drinks in the traditional way takes much time and boiling will also consume a lot of energy. As the first step, a detailed project proposal was drafted for production of instant herbal drinks.

A women's group of 05 was selected from very low income families and the technology was introduced. Again several community discussions were held on the technology and also to introduce the production for the communities. Once the group has started their production with the direct Involvement of Ministry of Science and Technology, they were able get the GMP certificate as their first step of quality assurance. The acquiring of GMP certificate has helped them to increase the demand for their production and they have secured orders from many leading supermarkets.

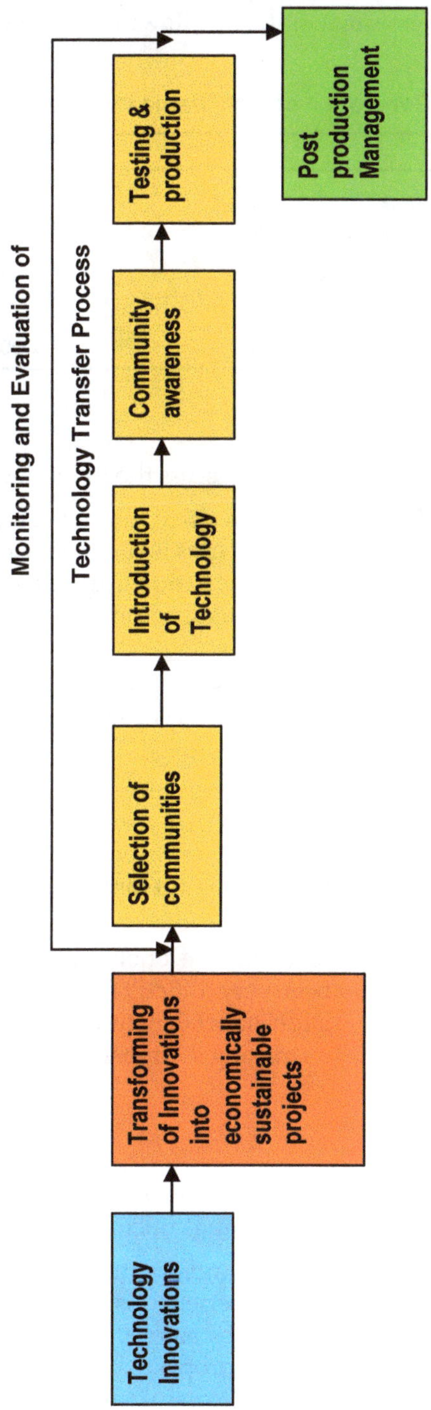

Figure 20.21: Steps in Technology Transfer of Innovations

The S&T Interventions has made an impact on improving their income and helping them for a better economical status. Following is the average income generation of the group.

Table 20.5: Income Generation of the Vidatha Women's Group

Description		*Amount Rs.*
Monthly cost	2750.00X 20 days	55,000.00
Daily income	3.00 X 1500	4,500.00
Monthly income	4,500.00 X 20	90,000.00
Monthly profit		35,000.00
Profit for one member	35,000.00/5	7,000.00

6.0 Challenges

In conducting these programmes we came across the following challenges.

- *Attitude*: The cultural attitudes that women are not capable of handling sophisticated or even simple technologies is the main drawback in these programmes, thus cultural perceptions for various production methodologies have also prevented many from absorbing new efficient technologies comfortably.
- *Communication*: Most of the documents are produced in English. But, the information flow to the rural communities is very slow and many of them also lack the knowledge on English language.
- *Cultural restrictions*: Many women in the rural communities have strong cultural bond with traditional technologies and sometimes their caste by birth have restricted them in exploring new technologies. Adoption into different livelihood has become very difficult for them due to these cultural restrictions.
- *Capital investment*: Many of the S&T based economic productions need some capital investments and it has been very difficult for them to find this capital investment. We lack many micro-financial institutions that provide initial finances for these entrepreneurs making it impossible for them to start their businesses.

7.0 Conclusions

During our attempt for women empowerment through science and technology interventions we were able to drawn the following conclusions.

- The most important factor is the technology transfer has to simple and easily handled by women. Our experience is that small and medium technologies transferred for industries have shown a greater success in implementation. Identification of appropriate technologies for different communities will help them to start successful enterprises and also provide ability to empower themselves economically, socially and politically as well.

- Another key factor to address understands of the social context within which the technology is to be used. Some research institutions develop technologies spending a huge amount of money but when technology is transferred the initial cost of machinery is so high that it becomes impossible for women to start the business. The complications in the machinery also prevent them from starting an industry using the technological interventions. Such technologies become useless in improving income generation specifically among women.
- The innovations should always be done based on the social needs. The technology introduced should be utilised in the existing condition of a particular area. In some instances the raw material cost and operation cost of innovative technologies are so high that, even though the women wanted to start the enterprises, the costing problem restricts them in their capacity. Continuous interrelationships with the community and community discussions and testing of models within community will help in overcoming these barriers.
- When empowering women through S&T Interventions, the technology transfer will not merely help in its success but other supporting sectors also have to be developed. For example if a woman is given a technology to add value and do a production by a raw material, the knowledge of business management, and quality control is the key to her success. The marketing and packaging knowledge will also help her to become a successful entrepreneur. Therefore it is important that a technology transfer programme become an integrated programme that incorporates all the supporting services as well.
- A proper micro-finance loan scheme is vital in women empowerment programmes since many rural women have no savings to start their own business. The Vidatha Programme together with a government policy was able to provide a lower interest loan with very convenient conditions for women. Out of the total loans 75 per cent has been taken by women and the repayment is as high as 99 per cent.
- Other important factor is creating the market opportunities for the production of these women. Vidatha programme has always attempted to provide better marketing opportunities for entrepreneurs through organizing trade fairs and community awareness programmes. In addition, all the Vidatha resource centres display the production and assist as much as possible in getting better marketing links.

Economic empowerment of women is the key for uplifting her social status. The economic independence will provide the opportunity for controlling of finance, management and decision-making.

As an outcome, women's views and ideas will be given due consideration in state policies which will also focus on women empowerment. Women having ownership of assets will become a part of decision making power within households as well as within society. Within rural Vidatha societies more than executive posts are held by women. Therefore women empowerment should address uplifting of economic, social and political aspects to create a social stability for women within the highly competitive society which will pave the way for a holistic social development as well.

References

Banerji.Anita and Sen Kumar Raj, 2000, Women and economic Development, Deep and Deep Publications, New Delhi, pp. 266.

Country Gender Assessment, Sri Lanka, south Asia regional department and regional and sustainable Development department, Asian development Bank report, Manilla, Philippines, 2004, pp. iii.

Economic and social Statistics of Sri Lanka a, Central Bank of Sri Lanka, 2006, pp. 127.

The Sri Lankan women-Partner in progress, report from department of census and statistics, Sri Lanka, 2007, pp.01.

List of Abbreviations

GMP: Good Manufacturing Practices
ITI: Industrial Technological Institute
S&T: Science and Technology
SLS: Sri Lanka Standards
SLSI: Sri Lanka Standards Institutions

Chapter 21

Production of Low-Cost Complementary Foods Using Locally Grown Foods in Tanzania: An Economic Empowerment of Women

Rose Rita Kingamkono
Director, Research Coordination and Promotion,
Tanzania Commission for Science and Technology, Dar es Salaam, Tanzania,
E-mail: rkingamkono@costech.or.tz

ABSTRACT

Childhood malnutrition is rampant in Tanzania and peaks during the complementary feeding period (6-24 months old children). The causes are complex and relate to social-economic, cultural and technological factors and skills. The immediate causes are low feeding frequency, high dietary bulk, poor nutrient bio-availability and microbiological contamination of foods fed to young children. Among the major constraints hindering caretakers to follow nutrition instructions are ignorance of preparation techniques, lower prestige of self-prepared foods and in particular lack of time and fuel. High quality industrially processed complementary products are now available in the market, but too expensive for the poor population. Production of low-cost nutritious cereal-legume blend at cottage industry using less time and fuel is possible. Such products are safe, convenient, competitive to industrially processed blends, and young children can consume them easily. The proposal intends to produce complementary foods using locally grown foods in selected districts in Tanzania where a varicty of cereals, legumes and oilseeds are grown. Particularly, where underutilized drought resistant food crops such as sorghum, millets, cowpeas, ground nuts and other legumes are abundant, childhood malnutrition is rampant and the population is poor. This will be a transfer of technology project aimed at adding value to the agricultural produce consequently contributing to rural development. It will adopt a market oriented approach empowering women economically through production and selling of low-cost complementary foods while improving the diet of their young children. Innovative processes will be employed to ensure high quality and diversified products availability. Using these procedures, cereal-legume blends become nutritionally adequate and safe for young child feeding. They have acceptable organoleptic properties and can stay unspoiled up to six months at room temperature. Science and technology processes will also be applied to ensure cheap and reliable supply of raw materials.

Keywords: *Childhood malnutrition, Complementary foods, Weaning foods, Child feeding, Women empowerment.*

1.0 Introduction

Childhood malnutrition is rampant in Tanzania and peaks during the complementary feeding period (6-24 months old children). Malnutrition imperils the viability of the society of Tanzania and survivors are left crippled, chronically vulnerable to illness, intellectually disabled, psychologically ill and with poor reproductive and productive performance. Thus malnutrition undermines Tanzanian's struggle for development, peace, equity and justice. The causes of malnutrition are complex and relate to social-economic, cultural and technological factors and skills. The immediate causes are high dietary bulk (low energy, protein and mineral density) of cereal-based diets used as complementary foods for young children in Tanzania (Mosha and Svanberg, 1983; Alderman *et al.,* 2006; Mensah and Tomkins, 2003).

Achieving a diet that is nutritionally adequate and safe is difficult among 6-to 24-month-old children who have a predominantly plant-based diet. There are several important challenges that caretakers have to overcome in order to provide safe and nutritious foods (Mensah and Tomkins, 2003). Complementary foods in Tanzania are basically a thin porridge prepared from cereal flours. Basically cereals are rich in energy, protein and minerals. However, with high water affinity when boiled the resultant porridge from cereal flours become too thick already at about ? 5 per cent W/V (Mosha and Svanberg, 2003). A young child, whose feeding capacity at this tender age is still developing, finds it difficult to intake enough food thus compromising nutrient intake (Gopaldas and John, 1991; Ashworth and Draper, 1992; Darling *et al.,* 1995; Bennet *et al.,* 1999).

To make it thin enough for comfortable feeding for young children, much smaller amounts of flour have to be used to prepare the porridge. In doing so the porridge becomes diluted to the extent that a child will have to consume large volumes per day to meet daily nutrient requirements.

But intake of such large volumes is limited by the small child gastric capacity per feeding. A porridge containing 5 per cent dry matter provides only 0.2 kcal/g, which is much lower than that provided by breastmilk (0.67kcal/g) (Mensah and Tomkins, 2003). This phenomenon, calls for higher feeding frequency which in turn is limited by high women workload who in Africa is the sole young children caretaker. In most communities, infant food is prepared once a day and stored in ambient temperature for feeding over prolonged periods. But this practice provides an opportunity for the food to be contaminated with enteropathogens risking infants to diarrhoea diseases (Kingamkono *et al*, 1994; 1998; 1999; Mensah and Tomkins, 2003).

As the child grows older other food commodities may be added to supplement the cereal porridge. In this case a variety of oilseed and pulses such as cowpeas, groundnuts (peanuts), pigeon peas are used. Hardly any animal foods and products are added. Indeed, cereal-legume blends provide complementary effect to amino acids composition of the mixture. Legumes bring in lysine which is limited in cereals, and cereals bring in sulfur-containing amino acids (methionine) limited in legumes subsequently improving the protein biological value and to some extent improving dietary bulkiness because legumes have lower starch content and higher minerals and protein content than cereals. However, the bioavailability of nutrients from cereal-based complementary food is interfered with by high levels of anti-nutritional factors inherent in cereals and legumes such as phytic acid, tannins, lactins, oxalic acid which impair the absorption or bioavailability of nutrients including iron, zinc, calcium and manganese (Gibson *et al.* 1998 and 2006; Mensah and Tomkins, 2003). According to Hurrell (2006) iron absorption may be as low as 2–3 per cent from porridge based on whole-grain cereals and legumes. In addition, many plants contain inhibitors of digestive enzymes and may damage the intestinal mucosa.

High quality industrially processed fortified infant cereals/ready-to-eat complementary products are commercially available in the market. These have been seen to significantly increase intakes of micronutrients such as calcium, iron, zinc, vitamin A, thiamine, riboflavin, niacin, vitamin B_6, vitamin B_{12} and vitamin C in infants, compared with infants who consume home-made prepared complementary foods (Faber, 2005). But the industrial processed ready-to-consume foods are too expensive for the majority of the poor population. A combination of income growth and nutrition interventions are therefore critical (Alderman *et al.*, 2006) to enable poor population communities to have affordable and convenient complementary foods which are nutritionally adequate and safe to complement the breast feeding.

It is possible to produce low-cost nutritious cereal-legume complementary foods using traditional technologies available for food preparation at household level. Technologies of particular importance in this case are: soaking, germination/malting, fermentation, dry roasting or toasting.

These are mainly carried out by women who spend a considerable amount of their time in food preparation. The blends produced have also been evaluated for their nutritional value and nutritional impacts in young child feeding in developing countries with promising results (Mensah and Tomkins, 2003). These household technologies are low-cost (Owino *et al.*, 2007) methods of food preservation which improve food safety in the household and contribute to the improvement of the nutritional value and digestibility of raw food materials (Mensah and Tomkins 2003). They require less time and fuel to prepare, young children accept them well and consume them easily, resulting in increased food consequently with significant improvement of energy and protein intake (Paul *et al.*, 2008) as well as significant improvement of child growth rate compared to cross-sectional infants who were not in the intervention (Lartey *et al.*, 1999; Gibson *et al.*, 2006).

2.0 Project Description and Rationale

Over the last decade, a market for cereal-legume blended flours has emerged specifically in some urban areas of Tanzania indicating demand for such convenient foods. Most of these are produced by small scale women entrepreneurs with no professional guidance on mixing ratios and treatment resulting in products that are variable and sub-standard in nutrition and safety quality.

This proposal intends to address the problem of food and nutrient intake in infants and young children through application of science and technological processes to ensure high quality of cereal-legume blended complementary foods in selected districts in Tanzania where a variety of cereals, legumes and oilseeds are grown. Where underutilized drought resistant food crops such as sorghum, millets, cowpeas, ground nuts and other legumes are grown, childhood malnutrition is rampant and the population is poor. To ensure regular and cheap supply and reliable source of raw materials, community members will be encouraged to cultivate these crops and sell them to the project.

Sorghum and millets are the second and fourth most widely cultivated cereal crops in Tanzania, producing over one million tons per year when sorghum and millets are combined (www.tanzania.go.tz/). However, much of it is consumed at household level and only less than 2 per cent of the harvest enters the formal market. On the other end, various legumes such as beans (*Phaceola vulgaris*); Cowpea (*Vigna unguiculata*), Soybean (*Glycine max*), pigeon pea (*Cajanus cajan*), chickpea (*Cicer arietinum*) and ground nuts (*Arachis hypogea*) are commonly integrated into the cereal farming systems throughout the country. According to the report of the State of the Economy in Tanzania, production of legumes in 2007 stood at 1.2m tons (www.tanzania.go.tz/).

The proposed project is a community-based pilot intervention involving creation of awareness, mobilization, empowerment and capacity building (teaching) community members on how to process quality cereal-legume blended flours suitable for young child feeding. The project will simultaneously address both childhood malnutrition and poverty among community members particularly women and youth who are usually under-privileged. It will adopt a market-oriented approach with the objective of empowering women and youth economically through production and selling of low-cost complementary foods while improving the diet of their children using locally grown and/or available foods.

Processing will involve cleaning, washing, steeping, germination and/or fermentation, drying, roasting, mixing, milling, parking and labelling. Using these procedures, cereal-legume blends become nutritionally adequate and safe for young child feeding (Mensah and Tomkins, 2003). They also have acceptable organoleptic properties and can stay unspoiled up to 6 months (Owino *et al.* 2007). Soaking and germination/malting are known to improve mineral and protein bioavailability resulting from significant reduction of anti-nutritional factors inherent in cereals and legumes (Ayyanagari *et al.*, 1987; Hotz and Gibson, 2001[a&b]; Mensah and Tomkins, 2003). Due to activated amylase activities, germination allows as much as three times more flour to be used to prepare porridge of the same consistency as normal non-starch-digested porridge, thus allowing a higher nutrient density in porridge of a given viscosity (Mosha and Svanberg 1983, Darling *et al.* 1994) which overcomes the dietary bulkiness problem. Fermentation on the other hand, allows preparation of porridge that can be stored and fed to the child for the whole day without risking the child getting diarrhoea (Kingamkono *et al.*, 1994, 1998, 1999). Roasting or toasting also reduces anti-nutritional factors specifically protease inhibitors and lactins, and results in diminishing effect of porridge viscosity due to starch dextrinization although it could affect protein solubility, but this could be counterbalanced by combining germination and toasting (Kingamkono, 1996).

3.0 Primary Benefits of the Project

The project falls in the agricultural sector, particularly with reference to transfer of technology for processing or adding value to the agricultural produce and thus contributing to rural development.

The project will adopt a market oriented approach with the objective of empowering women economically through production and selling of low-cost complementary foods while improving the diet of their young children using locally available foods. When well popularized, the products to be produced will find a way into the normal market competing with the industrially processed brands.

4.0 Secondary Benefits of the Project

- Agriculture and rural economic development by adding value to agricultural produce and product diversification by processing and marketing them.
- Provide a market for the drought resistant food crops that are otherwise underutilized.
- Much of the added value of the small-scale produced products will return to rural people leading to greater participation in rural development.
- Households buying the product will economize a substantial sum of money and save foreign currency drain.
- Create employment through backward and forward linkages on agricultural producers on one hand and distributors and consumers on the other hand.
- Reduce time for hard-to-cook cereals and legumes and therefore reduce the amount of fuel wood and charcoal needed for cooking

5.0 Project Strategies

- Development of small-scale community based prototypes for different processes (cleaning, washing, steeping, germination and/or fermentation, drying, roasting, mixing, milling, parking and labelling
- Development of standard operational procedures
- Product development, process designs and optimization
- Piloting products and business incubation
- Seed multiplication, distribution, training on improved agricultural practices and post harvest technologies

- Identify and/or facilitate establishment of small women or youth economic groups in each of the proposed district
- Train the women or youth groups on production of low-cost improved complementary foods
- Advocate and popularize the produced products through mass media, conducting demonstrations, develop and distribute IEC materials and organize for exhibitions during national trade fairs and public health events
- Train economic group members on management and entrepreneurship skills and bookkeeping
- Train economic group members to give nutritional education while they earn an income through retailing of the product
- Social marketing to popularize the product for wider utilization and linking the production units with potential markets
- Facilitate establishment of a cooperative of producers and distributors of low-cost complementary foods to maintain economies of scale in activities such as marketing, advertisement and quality control service
- Act as 'mother unit' to oversee and provide service to smaller 'daughter units' under a 'licensing' or 'franchising' contract and do the social marketing, monitoring and evaluation and quality assurance
- Members of the women economic groups jointly owning and managing the production units. They will be responsible for running it, provide labour, ensure constant supply of raw material, feeding their children, selling the products and providing nutrition education as they sell the products

6.0 Implementation Arrangement

- Creating active dialogues with district and community officials and leaders to acquire community consent and agree on the implementation of the project in their villages.
- Conducting baseline needs-assessment involving both qualitative and quantitative situational analysis.
- Feedback seminar and project planning with community members.
- Development and production of Information Education and Communication (IEC) and training modules and materials.
- Conducting Training of Trainers (TOT) seminar for District Extension Workers as trainers of community members. District extension workers to be involved in this regard will include: nutritionists, community development officers, agriculturists, MCH personnel and other representatives from NGOs operating in the districts whose activities are related to improvement of health and nutrition of the people. Leaders of community-based organizations such as women and youth economic groups will also be invited to participate in the TOT seminar so that they act as trainers of their fellow group members.

- Training community members by the district trained team.
- Continuous training of community members and income generating groups on entrepreneurship skills.
- Continuous facilitation linking individuals and economic groups with potential financial institutions for credit facilities.
- Continuous advocacy for popularization of project products through production of articles for newspapers, radio and TV programmes and collaboration with cultural groups and schools in the project area in development of drummer, songs and poems.
- Facilitate participation of income generating groups and individuals in trade fair and exhibitions as the need arises.

7.0 Project Core Team Institutions

- Tanzania Commission for Science Institution-Coordinating Institution
- Tanzania Food and Nutrition Centre
- Ministry of Agriculture and Food Security
- Ministry of Trade and Industry
- Tanzania Industrial Research Development Organization
- Sokoine University of Agriculture
- University of Dar es Salaam
- Agricultural Research Institutions
- Local government/district councils
- Community based organizations/civil societies

8.0 Conclusions

Childhood malnutrition is rampant in Tanzania and peaks during the complementary feeding period. The causes are complex and relate to social-economic, cultural and technological factors and skills. The immediate causes are low feeding frequency, high dietary bulk, poor nutrient bio-availability and microbiological contamination of foods fed to young children. It is possible to produce low-cost nutritious cereal-legume blend at cottage industry using less time and fuel. Such products are safe, convenient, competitive to industrially processed blends, and young children can consume them easily which is the intention of this proposal and will utilize a variety of cereals such as sorghum, millets, cowpeas, groundnuts and other legumes and oilseeds. It will be a transfer of technology project aimed at adding value to the agricultural produce consequently contributing to rural development. It will adopt a market oriented approach empowering women economically through production and selling of low-cost complementary foods while improving the diet of their young children.

References

Alderman, H., Hoogeveen, H., Rossi, M., 2006. Reducing Child Malnutrition in Tanzania: Combined Effects of Income Growth and programme Interventions. Economics of Human Biology, 4(1), pp. 1-23.

Ashworth, A., Draper, A., 1992. The potential of Traditional Technologies for Increasing the Energy Density of weaning foods: A critical Review of Existing Knowledge with Particular Reference to Malting and Fermentation. WHO/CDD/92.4 e. Geneva: World Health Organization.

Ayyangari, R., Rao, B.S.N., Roy, D.N., 1989. Lectins, Trypsin Inhibitors, BOAA and Tannins in Legumes and Cereals and the Effects of Processing. Food Chem, 34, pp. 229–38.

Bennet, V.A., Morales, E., Gonzalez, J., Peerson, J.M.L., de Romana, L.G., Brown, K.H., 1999. Effect of Dietary Viscosity and Energy Density on total daily energy consumption by young Peruvian children. Am J Clin Nutr, 70, pp. 285–91.

Darling, J., Kitundu, J.A., Kingamkono, R.R., 1994a. Weaning Food Technologies. Postgraduate Doctor Publications, 17, pp. 344-350.

Darling, J., Kitundu, J.A., Kingamkono, R.R., Msengi, A.E., Mduma, B., Sullvan, K.R., Tomkins, A., 1994b. Improved Energy Intakes using Amylase-Digested weaning foods in Tanzanian children with acute diarrhoea. Journal of Pediatric Gastroenterology and Nutrition, 21, pp. 73-81.

Faber, M., 2005. Complementary foods consumed by 6-12-month-old rural infants in South Africa are inadequate in micronutrients. Public Health Nutrition, 8(4), pp 373-381.

Gibson, R.S., Ferguson, E.L., Lehrfeld, J., 1998. Complementary foods for infant feeding in developing countries: their nutrient adequacy and improvement. Eur J Clin Nutr, 52, pp. 764–70.

Gibson, R.S., L. Perlas and C. Hotz., 2006. Improving the Bioavailability of Nutrients in Plant Foods at the household level. Proceedings of the Nutrition Society, 65, pp. 160-168. Cambridge University Press

Gopaldas, T., John, C., 1991. Evaluation of a Controlled 6 months Feeding trial on Intake by Infants and Toddlers fed High Energy-low bulk Gruel versus High energy-High bulk Gruel in addition to their Habitual Home diet. *J Trop Paediatr*, 38, pp. 278–83.

Hotz, C., Gibson, R.S., 2001a. Assessment of Home-based Processing Methods to reduce the Phytate content and Phytate/zinc ratio of white maize (*Zea mays*). *J Agric Food Chem*, 49, pp. 692–8.

Hotz, C., Gibson, R.S., 2001b. Temple L. A Home-based Method to reduce Phytate content and increase Zinc Bioavailability in maize-based complementary diets. *Int J Food Sci Nutr*, 52, pp. 133–42.

Hurrell, R.F., 2003. Influence of Vegetable Protein Sources on Trace Element and Mineral Bioavailability. The American Society for Nutritional Sciences *J. Nutr*, 33, pp. 2973S-2977S

Kingamkono, R.R., 1996. Effect of Dehulling and Germination on Nutritive value of Bulrush millet. Journal of Food and Nutrition of Tanzania, 7(1), pp. 3-7.

Kingamkono, R.R., Sjögren, E., Svanberg, U., 1999. Enteropathogenic Bacteria in Faecal Swabs of Young children fed on lactic acid Fermented Cereal gruel. Epidemiology and Infections, 122, pp. 23-32.

Kingamkono, R.R., Sjögren, E., Svanberg, U., 1998. Inhibition of Enterotoxin Production by, and Growth of Enteropathogens in lactic Acid-fermenting cereal gruel. World Journal of Microbiology and Bacteriology, 14, pp. 661-667.

Kingamkono, R.R., Sjögren, E., Svanberg, U., Kaijser, B., 1995. Inhibition of Strains of Enteropathogens in a lactic Acid-fermenting gruel. World Journal of Microbiology and Bacteriology, 11, pp. 299-303.

Kingamkono, R.R., Sjögren, E., Svanberg, U., Kaijser, B., 1994. pH and Acidity in Lactic-fermenting cereal gruel: Effect on viability of Enteropathogenic Microorganisms. World Journal of Microbiology and Bacteriology, 10, 664-69.

Lartey, A., Manu, A., Brown, K.H., Peerson, J.M., Dewey K.G., 1999. A randomized, Community-based trial of the Effects of Improved, Centrally Processed Complementary foods on Growth and Micronutrient Status of Ghanaian infants from 6 to 12 month of age. Am J Clin Nutr, 70, pp. 391–404.

Mensah, P., A Tomkins., 2003. Household-level Technologies to Improve the Availability and Preparation of Adequate and Safe Complementary foods. Food and Nutrition Bulletin, 24(1), pp. 114-126.

Mosha, A.C., Svanberg, U., 1983. Preparation of Weaning foods with High Nutrient Density using flour of Germinated cereals. Food Nutr Bull, 5, pp. 10–4.

Owino, V.O., Sinkala, M., Amadi, B., Tomkins A.M., Filteau, S.M., 2007. Acceptability, Storage Stability and Costing of α-Amylase-Treated maize-Beans-Groundnuts-Bambaranuts complementary blend. Journal of the Science of Food and Agriculture, 87(1), pp.1021-1029.

Paul, K.H., Dickin, K.L., Ali, N.S., Monterrosa, E.C., Stoltzfus, R.J., 2008. American Society for Nutrition J. Nutr, 138, pp. 1963-1968

State of the Economy in Tanzania 2007. Ministry of Planning and Empowerment. http://www.tanzania.go.tz/haliyauchumitanzania1/2007/index.html (accessed 13 November 2008)

Svanberg, U., Lorri, W., 1997. Fermentation and Nutrient availability. Food Control, 8, pp. 319–27.

Svanberg, U., Lorri, W., Sandberg, A.S., 1993. Lactic fermentation on non-tannin and High-tannin cereals: Effects on *in vitro* Estimation of Iron availability and Phytate Hydrolysis. J Food Sci, 58, pp. 408–12.

Chapter 22

Empowering Women through Science and Technology Interventions: The Case for Zambia

Atridah Mulonga
National Science and Technology Council (NSTC), Lusaka, Zambia
E-mail: mulongaatridah@yahoo.com, mulonga@nstc.org.zm

ABSTRACT

Zambia is a developing country with a population approximated at 11.9 million, out of which women constitute 51 per cent. The main economic sectors are mining and agriculture. The scientific and technological base is not developed enough to make a significant contribution to the socio-economic well being of the people. Participation of women in science and technology is low despite them being the majority in the population. Gender issues in science and technology are becoming of concern, Government and non-governmental organisations have embarked on projects to empower women through science and technology. Government has put in place Gender, Science and Technology (S&T), Education, Information and Communication Technology (ICT) policies that ensure that concerns of women are integrated at all levels of S&T development process. These policies also address the enhancement of women's participation in S&T, with a view to encouraging them to undertake scientific and technological ventures that will help reduce poverty among them. The Zambia Association for Research and Development runs a project, which empowers women through Information Technology literacy. The training has benefited many disadvantaged women. The Gender Research in Africa into ICTs for Empowerment is carrying out a research project on the use of cell phones and their contribution to women's independence and empowerment. The project aims to encourage greater ICT use to improve women's lives. The Ministry of Science, Technology and Vocational Training and Ministry of Education operate bursary schemes for undergraduate and postgraduate female students. The UNESCO Commission in Zambia has been carrying out 'Science and Mathematics Camps' for girls in schools during which they receive intensive tuition in mathematics and science subjects. The Forum for African Women Educationalists of Zambia holds 'Science, Mathematics and Technology (SMT) quizzes for girls in schools. These are strategies to motivate girls to participate in SMT subjects and encourage them to take up science-based careers.

Keywords: *Women, Empowerment, Science, Technology, Policies.*

1.0 Introduction

Zambia is a developing country located in Southern Africa, with an area of 752,612 square kilometres. The human population is estimated at 11.9 million and women constitute more than half of the population (Central Statistics Office 2003). The principal economic sectors are mining (especially the mining and refining of copper and cobalt) and agriculture. The major agricultural crops are maize, sorghum, millet and cassava, which are mainly for domestic consumption (Central Statistics Office, 2004).

Although it is endowed with abundant natural resources, Zambia has a low economic base resulting in a low socio-economic status of its people. Women in particular suffer a very low socio-economic status and this makes them vulnerable to poverty as well as social and cultural disadvantages compounded by gender imbalances (H. M. Consultancy, 2005). Gender-based disparities persist in favour of males in many areas. Women are heavily concentrated in informal sector occupations and within this sector, they are largely involved in areas marked by low capital, low profits and high competition, such as beverage and food-processing (NGOCC and ZARD 2004).

It has been clearly demonstrated by the developed world that a scientifically and technically skilled population is a prerequisite for economic and wealth sustainability and wellbeing. In Zambia however, the scientific and technological base, including human resources and institutions is not developed enough to make any significant contribution to the socio-economic wellbeing of the people.

The Science and Technology (S&T) sector remains relatively underdeveloped. Zambia and other developing countries must build up their science, technology and innovation capacity in order to make demonstrable progress in achieving sustainable national economic development and in achieving the Millennium Development Goals (MDGs). In order to do this effectively, there is need to increase the impact of science and technology by making it accessible to all the people of this country.

Despite women's predominance in the Zambian population, their participation in S&T still remains highly underplayed. However, gender issues in science and technology are increasingly becoming a focus of concern in Zambia. The Government, non-governmental organisations (NGOs) and other gender and human rights activists recognize the gender imbalances in social, economic, cultural and political spheres that have obstructed females from contributing effectively and benefiting from the science and technology development process. These groups have embarked on a number of projects to help curb this trend.

The Zambian government has shown commitment to the empowerment of women through ensuring gender equality and strengthening the capacity of women through special programmes aimed at raising their scientific, technical and innovative skills. The Government strives for the attainment of policies that reflect and recognize the role of women in science and technology development and which ensure that the concerns of women are integrated at all levels of the S&T development process. Zambia has a Gender Policy, a Science and Technology Policy, an Education Policy and an Information and Communication Technology (ICT) Policy, which address among other things, the enhancement of women's participation in the science and technology sector with a view to encouraging women to undertake various scientific and technological ventures that will help to reduce poverty among them.

A number of women organizations and movements have embarked on capacity building projects for women and girls. The projects aim at empowering women with scientific and technological skills that will enable them venture into activities that will improve their livelihoods. These efforts by Government and women organizations have not gone without challenges. Like many other developing countries, Zambia is faced with numerous obstacles that hinder full participation of women in science and technology for their own empowerment. Some of the obstacles result from cultural and social practices that obstruct women from exploring the resources and opportunities in S&T for their development.

2.0 Legal/Policy Frameworks Supporting the Empowerment of Women through Science and Technology

The Government of Zambia attaches great importance to issues of gender equality and the empowerment of women. In the quest to ensure gender equity, the government adopted the National Gender policy in 2000, which sets out guidelines for mainstreaming gender issues in all national development initiatives. To this effect, Zambia has undertaken a number of measures to realize gender mainstreaming in all sectors. For instance, Government has developed and is implementing the Fifth National Development Plan (FNDP) in which the promotion of gender equality and empowerment of women is a priority. Government has also put in place an enabling environment for gender equity through policy directives using all Government line Ministries to ensure that they allocate 30 per cent of their sector budgets to gender related activities. Already this policy directive is being implemented in 3 sectors namely Science &Technology, Education and Land.

2.1. Science and Technology Policy

The Government recognized that if both men and women are to participate effectively in the development and implementation of Science and Technology programmes it is important to have a Science and Technology (S&T) Policy that addresses gender concerns. The gender mainstreaming objectives of the policy are:

- To ensure that gender concerns are integrated at all levels of the S&T development process

- To increase the numbers of girls who take up science and technological subjects at all levels of learning
- To identify and eliminate factors which prevent females from progressing and excelling in science subjects.

Specific measures have been put in place to help meet these objectives and they include:

- Reviewing of the school and college curriculum on S&T to make it gender sensitive
- Establishing special funds and scholarships to award females for their education in the science and technical fields
- Promoting S&T subjects in girls schools by making these subjects compulsory
- Establishing and/or strengthening career counselling programmes to address problems which hinder girls progression in S&T
- Providing incentives to female teachers in S&T (Ministry of Science, Technology and Vocational Training, 1996).

In line with these strategies, the Government established the Ministry of Science, Technology and Vocational Training (MSTVT) Bursary Scheme to enable vulnerable students to access science and technology skills training. The objectives of this scheme is to encourage female students to take up skills training in S&T courses in order to promote critical skills among females for their development. The Ministry also runs a special bursary scheme for postgraduate female students in science and technology fields. This scheme has supported a number of students and, in its 5 years of implementation, it has produced more than 10 graduates. They are now holding high profile positions in the S&T sector.

The National Science and Technology Council (NSTC) give a 'Girl Child Scientist Award' every year. The award is given to a promising young girl scientist, identified at the S&T exhibitions that take place during the annual National Science and Technology Week and Forum. The award is aimed at encouraging scientific and technological innovation among young girls and to motivate them to pursue S&T related careers.

2.2 Education Policy

Government has reserved 30 per cent of bursaries for female students, over and above the general bursaries scheme, which is provided to all vulnerable students. With the implementation of this action in the education sector, Zambia has almost achieved gender parity at primary level (Ministry of Education, 2003).

2.3 Information and Communication Technology (ICT) Policy

The Information and Communication Technology (ICT) sector in Zambia has set a goal to increase participation and empowerment opportunities for women in national development through ICTs. The specific objectives are:

- To promote ICT as an alternative career for women in the formal educational system and beyond; to empower and encourage creativity and innovation among women;
- And to provide a forum for collaboration and exchange of ideas on matters affecting women.

The adopted Zambian ICT policy addresses gender and women issues and outlines a number of strategies to be implemented by Ministries responsible for Gender portfolios, in conjunction with traditional leadership and civil society. These strategies include: setting up special training programmes for women in ICTs; developing and promoting ICT awareness and literacy programmes among women; promoting the use of ICT's as tools to eliminate all forms of inequalities between the sexes; encouraging the participation and integration of women in organs and processes of National policy formulation, planning and decision making through affordable information access mechanisms; promoting innovative start-up (incubator) projects for women as ICT entrepreneurs; facilitating and encouraging the development of electronic networks and systems for associations and organizations engaged in the advancement of women issues; promoting and using ICTs in addressing social issues such as health among women; promoting electronic regular consultative processes and exchange of information among women and providing infrastructure and affordable access to ICT tools and services in rural and urban areas (Ministry of Communications and Transport, 2005).

3.0 Projects Aimed at Empowering Women through Science and Technology in Zambia

In line with the objectives of many policies, Government institutions, non-governmental organisations and the private sector have engaged in various ventures to empower women through Science and Technology. Their aim is to improve the livelihood of women in Zambia. One area that has advanced in this quest is Information and Communication Technology (ICT). Two projects aimed at empowering women in the field of ICTs are outlined below.

3.1 Women's Information and Technology (I.T) Literacy Training Programme: Changing Lives through Basic I.T. Training for the Underprivileged

The Zambia Association for Research and Development (ZARD) is a non-governmental organisation that works to empower women and uplift their status and living standards through participatory action-oriented and gender sensitive research, advocacy, training, networking, publishing and dissemination of information. With support from the International Institute for Communication and Development (IICD), ZARD is implementing a programme called Women's Information for Development Network (WIDNet). An important component of this programme is the basic Information Technology (I.T.) Training for Women. ZARD opened a Women's I.T. Literacy Training Centre for women and girls from member organisations and from low-income communities. The aim was to empower these women and girls with basic I.T. skills to enable them use ICT tools to access new knowledge and education, to study and access job opportunities, to establish new business contacts and to communicate and network for an improved livelihood.

Member organisations are all involved in women's rights, gender and development. They include organisations such as Women for Change (WFC), Zambia National Women's Lobby (ZNWL), Society for Women and AIDS in Zambia (SWAAZ), Women Entrepreneurs Development Association of Zambia (WEDAZ), to mention but a few. The training of staff in these organisations involves lectures, training manuals, demonstration and practical. ZARD usually conducts needs assessment and prepares demand driven training packages. These usually include Microsoft Office applications.

Training of staff has been done in more than 10 organisations since 2000. Those that have received training have since enhanced their I.T. skills, giving them efficiency and confidence in their jobs (Lungu, 2008). I.T. training is also done for individuals, usually the less educated from low-income communities. These usually come from the various HIV/AIDS Awareness focal point areas that ZARD has in the low class townships of Lusaka. Modules covered during the training for this disadvantaged class of women include basic windows, word processing, spreadsheets and use of internet, *i.e.* e-mails and search engines. The training strategy includes lectures, training manuals, demonstration and practical. The trainees are allocated a computer each during the training.

Since the inception of the programme in July 2006, ZARD has trained over 60 women and girls in basic I.T. skills. Follow-up on some of the trained women revealed that a number of them have since found jobs where they are applying their acquired skills *e.g.* typists, office assistants, till operators etc. Others are using the skills to manage their businesses, *e.g.* basic bookkeeping and maintaining records of their operations. Those who have found employment have said that they had advantage over other candidates for the jobs because of their I.T. literacy (Lungu, 2008).

Some of the challenges faced in implementing this programme include the negative attitude on the part of the low class, less educated women towards I.T in general. Most do not visualize the benefits of being empowered with I.T. skills. A lot of sensitization needs to be done in this regard. Another challenge is the language barrier. Most individual trainees have limited understanding of the English language, making it difficult for them to understand some commands associated with the use of computers. Others still, have a challenge of reading and writing (Lungu, 2008).

It has been concluded from this project that poor women from low-income communities in developing countries can use information technology to conquer the limitations that presently inhibit their access to resources and enable them to exercise essential rights for the development and improvement of their lives.

3.2 Zambian Mobile Cellular Phone Service Provision and the Advancement of Women in Zambia: Making Life Better for Zambian Women through Mobile Cellular Phones

The International Development Research Centre (IDRC)–Canada, and the Research for the Future–South Africa, through the Gender Research in Africa into

ICTs for Empowerment (GRACE), support a research project focused on the use of cell phones and their contribution to women's independence and empowerment. The general aim of the GRACE Research Zambia Project on Mobile Cellular Phone Service Provision and the Advancement of Women in Zambia is to explore how the provision of cell phone services affects the advancement of women in Zambia by examining the experiences of cell phone service providers, policy makers and women's empowerment organisations; and to explore the social, economic and cultural effects of the service on the country with a specific focus on the status of women.

The research questions and methodologies emphasise qualitative research techniques and appreciative inquiry approach. The major focus of the research has been Government, interest groups, civil society, mobile cellular phone service providers and ordinary mobile cellular phone users. The research also investigated the institutions that monitor gender issues and the wellbeing of women in relation to ICTs.

The research looked at the slogan for one of the Mobile Telephone services provider, Celtel (now Zain), which was 'Making life better' and investigated what this meant for the Zambian women. It tried to answer such questions as: does it mean that women have more freedom to communicate; do they have more control over the content of their conversations on the tool; do they have communication access to a greater number of people; do they have more freedom to express themselves? (Abraham, 2008).

The research revealed that women use cellular phones mainly for business activities, information gathering, and communication with the family and for emergencies. The research further revealed that women who use the tool conduct their businesses more efficiently than those who do not. The availability and use of this information technology reduces the amount of time needed for many transactions, increases the efficiency for many small businesses and enhances the possibility that such businesses can thrive and contribute to the well-being of Zambian women. Through their phones, users are assured of access to other users regardless of distance. Women's empowerment advocates have used this potential to channel advocacy messages to users of mobile phones. As a result, networks have formed that connect individual users on the basis of advocacy issues of common interest. These networks have turned into virtual communities where users can channel their own matters of interest to the general lists in a bid to seek support (Abraham, 2008).

In one district, a Women's Association runs a mobile phone project, which strategically bases its use of mobile phones on three principles: Advocacy, Livelihood and Entrepreneurship. Utilising these principles the association carries out advocacy work for women's progressive change. They, in collaboration with their community radio station, run a phone-in radio programme where they invite local leaders and create a forum where developmental questions can be asked and answered. They also use phones for network linkages around development issues for women's empowerment in their community. To improve their personal livelihood and sustain their use of the mobile phone, the women generate income through the entrepreneurial provision of communication services to their community members (Abraham, 2008).

The study also found that while communities can be created through use of mobile phones, the economic status of women predetermines their position in relation to the virtual networks sustained by mobile phone communication. Only women who can afford call units are privileged members of the mobile community with the ability to initiate conversations and the power to choose when to communicate. Those who cannot afford call units may be included at the discretion of those who do. This presents negative effects of such networking and exacerbates inequalities.

The Research is in its final stages and the results continue to form a pattern that suggests many possibilities for constructive uses of ICTs for the benefit of women and the need for policy intervention to regulate corporate influence on the quality and the nature of services. The major challenge of this research project is that there is no gender specific data on the subject of mobile cellular phones in the country coupled by the fact that the service providers do not take any interest in observing differentials in use between men and women (Abraham, 2008).

The on-going project will contribute to debates focusing on women's empowerment and ICTs through finding its own understandings of what "empowerment" and "gender" may mean in multiple African ICT contexts. Lessons learnt will be shared with policy makers and educators in the form of contextualized and local-specific recommendations. These recommendations will help establish a 'network of gender and ICT researchers in Africa'.

4.0 Interventions Aimed at Encouraging Girls to Take Up Science and Technology Subjects

Government has partnered with a number of organisations in ensuring gender equity in the area of education. Some of these organisations have ventured into programmes aimed at encouraging girls in schools to take up science and technology related subjects. These organisations include the Forum for African Women Educationalists of Zambia (FAWEZA) and the Zambia National Commission for UNESCO (ZNCU).

As a result of poor performance that the girl child has demonstrated in Science, Mathematics and Technology (SMT) subjects and a bias to food and nutrition and other art related subjects, FAWEZA sought to support school based SMT in order to promote gender sensitive and innovative teaching approaches which, when applied, will not only motivate girls taking SMT subjects but also encourage other girls to take up these subjects. This support has been through sponsorship of activities such as National Pupil SMT Quiz and SMT Teachers Exposition. Through funding from the Embassy of the Kingdom of the Netherlands FAWEZA held a National SMT quiz for pupils in high schools under the theme "Enhancing Gender Equity in SMT". The intended purpose was to motivate girls' participation in SMT subjects, promote the learning of SMT subjects, and encourage girls to work hard in SMT if they are to succeed in Science based careers, promote interaction among the pupils and encourage networking among the pupils (FAWEZA, 2008). Questions were prepared by the Junior Engineers, Technicians and Scientists (JETS) Secretariat. The questions ranged from general knowledge, Science, Mathematics and Technology.

In 1999, the UNESCO through the Zambia National Commission for UNESCO (ZNCU) instituted a special project on "Improving Access of Girls to Science and Technology and Vocational Education". This came as a result of realization from studies, that there is low participation of girls in science in institutions of higher learning, which has led to low participation of females in science, technological and vocational careers. As a strategy, the ZNCU has been conducting Science and Mathematics camps/clinics for girls in different provinces around the country, in order to promote the teaching and learning of science and technological subjects. The objectives of these camps are to provide the girls with intensive extra tuition in science and mathematics so as to bring them to the level where they will have confidence to compete favourably with boys, to expose them to and get them to interact with female role models in science and technology so as to encourage girls to pursue science and technology related careers, and to expose girls to situations of practical application of science in industry. The camps/clinics have been successful and the objectives have largely been met (Zambia National Commission for UNESCO, 2007).

5.0 Recommendations and Conclusions

Empowerment of women through S&T in Zambia has been faced with a challenge of limited access to facilities and limited utilization of S&T facilities by most women especially those in rural areas and those from low-income communities. Other barriers include lack of appropriate content, bad perception about and low interest of women in S&T, especially among girls at basic education level; and many socio-cultural factors. To change this state of matters, it would be recommended to Government and all stakeholders involved in the welfare of women to consider the following among other strategies:

- Develop and support more science and technology projects for women to venture in either as individuals or as women groups;
- Give grants to women to start up scientific and technological businesses;
- Award scholarships for women to study science and technology courses;
- Influence the development of school curricular that support enhanced participation of females in mathematics and science subjects;
- Develop and implement S&T sensitisation programmes for women.

Science and technology offers many opportunities for the development and economic empowerment of women. Success in S&T programmes for women, especially those from low-income communities and in rural areas, will create a grass-roots effect, which will bring about widespread empowerment and development for women. Eventually, this can lead to a strong S&T entrepreneurial sector that can contribute to the economic empowerment of Zambia as a whole. Increased sensitisation on the benefits of S&T is essential to support a Zambian knowledge-based society. In promoting science and technology, it is important that the community is able to relate S&T to everyday life and appreciate how science and technology improves our economic, social and cultural wellbeing.

Acknowledgements

I am grateful to the staff at the Ministry of Science Technology and Vocational Training (MSTVT) and the National Science and Technology Council (NSTC) for their support and encouragement. I would like to express my sincere gratitude to Chimika Lungu of the Zambia Association for Research and Development (ZARD) and Kiss Abraham of Gender Research in Africa into ICTs for Empowerment (GRACE) for their valuable contributions to this paper. I am grateful to the Centre for Science and Technology of the Non-Aligned and other Developing Counties (NAM S&T Centre), the Iranian Research Organisation for Science and Technology (IROST) and the Iranian Association of Women's Studies (IAWS) for allowing me the opportunity to present this paper at the International Workshop on Empowerment of Women through Science and Technology Interventions.

References

Abraham, K., 2008. The Names in your Address Book. Gender Research in Africa into ICTs for Empowerment (GRACE), Lusaka, Zambia.

Central Statistics Office, 2003. Zambia Demographic and Health Survey. Government of the Republic of Zambia, Lusaka, Zambia.

Central Statistics Office, 2004. Selected Socio-economic Indicators. Government of the Republic of Zambia, Lusaka, Zambia.

FAWEZA, 2008. Science, Mathematics and Technology (SMT) Quiz. Forum for African Women Educationalists of Zambia, Lusaka, Zambia.

H. M. Consultancy Services Limited, 2005. Zambia: Country Gender Profile. JICA, Lusaka Zambia

Lungu, C., 2008. Women's Literacy Training Programme. Report. Zambia Association for Research and Development (ZARD), Lusaka, Zambia.

Ministry of Education, 2003. Strategic Plan 2003 – 2007. Government of the Republic of Zambia, Lusaka, Zambia.

Ministry of Communications and Transport, 2005. National Information and Communication Technology Policy. Government of the Republic of Zambia, Lusaka, Zambia.

Ministry of Science, Technology and Vocational Training, 1996. National Science and Technology Policy. Government of the Republic of Zambia, Lusaka Zambia.

NGOCC and ZARD, 2004. Ngo-Beijing + 10 Shadow Report on the Situation of Women in Zambia during the Period 1999 – 2004. Lusaka, Zambia.

http://www.ngocc.org.zm

Zambia National Commission for UNESCO, 2007. Zambia and UNESCO. Report. ZNCU-UNESCO, Lusaka, Zambia.